Multitrophic Interactions for Sustainable Agroforestry

Harnessing the Power of Beneficial Insects and Ecological Dynamics

Multitrophic Interactions for Sustainable Agroforestry

Harnessing the Power of Beneficial Insects and Ecological Dynamics

Hanuman Singh Jatav, PhD
Vishnu D. Rajput, PhD
Ipsita Samal, PhD
Tanmaya Kumar Bhoi, PhD
Editors

AAP | APPLE ACADEMIC PRESS

First edition published 2027

Apple Academic Press Inc.
1265 Goldenrod Circle, NE,
Palm Bay, FL 32905 USA

760 Laurentian Drive, Unit 19,
Burlington, ON L7N 0A4, Canada

CRC Press
2385 NW Executive Center Drive,
Suite 320, Boca Raton FL 33431

4 Park Square, Milton Park,
Abingdon, Oxon, OX14 4RN, UK

Library and Archives Canada Cataloguing in Publication

CIP data on file with Canada Library and Archives

Library of Congress Cataloging-in-Publication Data

CIP data on file with US Library of Congress

ISBN: 978-1-77964-124-3 (hbk)
ISBN: 978-1-77964-125-0 (ebk)

DOI: 10.1201/9781779641250

About the Editor

Hanuman Singh Jatav, PhD
Assistant Professor of Soil Science and Agricultural Chemistry, Sri Karan Narendra Agriculture University, Jobner, Jaipur, Rajasthan, India

Hanuman Singh Jatav, PhD, is an Assistant Professor of Soil Science and Agricultural Chemistry at Sri Karan Narendra Agriculture University in Jobner, Jaipur, Rajasthan, India. He specializes in sewage sludge, biochar, micronutrients, heavy metals, and soil fertility management, having taught for over six years and conducted research for 12 years. He earned his BSc (Honors) in Agriculture from Maharana Pratap University of Agriculture and Technology in Udaipur, Rajasthan, India, and his MSc (Agriculture) and PhD in Soil Science and Agricultural Chemistry from the Institute of Agricultural Sciences at Banaras Hindu University in Varanasi, Uttar Pradesh, India, under the supervision of Prof. Satish Kumar Singh (FISSS, FISAC). Dr. Jatav is dedicated to advancing environmental sustainability through his research, having published over 120 papers including research papers, review papers, books, book chapters, and technical papers in various national and international journals and publishers. He serves as an academic editor and member of subject expertise in many internationally acclaimed journals, and is an active member of more than 20 national and international scientific societies. Dr. Jatav has been recognized for his contributions, receiving a Certificate of Gratitude as an Ambassador by IUCN: International Union for Conservation of Nature-Gland, Switzerland, at the World Conservation Congress during the Global Youth Summit. He has also been honored as Outstanding Assistant Professor of Sri Karan Narendra Agriculture University, Jobner, Jaipur, Rajasthan, India.

Vishnu D. Rajput, PhD
Leading Researcher (Associate Professor), Head of International Laboratory of Nanobiotechnology, Southern Federal University

Vishnu D. Rajput, PhD, is a Leading Researcher (Associate Professor) and Head of the International Lab on Nanobiotechnology & Soil Health at Southern Federal University, Rostov-on-Don, Russia. His current research focuses on the toxic effects of bulk and nanoforms of metals, including bioaccumulation, bio/geo-transformations, uptake, translocation, and their impact on plant physiology, morphology, anatomy, and the ultrastructure of cellular and subcellular organelles. Dr. Rajput's work also involves investigating cytomorphometric modifications and DNA damage induced by these metal forms. Dr. Rajput has made significant contributions to the field, particularly in understanding how nanoparticles and heavy metals interact with plants, soil, microbial communities, and the broader environment. He has explored remediation approaches using biochar/nano-biochar-based sorbents and nanotechnological interventions. With extensive experience and experimental work, he has provided comprehensive insights into the state of research in environmental science related to these interactions and remediation technologies. He has an impressive publication record, with a total of 351 scientific publications, including 204 peer-reviewed full-length articles, 18 books, 70 chapters (Scopus indexed), and 44 conference articles. Dr. Rajput serves on the editorial boards of various high-impact journals and leads several grants, including Mega-Grant and BRICS, and participates as a performer/co-PI in numerous national and international projects. Dr. Rajput has been recognized for his outstanding contributions to academic, creative research, and publication activities. He has received a Certificate for Appreciation 2019, Certificate of Honor 2020, Diploma Award 2021, and Letter of Gratitude 2022 and 2023 from Southern Federal University, Russia. Additionally, he has been granted "Highly Qualified Specialist" status by the Russian government and is listed among the top 2% of Russian scientists.

Ipsita Samal, PhD
Scientist (Entomology), Indian Council of Agricultural Research–National Research Centre on Litchi, Mushahari, Muzaffarpur, Bihar, India

Ipsita Samal, PhD, is currently working as a Scientist (Entomology) at the ICAR-National Research Centre on Litchi, Mushahari, Muzaffarpur, Bihar, India. She completed PhD (Entomology) from the Indian Agricultural Research Institute, New Delhi. She has obtained her MSc degree from the *Indian Agricultural Research Institute* (IARI), New Delhi, and graduated from the College of Agriculture, Bhawanipatna, OUAT, Odisha. Her scientific interests are in field of entomology, host plant resistance, integrated pest management aspects of litchi, and understanding biochemical-molecular basis of resistance in insect pests. She is an awardee of a Dr. S. Pradhan memorial award (from ICAR-IARI, New Delhi), ICAR-JRF, ICAR-NET, and IARI-PhD in entomology. She has published several articles, research papers, and book chapters and presented her work at various national and international conferences.

Tanmaya Kumar Bhoi
Scientist, Arid Forest Research Institute, Jodhpur, Rajasthan, India

Dr. Tanmaya Kumar Bhoi, PhD, is a distinguished scientist with a specialization in the intricate realm of insect science. His extensive expertise encompasses research in the areas of host plant resistance, biological control, and integrated insect pest management. With a dedicated focus on advancing agricultural practices, he has made contributions that have significantly impacted the field of entomology. His research delves into the mechanisms of host -plant resistance, aiming to enhance crop resilience against insect pests through innovative strategies. Additionally, his work in biological control emphasizes sustainable and environmentally friendly approaches to manage insect populations, mitigating the reliance on chemical interventions. His commitment to advancing knowledge is underscored by a prolific publication record, with numerous scholarly contributions both at the national and international levels. These publications reflect not only the depth of his research but also its broader impact on the scientific community and agricultural practices. He is multifaceted expertise positions him as a key figure in the ongoing efforts to develop effective and sustainable solutions for insect pest management, contributing significantly to the advancement of entomological science and its practical applications in forestry.

Contents

Contributors .... *xi*

Abbreviations .... *xv*

Acknowledgments .... *xvii*

Preface .... *xix*

1. **Exploring the Role of Beneficial Insects in Sustainable Pest Management and Biodiversity Conservation 1**
   Chada A. Reddy, Mahesh Pathak, Kasturi Sarmah, and Sourav Sen

2. **Biofertilizers: Key Contributors to Sustainable Agroforestry 23**
   Vipula Vyas, Sunil Choudhary, Tanmaya Kumar Bhoi, Kuldeep Sharma, and Sangeeta Singh

3. **Conservatory Measures for Natural Enemies in Pest Control in Agroforestry: Sustainable Multitrophic Interactions 51**
   Satyabrata Sarangi, Suman Samilita Dash, and Aradhana Panda

4. **Evaluation of Plant–Microbe Interactions for Sustainable Agroforestry 85**
   Sunil Choudhary and Vipula Vyas

5. **In Multitrophic Agroforestry, Integrated Pest and Disease Management 119**
   Manoj K. Tripathy, Sashanka S. Dash, Subash C. Mahapatra, and Sajiya Quadri

6. **Multitrophic Pollination Networks: Enhancing Biodiversity Conservation and Crop Yield Through Integrated Ecological Interactions 149**
   Jasti Sri Vishnu Murthy, Alagesan Keerthana, Logeswaran K, Kanchan Kadawla, Feby Atee, Bindu Gudi Ramakrishna, Deepak Kumar Mahanta, Anil Kumar, and Vinod Kumar Dubey

7. **Indigenous Wisdom and Traditional Knowledge in Plant Protection Through Multitrophic Dynamics 175**
   Souvik Acharya

8. Harnessing Traditional Knowledge for Plant Protection: Exploring Multitrophic Dynamics in Agroecosystems 191
Ankita Mukhopadhyay Biswas, Pabitra Kumar Ghosh, and Swarnali Duary

9. Integrated Pest and Disease Management (IPDM) in Multitrophic Agroforestry 211
Parthasarathy Seethapathy, Jagadeesh Bathula, Mohammad Abdul Waseem, and Sivaraj Paramasivam

10. Symbiotic Relationships: Exploring the Role of Insects in Agroforestry Systems 243
Deepak Kumar Mahanta, J. Komal, Tanmaya Kumar Bhoi, and Ipsita Samal

Index 271

Contributors

Souvik Acharya
Faculty Centre for Integrated Rural Development and Management,
Ramakrishna Mission Vivekananda Educational and Research Institute, Kolkata, West Bengal, India

Feby Atee
Department of Entomology, Tamil Nadu Agricultural University, Coimbatore, Tamil Nadu, India

Jagadeesh Bathula
Department of Natural Resource Management and Conservation,
Forest College and Research Institute, Mulugu, Telangana, India

Tanmaya Kumar Bhoi
Forest Protection Division, Indian Council of Forestry Research and Education-Arid Forest Research Institute (ICFRE-AFRI), Jodhpur, Rajasthan, India

Ankita Mukhopadhyay Biswas
Department of Agriculture, Brainware University, Barasat, Kolkata, West Bengal, India

Sunil Choudhary
S.B.R.M. PG Government College, Nagaur, Rajasthan, India

Suman Samilita Dash
Department of Entomology, Odisha University of Agriculture and Technology, Bhubaneswar, Odisha, India

Sashanka S. Dash
Department of Entomology, College of Agriculture, OUAT, Bhubaneswar, Odisha, India

Vinod Kumar Dubey
Department of Genetics and Plant Breeding, College of Agriculture, Acharya Narendra Deva University of Agriculture and Technology, Kumarganj, Ayodhya, Uttar Pradesh, India

Swarnali Duary
Department of Agriculture, Brainware University, Barasat, Kolkata, West Bengal, India

Pabitra Kumar Ghosh
Department of Agriculture, Brainware University, Barasat, Kolkata, West Bengal, India

Kanchan Kadawla
Department of Entomology, College of Agriculture,
Chaudhary Charan Singh Haryana Agricultural University, Hisar, Haryana, India

Alagesan Keerthana
Department of Entomology, Tamil Nadu Agricultural University, Coimbatore, Tamil Nadu, India

J. Komal
Basic Seed Multiplication and Training Centre, Central Silk Board, Kharsawan, Jharkhand, India

Anil Kumar
Sugarcane Research Institute, Department of Entomology, Post Graduate College of Agriculture, Dr. Rajendra Prasad Central Agricultural University, Pusa, Bihar, India

K. Logeswaran
Division of Entomology and Nematology, Indian Institute of Horticultural Research, Bengaluru, Karnataka, India

Deepak Kumar Mahanta
Forest Entomology Discipline, Forest Protection Division, Indian Council of Forestry Research and Education-Forest Research Institute (ICFRE-FRI), Dehradun, Uttarakhand, India

Subash C. Mahapatra
AICRP on Agroforestry, OUAT, Bhubaneswar, Odisha, India

Jasti Sri Vishnu Murthy
Department of Agricultural Entomology, College of Agriculture, Kerala Agricultural University, Thrissur, Kerala, India

Aradhana Panda
Department of Entomology, School of Agriculture, DRIEMS University, Tangi, Cuttack, Odisha, India

Sivaraj Paramasivam
Department of Agricultural Extension, Amrita School of Agricultural Sciences,
Amrita Vishwa Vidyapeetham, Coimbatore, Tamil Nadu, India

Mahesh Pathak
Department of Entomology, School of Crop Protection,
College of Post Graduate Studies in Agricultural Sciences, Central Agricultural University (Imphal), Umiam, Meghalaya India

Sajiya Quadri
Department of Entomology, College of Agriculture, OUAT, Bhubaneswar, Odisha, India

Bindu Gudi Ramakrishna
Department of Agricultural Entomology, College of Agriculture, Kerala Agricultural University, Thrissur, Kerala, India

Chada A. Reddy
SRF, PMD, NIPHM, Hyderabad, India

Ipsita Samal
ICAR-National Research Centre on Litchi, Mushahari, Ramna, Muzaffarpur, Bihar, India

Satyabrata Sarangi
Department of Entomology, Odisha University of Agriculture and Technology, Bhubaneswar, Odisha, India

Kasturi Sarmah
Department of Agricultural Entomology, Centre of Plant Protection Studies,
Tamil Nadu Agricultural University, Coimbatore, Tamil Nadu, India

Sourav Sen
Assistant Professor, Department of Agriculture, Jharkhand Rai University, Ranchi, Jharkhand, India

Kuldeep Sharma
Forest Protection Division, ICFRE-Arid Forest Research Institute (ICFRE-AFRI), Jodhpur, Rajasthan, India

Sangeeta Singh
Forest Protection Division, ICFRE-Arid Forest Research Institute (ICFRE-AFRI), Jodhpur, Rajasthan, India

Parthasarathy Seethapathy
Agrotechnology Division, CSIR - Institute of Himalayan Bioresource Technology, Palampur, Himachal Pradesh, India

Manoj K. Tripathy
Department of Entomology, College of Agriculture, OUAT, Bhubaneswar, Odisha, India

Vipula Vyas
Forest Protection Division, ICFRE-Arid Forest Research Institute (ICFRE-AFRI), Jodhpur, Rajasthan, India

Mohammad Abdul Waseem
Department of Basic and Social Sciences, Forest College and Research Institute, Mulugu, Telangana, India

Abbreviations

AM	arbuscular mycorrhizal
AMF	arbuscular mycorrhizal fungi
ANI	average nucleotide identity
AOD	acute oak decline
BNF	biological nitrogen fixation
BSF	black soldier fly
BSFL	black soldier fly larvae
CBB	coffee berry borer
CBC	conservation biological control
CLM	coffee leaf miner
CLR	coffee leaf rust
CR	Congo red
DED	Dutch elm disease
DMTS	dimethyl trisulfide
EI	ecological intensification
ERIC	enterobacterial repetitive intergenic consensus
ES	ecosystem services
IPDM	integrated pest and disease management
IPM	integrated pest management
IPPP	indigenous plant protection practice
ISR	induced systemic resistance
ITK	indigenous technical knowledge
IWTK	indigenous wisdom and traditional knowledge
KSB	K-synthetic bacteria
MAF	multitrophic agroforestry
MLSA	multilocus sequence analysis
NE	natural enemy
PAGE	polyacrylamide gel electrophoresis
PGP	plant growth promoting
PGPR	plant growth-promoting rhizobacteria
PSB	phosphate-solubilizing bacteria

PSO	phosphate-solubilizing organism
RHH	resource heterogeneity hypothesis
RNB	root nodule bacteria
TCD	thousand cankers disease
VOCs	volatile organic compounds
YEMA	Yeast Extract Mannitol Agar

Acknowledgments

I, Dr. Hanuman Singh Jatav, am deeply grateful and want to express my heartfelt thanks to those who have contributed to the compilation of this book, *Multitrophic Interactions for Sustainable Agroforestry: Harnessing the Power of Beneficial Insects and Ecological Dynamics.* I would like to extend my sincere gratitude to Prof. Balraj Singh, Hon'ble Vice-Chancellor of Sri Karan Narendra Agriculture University, Jobner, Jaipur, for his motivation to undertake this project. I am also deeply indebted to Dr. Amitava Rakshit, Associate Professor at Banaras Hindu University, Varanasi, for his lifelong motivation and guidance. I would also like to thank Prof. K. K. Sharma (HOD, Soil Science and Agricultural Chemistry) and Dr. S. K. Dadhich for their direct and indirect support throughout this endeavor.

I would like to extend my gratitude to Prof. Harphool Singh (Dean, College of Agriculture [COA], Fatehpur, Sikar, Rajasthan) and Prof. S. R. Dhaka (Dean, COA, Bharatpur, Rajasthan) for their guidance and support. I am also thankful to Dr. Sanjay K. Attar, Dr. Muddser A. Khan, Dr. K. C. Verma, Dr. Kailash Chandra, Dr. Mujahid Khan, Dr. Lokesh Kumar Jat, Dr. Champa L. Khatik, Dr. Shubita Kumawat, Dr. Ramu Meena, Dr. Ajeet Singh, and Dr. Vishnu D. Rajput for their direct and indirect support.

As a student of Banaras Hindu University, Varanasi, I am grateful to Prof. A.K. Ghosh, Prof. Nirmal De, Prof. Surendra Singh (FISSS), Prof. Priyankar Raha, Prof. P.K. Sharma, Dr. Y.V. Singh, and Dr. Ashish Latare for their support and guidance during my studies. I would also like to thank my school teachers, Shri Surendra Sarthak Ji, Shri Ghanshyam Ji, Shri Harkesh Ji, and Shri Hanuman Ji, for their early inspiration.

I am also grateful to my friends, Dr. Sunil Kumar, Dr. Dinesh Jinger, Dr. Sonu Kumar, Dr. Surendra Singh Jatav, Dharamveer Verma, and Rajat, for their encouragement and support. I would like to thank Prof. Bijay Singh and Prof. Satish Kumar Singh for their regular efforts in compiling and critically evaluating this book. I am also thankful to Dr. Vishnu D. Rajput for his continued support in checking the content to make it worthy.

I extend my sincere gratitude to my family members who have always supported me in difficult situations. The blessings of Lord Hanuman Ji, Goddess Maa Sharde, and my parents, Smt. Supedi Devi and Shri Babulal

Jatav, have been a constant source of patience and power to overcome the challenges that came my way during the completion of this project. I can only pray for their continued blessings.

—Dr. H. S. Jatav
Assistant Professor
Soil Science and Agricultural Chemistry,
COA, Shekhawati, Sikar, Fatehpur, Rajasthan
(Sri Karan Narendra Agriculture University, Jobner, Rajasthan)

Preface

This book, *Multitrophic Interactions for Sustainable Agroforestry: Harnessing the Power of Beneficial Insects and Ecological Dynamics*, is a comprehensive exploration of the intricate relationships between beneficial insects, ecological dynamics, and sustainable agroforestry practices. This book delves into the multifaceted world of multitrophic interactions, highlighting their crucial role in maintaining ecological balance and promoting sustainable agricultural practices.

Agroforestry, as a land-use strategy, offers a unique opportunity to integrate trees and crops in interacting combinations, thereby fostering a diverse array of ecological interactions. By embracing these interactions, we can harness the power of beneficial insects and ecological dynamics to enhance ecosystem resilience, improve crop yields, and mitigate climate change.

The chapters in this book bring together cutting-edge research and practical insights from experts in the field. From the role of beneficial insects in pollination networks to the ecological basis of multitrophic relationships, each chapter provides a nuanced understanding of the complex dynamics at play in agroforestry systems. The book is structured to provide a comprehensive overview of the current state of knowledge in multitrophic interactions and their applications in agroforestry. It begins by exploring the historical evolution of agroforestry and its potential to address climate change, followed by in-depth discussions of the ecological and economic benefits of multitrophic interactions. Throughout the book, the authors emphasize the importance of integrating ecological and economic considerations to develop sustainable agroforestry practices. They highlight the need for a holistic approach that balances the needs of farmers, ecosystems, and the environment.

The book is a valuable resource for researchers, policymakers, and practitioners seeking to develop and implement sustainable agroforestry practices. It offers a wealth of information and insights that can inform policy and decision-making, ultimately contributing to a more resilient and sustainable food system. Thus, this book is a testament to the power of multitrophic interactions in agroforestry and their potential to drive sustainable development. We hope that it will inspire a new generation of researchers and practitioners to continue exploring the frontiers of agroforestry and its role in shaping a more sustainable future.

CHAPTER 1

Exploring the Role of Beneficial Insects in Sustainable Pest Management and Biodiversity Conservation

CHADA A. REDDY,[1*] MAHESH PATHAK,[2] KASTURI SARMAH,[3] and SOURAV SEN[4]

[1]*SRF, PMD, NIPHM, Hyderabad, India*

[2]*Department of Entomology, School of Crop Protection, College of Post Graduate Studies in Agricultural Sciences, Central Agricultural University (Imphal), Umiam, Meghalaya, India*

[3]*Department of Agricultural Entomology, Centre of Plant Protection Studies, Tamil Nadu Agricultural University, Coimbatore, Tamil Nadu, India*

[4]*Department of Agriculture, Jharkhand Rai University, Ranchi, Jharkhand, India*

*Corresponding author

ABSTRACT

Agroforestry systems are increasingly recognized as sustainable land management approaches that enhance food security, mitigate climate change, and conserve biodiversity. Beneficial insects—pollinators, natural enemies, decomposers, and scavengers—play a pivotal role in maintaining these systems. Pollinators such as bees enhance fruit and nut production by ensuring successful crop fertilization, while plant diversity in agroforestry landscapes further strengthens pollination services. Natural enemies, including predators and parasitoids, regulate pest populations, reducing reliance on chemical pesticides and supporting ecological balance.

Multitrophic Interactions for Sustainable Agroforestry: Harnessing the Power of Beneficial Insects and Ecological Dynamics. Hanuman Singh Jatav, Vishnu D. Rajput, Ipsita Samal, and Tanmaya Kumar Bhoi (Eds).

DOI: 10.1201/9781779641250-1

Decomposers like dung beetles accelerate nutrient cycling through organic matter breakdown and soil enrichment, improving fertility and crop productivity. Scavenger insects also contribute by cleaning the environment, reinforcing ecosystem health. Together, these insect groups act as "ecosystem guardians," fostering resilience, sustainability, and biodiversity within agroforestry landscapes. However, challenges such as habitat loss, pesticide use, and climate change threaten their populations. To safeguard their services, strategies such as promoting habitat diversity, minimizing chemical inputs, and creating insect-friendly environments are essential. By integrating conservation practices into agroforestry systems, farmers can harness the full potential of beneficial insects, ensuring ecological stability and long-term productivity. Ultimately, protecting these unsung heroes is key to achieving sustainable and resilient agroforestry systems..

1.1 INTRODUCTION

Agroforestry is a sustainable land management system that integrates the cultivation of trees and shrubs with crops or livestock that has gained prominence as a crucial solution to tackle pressing global challenges, including food security, climate change, and biodiversity conservation [1, 2]. This innovative system integrates trees and shrubs into agricultural landscapes, creating a dynamic and diverse environment that offers a myriad of ecological, economic, and social benefits [3]. Within the intricate tapestry of agroforestry systems, a spotlight is cast on the indispensable role played by beneficial insects. These organisms, encompassing pollinators, natural enemies, decomposers, and insect scavengers, emerge as unsung heroes in the pursuit of sustainable agriculture. The intricate dance of these insects within the agroforestry ecosystem significantly contributes to the maintenance and enhancement of ecological balance [4].

In this chapter, the primary objective is to demystify the intricate connections existing within agroforestry systems, specifically focusing on the symbiotic relationships between beneficial insects and these agricultural setups. The central theme revolves around understanding the pivotal mechanisms and advantages that underscore the invaluable role of these insects as custodians of agroforestry ecosystems.

Pollinators, such as bees and butterflies, are instrumental in facilitating the reproduction of flowering plants, thereby ensuring a bountiful harvest for crops. The presence of diverse plant species within agroforestry systems provides a rich and continuous source of nectar and pollen, creating an ideal habitat for these essential pollinators [5]. A significant aspect of this exploration centers on the ecological services rendered by pollinators, with a particular emphasis on their profound impact on the production of fruits and nuts. By delving into the dynamics of these interactions, the chapter seeks to unravel the synergies between pollinators and the diverse array of plant species coexisting within agroforestry environments. This understanding is crucial for elucidating how the health and productivity of these ecosystems are intricately linked to the presence and activities of these pollinating insects.

Natural enemies, including predatory insects and parasitoids, act as a biological control mechanism, keeping pest populations in check [6]. By fostering a natural balance between pests and their predators, agroforestry systems reduce the reliance on chemical pesticides, promoting environmentally friendly and sustainable pest management practices. Moreover, the chapter delves into the fundamental role played by beneficial insects in the realm of biological pest control. Here, these insects function as both predators and parasitoids, contributing significantly to the regulation of pest populations within agroforestry systems [7]. Their predatory activities help in keeping pest numbers at manageable levels, thus mitigating the need for excessive chemical interventions. This not only promotes ecological sustainability but also underscores the potential for a more balanced and resilient agroecosystem.

By comprehensively examining the multifaceted relationships between beneficial insects, pollinators, and the intricate web of plant life within agroforestry systems, this chapter aims to provide insights that extend beyond the conventional understanding of agricultural practices. It emphasizes the importance of fostering and conserving these beneficial insect populations as a sustainable and effective strategy for enhancing the productivity and resilience of agroforestry ecosystems. Ultimately, this holistic approach advocates for a harmonious coexistence between agriculture and nature, recognizing the integral role played by these insects in maintaining the delicate balance of these intricate ecosystems. The statement emphasizes the multifaceted contributions of beneficial insects, with a specific focus on decomposers such as soil arthropods, flies, and beetles, extending their significance beyond traditional pest control measures [8].

First, it underscores the nutrient cycling capacity of these insects, particularly the decomposers. In the case of dung beetles, they play a crucial role in breaking down organic matter, facilitating its decomposition [9]. This process is not only vital for waste management but also contributes significantly to nutrient cycling within ecosystems. The breakdown of organic materials by dung beetles leads to the release of essential nutrients into the soil, a process known as nutrient mineralization. This enrichment of the soil is fundamental for agroforestry systems, as it enhances the availability of nutrients necessary for the growth and health of crops. Furthermore, the statement highlights the broader impact on soil health. The active involvement of beneficial insects in organic matter decomposition helps foster soil health by maintaining a balanced nutrient profile. This, in turn, promotes the sustainability and resilience of agroforestry landscapes [10].

The text also touches upon the role of insect scavengers, such as ants, blowflies, and dermestid beetles, emphasizing their importance in cleaning the ecosystem. Scavengers play a critical role in the removal of dead and decaying organic matter, preventing the accumulation of waste that could otherwise become a breeding ground for pests or diseases [11]. The interconnectedness of these organisms within the complex web of interactions in agroforestry landscapes is emphasized, illustrating how each contributes to the overall health and functionality of the ecosystem.

One key strategy discussed is the establishment of diverse agroforestry systems. By promoting diversity within these agricultural and forestry landscapes, the chapter suggests that it is possible to create more hospitable environments for beneficial insects. This diversity can include a variety of plant species and structures that offer suitable habitats and food sources for these insects, fostering a balanced and resilient ecosystem. Moreover, the chapter advocates for the reduction of pesticide usage as a crucial measure to support beneficial insect populations. Pesticides, while intended to protect crops, often have detrimental effects on nontarget species, including the very insects that play a vital role in maintaining ecological balance. By minimizing pesticide use, agroforestry systems can better support the well-being of beneficial insects [12]. A third important strategy highlighted is the creation of insect-friendly habitats. This involves intentionally designing and managing landscapes to provide favorable conditions for the reproduction, feeding, and overall thriving of beneficial insects. This could involve preserving natural habitats within or adjacent to agroforestry systems or implementing specific practices that encourage insect populations.

Collectively, these beneficial insects form an intricate web of interactions within agroforestry systems, fostering resilience and adaptability. Their contribution goes beyond mere pest control and pollination; they actively participate in the creation of a robust and self-sustaining ecosystem. This not only optimizes the overall sustainability of agroforestry landscapes but also enhances productivity by creating a harmonious environment where agriculture and nature coexist. The chapter delves into these challenges, aiming to provide a comprehensive understanding of the obstacles that beneficial insects encounter within agroforestry systems. It goes beyond merely identifying the issues and also explores potential solutions and opportunities for conservation and enhancement of these insect populations. Overall, this chapter serves as a comprehensive exploration of the multifaceted roles of beneficial insects in agroforestry systems, offering insights into the intricate dynamics that shape the sustainability and resilience of these integrated landscapes.

1.2 BENEFICIAL INSECTS

1.2.1 POLLINATORS

Pollinators play a crucial role in agroforestry, as approximately 85% of global flowering plants rely on animal pollination, primarily carried out by insects [13]. Various pollinators, including bees, honey bees, wasps, houseflies, beetles, butterflies, moths, and bats, contribute to the ecosystem's well-being. Bees, in particular, are crucial for pollinating the majority of agricultural crops. Unfortunately, these essential pollinators are experiencing a global decline in both abundance and diversity, attributed to threats such as land degradation, habitat fragmentation, the introduction of nonnative species, diseases, and the indiscriminate use of chemical pesticides. As per the findings of the Intergovernmental Science-Policy Platform on Biodiversity and Ecosystem Services (IPBES) report [14], a potential threat of extinction looms over 40% of global invertebrate pollinator species. The diminishing numbers of pollinators can have profound consequences for the overall ecosystem. To tackle this challenge, proactive measures are essential for managing and restoring their habitats. There is a pressing need for concerted efforts to enhance both the abundance and diversity of pollinators. Initiatives focused on conserving these vital species are critical to ensuring the ongoing health and sustainability of ecosystems.

Agroforestry has the potential to enhance pollinator habitats and services by integrating a variety of structural and functional elements into agricultural landscapes. The adoption of agroforestry practices offers four key advantages for pollinators: (1) the creation of suitable habitats, (2) increased accessibility to foraging resources and nesting sites, (3) improved site and landscape connectivity, and (4) a decrease in pesticide exposure. In order to fulfill their nutritional needs for pollen and nectar, pollinators require a diverse range of flowering plants throughout the foraging season. Numerous studies indicate that the woody species commonly employed in temperate agroforestry practices play a crucial role in providing floral resources for pollinators. Tree and shrub species such as maple (*Acer* spp.), horse chestnut (*Aesculus* spp.), basswood (*Tilia* spp.), willow (*Salix* spp.), brambles (*Rubus* spp.), cherry and plum (*Prunus* spp.), and service berry (*Amelanchier* spp.) contribute abundant nectar with relatively high sugar content [15]. Pollen serves as a significant protein source employed by native bees, honey bees, and specific wasps for nourishing their offspring or provisioning their eggs. Additionally, certain adult and larval flies, along with beetles, integrate pollen into their diet. Numerous notable woody species, such as willow, maple, cherry, plum, brambles, chestnut (*Castanea* spp.), and ash (*Fraxinus* spp.), supply pollen that is rich in elevated concentrations of amino acids, sterols, trace minerals, and other nutritionally essential compounds, offering benefits to bees and a diverse array of pollinators [16]. Certain bee species exhibit a specialization in pollen consumption known as oligolecty, relying exclusively on specific shrubs and trees within distinct plant families. Examples include willows, dogwoods (Cornaceae), heaths such as blueberry and huckleberry (Ericaceae), buckthorns such as New Jersey tea (Rhamnaceae), and roses (Rosaceae) [17, 18]. Additionally, apple trees (*Malus* spp.), hawthorn shrubs (*Crataegus* spp.), and elderberry plants (*Sambucus* spp.) have been recognized as attractive sources for pollinators seeking food [19, 20]. In addition to gathering pollen and nectar, bees are actively involved in obtaining resins and floral oils from various trees and plants. This behavior plays a crucial role in the construction of nests and the provision of resources for their offspring, as emphasized by studies conducted by Policarová et al. and Portman et al. [21, 22]. Some ground-nesting bee species utilize tree resins in the establishment of their nests, as documented by Wcislo and Cane [23]. Additionally, honey bees make use of resin in the formation of propolis, a substance employed to seal openings in their hives. Tree species renowned for abundant resin sources include pine (*Pinus* spp.), birch (*Betula* spp.), elm (*Ulmus* spp.), alder (*Alnus* spp.), beech (*Fagus* spp.), and horse chestnut, as reported by Greenaway et al. [24].

The impact of various weather conditions, such as ambient temperature, wind speed, and precipitation, significantly influences the process of pollination [25, 26]. Agroforestry techniques significantly influence essential meteorological factors, specifically temperature and wind speed. The inclusion of trees in windbreaks has the capacity to diminish air circulation and alter temperatures in cultivated regions. In locations situated downwind of windbreaks, there is a discernible elevation in daytime air temperatures by several degrees [27]. This temperature increase holds the potential to enhance pollinator activity and pollination, particularly in areas devoted to the cultivation of vegetables and fruits, where air temperatures might fall below the optimal range during the pollination period. Additionally, windbreaks provide protection against lower winter temperatures and act as a shield for beehives against strong winds [28]. Agroforestry techniques, exemplified by the creation of hedgerows, can function as sites for butterfly and moth species to lay eggs, host larvae, and provide overwintering habitats [29]. Furthermore, they offer suitable ground-nesting habitats for bees, including ground-nesting species such as bumblebees [30].

Agroforestry practices have the potential to alter the pathways through which pesticides are exposed, consequently impacting the risks posed to pollinators. Current literature emphasizes two main avenues by which agroforestry practices can influence pesticide exposure: (1) the reduction of spray drift and (2) the mitigation of runoff. The implementation of windbreaks and linear agroforestry plantings plays a pivotal role in diminishing wind speeds and capturing particles, resulting in a significant 80–90% reduction in direct exposure of pollinators to pesticide droplets [31, 32]. In addition to mitigating spray drift, pesticides, whether in liquid form, as dust for seed coating, or attached to soil particles, have the potential to be transported off target via runoff [33]. Pavlidis [34] documented a noteworthy reduction ranging from 40 to 100% in the runoff of pesticides, including herbicides, with the adoption of agroforestry systems.

1.2.2 NATURAL ENEMIES

In intricate agroforestry ecosystems that host a diverse array of arthropod taxa, each playing various roles in providing ecosystem services or disservices and engaging in interactions such as predation, competition, and parasitism. Natural enemies or biological control agents play a crucial role in managing pest species. These organisms, which can be insects or plant diseases, serve as effective tools in pest control. In the realm of insect pests, biological control agents encompass predators, parasitoids, and pathogens. Meanwhile, those combating plant diseases are commonly referred to as antagonists.

Predatory insects, such as lacewings, ladybugs, and praying mantis, offer valuable contributions by directly preying on pests such as aphids. Ladybugs, exemplified by *Coleomegilla maculata*, have been acknowledged across various cultures for their predatory behavior spanning centuries. The term "parasite" is often assigned to insects that parasitize others, typically being smaller with a shorter life cycle than their hosts. Although they generally avoid killing their hosts, parasitoids differ by ultimately sterilizing, killing, and sometimes consuming the host. Once the parasitoid concludes its life cycle, it transforms into a free-living insect, no longer reliant on the host. Hymenopterans or Dipterans orders house the majority of parasitoids, comprising up to 25% of all insects. Parasitoids target various pest life stages, including egg parasitism [35]. Pathogens, encompassing bacteria, fungi, nematodes, protozoa, and viruses, constitute microorganisms capable of infecting and terminating the host. In certain conditions, such as prolonged high humidity or dense pest populations, naturally occurring pathogens can significantly diminish populations of aphids, caterpillars, mites, and other invertebrates.

Agroforestry, characterized by cultivating trees alongside conventional crops, is gaining popularity, especially in tropical regions. Despite its increasing adoption, limited research has explored the impact of agroforestry systems on natural enemy populations. Conservation biocontrol aims to enhance the natural regulation of pest populations through their natural enemies, playing a pivotal role in achieving sustainable production, particularly in organic farming systems. The amalgamation of trees alongside crops or animals in agroforestry systems yields both economic and environmental advantages. These encompass carbon sequestration, enhanced soil fertility, diminished soil erosion, and heightened biodiversity [6].

The historical practice of employing natural elements for pest control dates back centuries. As early as AD 304, Chinese farmers employed a strategic approach by preserving nests of citrus ants (*Oecophylla smaragdina*) in orchards to effectively address pest outbreaks. The inception of the contemporary concept of classical biological control can be traced to 1888 in California's citrus orchards. During this period, introduced vedalia beetles (*Rodolia cardinalis*) demonstrated their efficacy in managing cottony cushion scale (*Icerya purchasi*) pests [36].

Promoting diversity in plant life within agroforestry ecosystems is essential for strengthening system stability and reducing the risk of significant insect pest outbreaks commonly associated with monocultures [7]. The "natural enemies hypothesis" posits that increased vegetational diversity enhances the size and impact of natural enemies that regulate herbivorous arthropod pests. According to this idea, the heightened abundance and diversity of arthropod predators and parasitoids result from the increased availability of suitable microhabitats and diverse food resources, such as nectar, pollen, and alternative hosts or prey.

The supply of these nutritional sources frequently prolongs the lifespan and enhances the reproductive capabilities of indigenous predators, bolstering their functional and numerical responses. Additionally, chemical cues emitted by affiliated plants could enhance the efficiency and overall performance of these natural enemies. Despite substantial support for the natural enemies hypothesis in various studies, some findings suggest neutral or negative responses, underscoring the unpredictable nature of tritrophic interactions involving plants, herbivores, and natural enemies. As a result, there exists no all-encompassing regulation or universally applicable theory that forecasts the population size or behavior of natural adversaries within agroforestry ecosystems characterized by diverse vegetation. This underscores the necessity for distinct assessments tailored to each specific system.

In tropical agroecosystems, agroforestry has demonstrated its impact on prey–predator communities and fruit infestation in cocoa and coffee plantations. While various mechanisms influence these dynamics, limited research has explored temperate agroforestry contexts. Notably, practices such as alley cropping, which maintain grassy strips along tree lines, create perennial habitats advantageous for biocontrol in orchards, as discussed by Simon et al. [37].

The role of grassy strips is vital in enhancing aphid biocontrol by increasing the activity–density of predatory arthropods in cultivated areas. However, research, exemplified by the work of Holland et al. [38], suggests that field margins may not significantly influence aphid predation, especially as airborne predators show reduced reliance on these margins. In a study conducted in a walnut tree–cereal agroforestry system in the South of France, no significant differences were observed in parasitoid communities and aphid parasitism when compared to control plots.

In tropical agroforestry environments, established trees directly influence predator communities by serving as a food source for parasitoid wasps, resulting in heightened parasitoid richness and abundance [39]. Moreover, mature trees can alter microclimates, impacting ground arthropod communities and the daily activities of predators—an essential factor in pest biocontrol.

While some argue that increased diversity in agroforestry systems may lead to greater insect diversity and reduced pest abundance compared to monoculture, the correlation between pest control and predator diversity is not universally positive. It depends on specific interactions, such as the presence of alternative food resources, intraguild interactions, and the functional characteristics of species, such as size or dietary preferences.

1.2.3 DECOMPOSERS

There is a slight difference among predators, scavengers, and decomposers, that is, predator kills and consumes their prey. Lion and zebra, vedalia beetle, cottony cushion scale, etc., are a few examples of predators. Scavengers are the organisms, which consume dead and decaying biomass, which include flesh fly and blow fly, etc., whereas decomposers are organisms that carry out decomposition or the breakdown of dead or decaying organisms, for example, dung beetles, black soldier fly (BSF), fungi, and bacteria.

1.2.3.1 DUNG BEETLE

These insects exhibit coprophagy, a behavior characterized by the consumption of the excrement of various organisms, with a particular preference for the feces of herbivores. The excrement, which contains partially processed grass and emits a pungent liquid, serves as the primary nourishment for mature beetles. Scientists classify dung beetles into different categories based on their behavior, specifically identifying them as rollers, tunnelers, or dwellers [9].

Rollers engage in a distinctive process of shaping a small portion of dung into a spherical form, rolling it away, and burying it. The resulting balls serve various purposes, acting as receptacles for females to deposit their eggs (referred to as brood balls) or as a source of sustenance for adult beetles. In contrast, tunnelers land on a dung pile and directly excavate into it, burying a portion of the dung. Dwellers, on the other hand, choose to remain on the surface of the dung pile, utilizing it as a location for laying eggs and nurturing their offspring. Although all three groups play significant roles, dung rollers are particularly crucial in dispersing dung to different locations.

For dung rollers, a male and female form a pair bond after a chance encounter at a dung pat. The male presents the female with a sizable brood ball, and if she agrees, they collaborate to move the ball, with the option for the female to ride atop it. Caution is necessary during this process, as other beetles may attempt to pilfer their valuable possession. Once a suitable location is identified, the pair buries the ball in a soft spot before engaging in mating. Following this, the male departs in search of additional mates, while the female remains behind to create one or more brood balls, each containing a single egg. She meticulously covers and seals each ball using a mixture of dung, saliva, and her own excrement before concealing them underground. Some dung beetle mothers invest up to 2 months caring for the hatched larvae, known as grubs, by cleaning them and disposing of their waste.

Butyric acid holds significance as a crucial volatile signal for dung beetles within the specified synthetic compounds, namely butyric acid, 2-butanone, skatole, indole, and various blends, comprising the dung bouquet [40]. Certain species of dung beetles are transitioning from their traditional dung-based diet to alternative sources such as carrion, decaying fruits, fungi, and more. Among these species, *Anoplotrupes stercorosus* stands out as the predominant dung beetle observed on vertebrate cadavers. In a study conducted by Weithmann et al. [41], it was revealed that *A. stercorosus* was particularly drawn to a combination of dimethyl trisulfide (DMTS), 3-octanone, and benzaldehyde compounds.

The advantages of dung beetles encompass the prevention of "dung pollution," the enhancement of soil aeration and structure, the augmentation of humus levels in the soil, and the inhibition of parasitic fly populations from breeding in the recent feces of mammals.

1.2.3.2 BLACK SOLDIER FLY: HERMETIA ILLUCENS (STRATIOMYIDAE; DIPTERA)

Both organic and inorganic wastes are found in the landfill garbage. More than half of the overall trash is made up of organic material. Waste encompasses various materials, including food, paper, and hazardous substances such as paint, packaging, grass clippings, furniture, clothing, bottles, batteries, asbestos, healthcare waste, sewage sludge, and radioactive materials [42]. The potential contamination of both groundwater and surface environments is a concern associated with these diverse waste components [43]. Due to improper management of landfill garbage, it causes serious risks to human health directly by the emission of toxic gases. On the other hand, a variety of microorganisms and arthropods have their natural habitats in landfills. It demonstrates that landfills and their surrounding areas are havens for a diverse range of arthropods. Effective waste site management in low- and middle-income countries has the potential to mitigate the population dynamics of various insect pests, thereby reducing associated health risks. It was found that black soldier fly larvae (BSFL) insect was able to digest several organic materials by releasing various digestive enzymes.

The female deposits a cluster of 500 eggs on decomposing organic matter. After hatching, the larvae consume the decaying material and relocate to a dry area for the pupation stage. This process has diverse applications, such as managing municipal organic waste [44], mitigating harmful microorganisms, facilitating biowaste management, and serving as livestock feed. Indonesia, with its extensive rice production, generates substantial postharvest waste. The BSF larvae exhibit the remarkable ability to digest raw rice straw, aided by various digestive enzymes [45]. Consequently, BSF larvae are utilized to convert straw into valuable biomass [46]. When applied to chicken manure, BSF larvae accelerate the inactivation of *Salmonella* and *Escherichia coli* [47], transforming the manure into a more nutritious substance and preventing cross-contamination of crops. The prepupae biomass of BSF comprises 32% proteins, 37% lipids, 19% minerals, and 9% chitin, making it a suitable option for livestock feed [48].

A Memorandum of Understanding (MoU) was executed between the Indian Council of Agricultural Research (ICAR) and the Central Institute of Brackishwater Aquaculture on January 20, 2021. The purpose of this collaboration is to launch a new startup initiative focused on the production of BSF meal, serving as an efficient and sustainable alternative to traditional fishmeal. The initiative is particularly relevant in metropolitan areas such as Chennai, where a significant amount of organic waste is generated daily, posing challenges in terms of disposal costs and contributing to harmful methane emissions. The startup's primary objective is to address this issue by collecting nutrient-rich kitchen waste from multistoried residential apartments, offering this service without any charge. The gathered waste will undergo a process of biological conversion facilitated by the BSF. This transformative process will yield a high-quality meal that is rich in proteins and lipids. Moreover, the byproduct, containing valuable inorganic nutrients, can be repurposed as an organic fertilizer suitable for agricultural and horticultural applications.

1.2.3.3 SCAVENGERS

An organism is called a scavenger when it mostly eats decaying biomass, such as animal or plant debris. Many scavengers are carnivorous, indicating they are animals that feed on meat. Scavengers typically eat animals that have either died naturally or have been killed by another carnivore, in contrast to most carnivores that hunt and kill their prey. They contribute to the food chain. The third trophic level is typically occupied by scavengers, which can be carnivores or omnivores—organisms that consume both plants and animals. Scavengers are crucial to the food chain. They prevent animal carcasses, sometimes known as carrion, from entering an ecosystem. Scavengers decompose these organic components, which are then recycled as nutrients into the environment. For instance, vultures eat dead animals, while termites eat the leftovers of dead plants. In the process of animal carrion decomposition, invertebrates such as flies and beetles, along with larger scavengers such as opossums, raccoons, and vultures, play a crucial role. Additionally, microbes such as bacteria and fungi are indispensable for the breakdown of carrion.

1.2.3.3.1 Insect Orders Involved in Scavenging

The important orders, which do scavenging are Coleoptera, Diptera, Isoptera, Hymenoptera, and Blattodea. Beyond these, there are many other orders, which do scavenging; they are butterflies and moths (order Lepidoptera); caddisflies (order Trichoptera); crickets, grasshoppers, and katydids (order Orthoptera); diplurans (order Diplura); dragonflies and damselflies (order Odonata); earwigs (order Dermaptera); fleas (order Siphonaptera); gladiator bugs (order Mantophasmatodea); hemipterans (order Hemiptera); homopterans (order Homoptera), ice bugs (order Grylloblattodea); lice (order Phthiraptera); mantis (order Mantodea); mayflies (order Ephemeroptera); megalopterans (order Megaloptera); psocids (order Psocoptera); scorpionflies (order Mecoptera); stoneflies (order Plecoptera); strepsipterans (order Strepsiptera); thrips (order Thysanoptera); true bugs (order Heteroptera); walking sticks (order Phasmida); webspinners (order Embioptera); and apterygote (subclass Apterygota).

What attracts insects toward carrion? [40]

The decomposition of cadaver tissue leads to the release of volatile organic compounds (VOCs), including indolic compounds such as skatole and indole, carboxylic acids like butanoic acid, and sulfur-rich compounds such as dimethyl disulfide and dimethyl trisulfide, all of which are characteristic byproducts of decay. These VOCs are often nitrogen-rich dimethyl disulfide and DMTS.

1.2.3.3.2 Stages of Decomposition

Corpse decomposition is a continuous process, which takes a few weeks to years, depending on the environmental conditions. It is divided into five stages; they are (1) fresh, (2) bloat or putrefaction, (3) decay or black putrefaction, (4) postdecay, and (5) dry or skeletonization [11, 49, 50]. Coupled with the broad stages of decomposition, putrefaction and autolysis are two stages of chemical breakdown. These two phases aid in the chemical disintegration process that breaks down the body's vital elements. The microbiome of the corpse disintegrates, followed by the necrobiome, which throughout time exhibits predictable changes.

1.2.3.3.3 Process of Decomposition

At the fresh stage even though the corpse first seems to be in fine shape, the bacteria that were feeding on the contents of the gut before death now begin to consume the intestine itself. When they eventually leave the intestine, they start to damage the internal organs that are close by. Natural digestive enzymes are produced by the body, which are mostly located in the colon, permeate the entire body and facilitate bodily breakdown. More specifically, the enzymes within it are released as a cell dies. These enzymes break down the cell and all of its connections to other cells. Then comes bloat stage where fluids are released into body cavities as a result of bacteria dissolving tissues and cells. They frequently breathe without oxygen (anaerobically) and generate a variety of gases as waste products, including hydrogen sulfide, methane, cadaverine, and putrescine. These gases may smell bad to humans, yet a variety of insects find them to be quite appealing. Pressure within the body is caused by the accumulation of gas as a result of the furious activity of the bacteria that are proliferating. By forcing fluids out of cells and blood vessels and into the body cavity, this pressure inflates the body. At the decay stage, the swollen body gradually diminishes, transforming into a flattened form with a smooth, creamy-textured flesh over time. The exposed sections of the body take on a dark hue, emitting a noxious odor of decay throughout the vicinity. During this phase, numerous bodily fluids seep out, saturating the surrounding soil. These fluids become a source of nutrition for other insects and mites. Insects play a significant role in consuming the majority of the flesh. Alternatively, in the absence of insects, bacterial decomposition becomes pivotal, eventually resulting in the body being consumed by bacteria. Then comes the postdecay stage where the body undergoes dehydration, and any remaining flesh is forsaken. The emergence of a pungent cheesy odor, attributed to butyric acid, attracts a fresh set of organisms associated with corpses. Concurrently, as the body undergoes fermentation, mold develops on the areas in direct contact with the ground. In the dry stage, the body undergoes a drying process and decomposes at an exceptionally sluggish rate. Over time, all the hair sheds, ultimately revealing only the skeletal remains.

1.2.3.3.4 Insect Activity at Each Stage of Decomposition

During the fresh stage, when a person passes away, flies are attracted to the body. Blowflies and house flies have the ability to deposit their eggs near wounds and various bodily openings such as the mouth, nose, eyes, anus, and genitalia because they lack the protective mechanisms found in a living organism. Typically, these eggs hatch and develop into maggots within a day, eventually maturing into adult flies. The entire life cycle of a fly, from egg to maggot to fly, is usually completed in 2–3 weeks, though this process may be prolonged at lower temperatures. Then starts the bloat stage, where they move throughout the organism, the juvenile maggots spread bacteria, emit digestive enzymes, and tear tissues using their mouth hooks. They collectively crawl as a cluster, exchanging body heat and digestive fluids. The cadaver undergoes accelerated decomposition, drawing in additional blowflies, beetles, mites, and flesh flies due to the emanating scents and bodily fluids. Subsequent predatory flies and beetles that arrive consume both the maggots and the decomposing tissue. Parasitoid wasps, which deposit their eggs within pupae and later maggot stages, become part of this ecological cycle. At the decay stage, the body is infested with multiple generations of maggots, including fully developed ones. These maggots leave the body and move into the soil, where they transform into pupae. When predatory beetles deposit their eggs in the corpse, the hatching larvae consume the decaying tissue. It is more common for parasitoid wasps to lay their eggs within pupae and maggots. At the postdecay stage, eating soft food alters the body in a way that makes it less attractive to maggot mouth hooks and more compatible with the chewing mouthparts of beetles. Beetles, particularly in their larval stage, feed on ligaments and skin as their source of sustenance. These larvae hatch from eggs laid by adults that previously fed on the body during its initial stages of decomposition. Despite the scarcity of moist flesh in the early stages of decay, the cheese fly scavenges any remaining moist tissue during this phase. Throughout this process, various predators and parasitoids, such as wasp and beetle larvae, continue to be present. At the dry stage, tineid moths and bacteria were the organisms that were capable of consuming human hair. Mites, in turn, feed on these microorganisms. The duration of their presence on the body is determined by the amount of hair and the species involved. Humans and pigs, having minimal hair compared to other animals, experience a relatively short duration of this phenomenon.

1.3 CHALLENGES AND OPPORTUNITIES

1. Lack of Awareness and Education:

    a. Challenge: Farmers and land managers may lack awareness of the crucial roles played by beneficial insects in agroforestry systems. This can result in unintentional actions that harm these populations.

b. Opportunity: Education and outreach programs can raise awareness about the importance of beneficial insects. Providing information on the ecological services they offer and promoting sustainable agricultural practices can lead to better insect conservation efforts.

2. Climate Change Impact:

   a. Challenge: Climate change can disrupt the delicate balance of agroforestry ecosystems, affecting the abundance and distribution of beneficial insects. Shifts in temperature and precipitation patterns may alter the availability of resources for these insects.
   b. Opportunity: Implementing adaptive management practices that consider the impacts of climate change can enhance the resilience of agroforestry systems. This may involve adjusting planting times, selecting climate-resilient plant species, and developing strategies to cope with changing conditions.

3. Pesticide Usage and Its Impact:

   a. Challenge: The reliance on chemical pesticides poses a significant threat to the health and survival of beneficial insects within agroforestry systems. These substances can harm nontarget species, including essential pollinators and natural enemies.
   b. Opportunity: Reducing pesticide usage is a key strategy to safeguard beneficial insect populations. Implementing alternative pest management practices that prioritize ecological sustainability can contribute to a healthier agroforestry ecosystem.

4. Habitat Fragmentation and Loss:

   a. Challenge: Urbanization and agricultural expansion often lead to habitat fragmentation, disrupting the natural habitats of beneficial insects. This can limit their movement, reproduction, and overall ability to contribute to ecosystem balance.
   b. Opportunity: Creating and preserving insect-friendly habitats within and around agroforestry landscapes can mitigate the

negative effects of habitat loss. Designing landscapes to support insect populations enhances their chances of thriving.

Indirect methods, such as applying mulch or intercropping to increase nonpest prey, can also bolster predator populations. Tiny areas of indigenous flora within and surrounding agricultural lands play a pivotal role as essential assets for sustenance and wintering shelters for various species. Findings indicate that engaging in conservation efforts at a broader landscape level can additionally prove advantageous for farmers by bolstering predator populations and improving pest control services [12].

1.4 CONCLUSION

In conclusion, agroforestry emerges as a sustainable and innovative land management approach with far-reaching implications for addressing global challenges such as food security, climate change, and biodiversity conservation. This chapter has focused on unraveling the intricate relationships within agroforestry systems, emphasizing the pivotal role of beneficial insect pollinators, natural enemies, decomposers, and scavengers. The synergies between these insects and the diverse plant life within agroforestry environments contribute significantly to ecological balance and sustainable agriculture.

The exploration of pollinators underscores their vital role in ensuring crop reproduction and harvest abundance, highlighting the interconnectedness of plant and insect species. Natural enemies, acting as biological control mechanisms, reduce the reliance on chemical pesticides, promoting environmentally friendly pest management. Additionally, the chapter emphasizes the often-overlooked decomposers and scavengers, such as dung beetles and ants, shedding light on their contributions to nutrient cycling, waste management, and overall soil health. Strategies such as promoting diversity within agroforestry systems, reducing pesticide usage, and creating insect-friendly habitats emerge as crucial for supporting and conserving beneficial insect populations. By adopting these approaches, agroforestry landscapes can foster a harmonious coexistence between agriculture and nature, optimizing sustainability, resilience, and productivity.

In essence, this chapter provides a holistic understanding of the multifaceted roles of beneficial insects in agroforestry systems. It not only identifies challenges but also explores solutions, offering insights into the dynamics shaping the integrated landscapes' sustainability. By recognizing the significance of these unsung heroes, the chapter advocates for a balanced and resilient approach to agroecosystems, where agriculture and nature thrive in tandem.

KEYWORDS

- **sustainable land management**
- **biodiversity conservation**
- **ecological balance**
- **predator**
- **scavengers**
- **coprophagy**
- **volatile organic compounds**
- **carrion**

REFERENCES

1. Raj, A., Jhariya, M. K., Pithoura, F. (2014). Need of agroforestry and impact on ecosystem. *Journal of Plant Development Sciences*, *6*, 577–581.
2. Sharma, P., Singh, M. K., Tiwari, P., Verma, K. (2017). Agroforestry systems: Opportunities and challenges in India. *Journal of Pharmacognosy and Phytochemistry*, *6*, 953–957.
3. Jose, S. (2009). Agroforestry for ecosystem services and environmental benefits: An overview. *Agroforestry Systems*, *76*, 1–10.
4. Torralba, M., Fagerholm, N., Burgess, P. J., Moreno, G., Plieninger, T. (2016). Do European agroforestry systems enhance biodiversity and ecosystem services? A meta-analysis. *Agriculture, Ecosystems and Environment*, *230*, 150–161.
5. Staton, T., Walters, R., Smith, J., Breeze, T., Girling, R. (2021). Management to promote flowering understoreys benefits natural enemy diversity, aphid suppression and income in an agroforestry system. *Agronomy*, *11*, 651.
6. Martin-Chave, A., Béral, C., Capowiez, Y. (2019). Agroforestry has an impact on nocturnal predation by ground beetles and Opiliones in a temperate organic alley cropping system. *Biological Control*, *129*, 128–135.
7. Ogol, C. K., Spence, J. R., Keddie, A. (1998). Natural enemy abundance and activity in a maize-leucaena agroforestry system in Kenya. *Environmental Entomology*, *27*, 1444–1451.

8. Roy, S., Roy, M. M., Bano, R., Saxena, P. (2017). Soil microarthropods: Biodiversity and role in grassland and agroforestry ecosystems. In: *Agroforestry: Anecdotal to Modern Science* (pp. 669–689).
9. Byk, A., Piętka, J. (2018). Dung beetles and their role in the nature. *Edukacja Biologiczna i Środowiskowa, 1*, 17–26.
10. Sollen-Norrlin, M., Ghaley, B. B., Rintoul, N. L. J. (2020). Agroforestry benefits and challenges for adoption in Europe and beyond. *Sustainability, 12*, 7001.
11. Verheggen, F., Perrault, K. A., Megido, R. C., Dubois, L. M., Francis, F., Haubruge, E., Forbes, S. L., Focant, J. F., Stefanuto, P. H. (2017). The odor of death: An overview of current knowledge on characterization and applications. *Bioscience, 67*, 600–613.
12. Landis, D. A., Wratten, S. D., Gurr, G. M. (2000). Habitat management to conserve natural enemies of arthropod pests in agriculture. *Annual Review of Entomology, 45*, 175–201.
13. Ollerton, J., Winfree, R., Tarrant, S. (2011). How many flowering plants are pollinated by animals? *Oikos, 120*, 321–326.
14. IPBES. (2016). *Assessment Report on Pollinators, Pollination and Food Production.* Inter Governmental Science-Policy Platform on Biodiversity and Ecosystem Services: Bonn, Germany.
15. Merckx, T., Feber, R. E., Mclaughlan, C., Bourn, N. A. D., Parsons, M. S., Townsend, M. C., Riordan, P., Macdonald, D. W. (2010). Shelter benefits less mobile moth species: The field-scale effect of hedgerow trees. *Agriculture, Ecosystems and Environment, 138*, 147–151.
16. Batra, S. W. T. (1985). Red maple (*Acer rubrum* l.), an important early spring food resource for honey bees and other insects. *Journal of the Kansas Entomological Society, 58*, 169–172.
17. Dötterl, S., Vereecken, N. J. (2010). The chemical ecology and evolution of bee–flower interactions: A review and perspectives. *Canadian Journal of Zoology, 88*, 668–697.
18. Fowler, J. (2016). Specialist bees of the Northeast: Host plants and habitat conservation. *Northeastern Naturalist, 23*, 305–320.
19. Minarro, M., Prida, E. (2013). Hedgerows surrounding organic apple orchards in north-west Spain: Potential to conserve beneficial insects. *Agricultural and Forest Entomology, 15*, 382–390.
20. Ponisio, L. C., Gaiarsa, M. P., Kremen, C. (2017). Opportunistic attachment assembles plant-pollinator networks. *Ecology Letters, 20*, 1261–1272.
21. Policarová, J., Cardinal, S., Martins, A. C., Straka, J. (2019). The role of floral oils in the evolution of Apis bees (Hymenoptera: Apidae). *Biological Journal of the Linnean Society*, 128, 486–497.
22. Portman, Z. M., Orr, M. C., Griswold, T. (2019). A review and updated classification of pollen gathering behavior in bees (Hymenoptera, Apoidea). *Journal of Hymenoptera Research, 71*, 171–208.
23. Wcislo, W. T., Cane, J. H. (1996). Floral resource utilization by solitary bees (Hymenoptera: Apoidea) and exploitation of their stored foods by natural enemies. *Annual Review of Entomology, 41*, 257–286.
24. Greenaway, W., Scaysbrook, T., Whatley, F. R. (1990). The composition and plant origins of propolis: A report of work at Oxford. *Bee World, 71*, 107–118.

25. Corbet, S. A. (1990). Pollination and the weather. *Israel Journal of Plant Sciences, 39*, 13–30.
26. Vicens, N., Bosch, J. (2000). Weather-dependent pollinator activity in an apple orchard, with special reference to *Osmia cornuta* and *Apis mellifera* (Hymenoptera: Megachilidae and Apidae). *Environmental Entomology, 29*, 413–420.
27. McNaughton, K. G. (1988). Effects of windbreaks on turbulent transport and microclimate. *Agriculture, Ecosystems and Environment, 22*, 17–39.
28. Haydak, M. H. (1958). Wintering of bees in minnesota. *Journal of Economic Entomology, 51*, 332–334.
29. Dover, J., Sparks, T. (2000). A review of the ecology of butterflies in British hedgerows. *Journal of Environment and Management, 60*, 51–63.
30. Morandin, L. A., Kremen, C. (2013). Hedgerow restoration promotes pollinator populations and exports native bees to adjacent fields. *Ecological Applications, 23*, 829–839.
31. Bischof, M., Coffey, T., Drennan, M. (2019). The effectiveness of riparian hedgerows at intercepting drift from aerial pesticide application. *Journal of Environmental Quality, 48*, 1481–1488.
32. Davis, B. N. K., Brown, M. J., Frost, A. J., Yates, T. J., Plant, R. A. (1994). The effects of hedges on spray deposition and on the biological impact of pesticide spray drift. *Ecotoxicology and Environmental Safety, 27*, 281–293.
33. Reichenberger, S., Bach, M., Skitschak, A., Frede, H. G. (2007). Mitigation strategies to reduce pesticide inputs into ground and surface water and their effectiveness; A review. *Science of the Total Environment, 384*, 1–35.
34. Pavlidis, G. (2017). Pollution control by agroforestry systems: A short review. *European Water, 59*, 297–301.
35. Sanda, N. B., Sunusi, M. (2014). Fundamentals of biological control of pests. *International Journal of Clinical & Biological Sciences, 1*, 1–11.
36. Garbach, K., Milder, J. C., Montenegro, M., Karp, D. S., DeClerck, F. A. J. (2014). Biodiversity and ecosystem services in agroecosystems. *Encyclopedia of Agriculture and Food Systems, 2*, 21–40.
37. Simon, S., Bouvier, J. C., Debras, J. F., Sauphanor, B. (2011). Biodiversity and pest management in orchard systems. *Sustainable Agriculture, 2*, 693–709.
38. Holland, J. M., Oaten, H., Southway, S., Moreby, S. (2008). The effectiveness of field margin enhancement for cereal aphid control by different natural enemy guilds. *Biological Control, 47*, 71–76.
39. Jarrett, C., Cyril, K., Haydon, D. T., Wandji, C. A., Ferreira, D. F., Welch, A. J., Powell, L. L., Matthiopoulos, J. (2023). Fewer pests and more ecosystem service-providing arthropods in shady African cocoa farms: Insights from a data integration study. *Journal of Applied Ecology, 61*, 304–315.
40. Wurmitzer, C., Blüthgen, N., Krell, F. T., Maldonado, B., Ocampo, F., Müller, J. K., Schmitt, T. (2017). Attraction of dung beetles to herbivore dung and synthetic compounds in a comparative field study. *Chemoecology, 27*, 75–84.
41. Weithmann, S., Von Hoermann, C., Schmitt, T., Steiger, S., Ayasse, M. (2020). The attraction of the dung beetle *Anoplotrupes stercorosus* (Coleoptera: Geotrupidae) to volatiles from vertebrate cadavers. *Insects, 11*, 476.

42. Samson, J., Berteaux, D., McGill, B. J., Humphries, M. M. (2011). Geographic disparities and moral hazards in the predicted impacts of climate change on human populations. *Global Ecology and Biogeography*, *20*, 532–544.
43. Taylor, R., A. Allen. (2006). Waste disposal and landfill: Potential hazards and information needs. In: *Protecting Groundwater for Health: Managing the Quality of Drinking-water Sources* (pp. 339–362).
44. Mertenat, A., Diener, S., Zurbrügg, C. (2019). Black Soldier Fly biowaste treatment–Assessment of global warming potential. *Waste Management*, *84*, 173–181.
45. Diener, S., Solano, N. M. S., Gutierrez, F. R., Zurbrugg, C., Tockner, K. (2011). Biological treatment of municipal organic waste using black soldier fly larvae. *Waste Biomass Valorization*, *2*, 357–363.
46. Manurung, R., Supriatna, A., Esyanthi, R. R., Putra, R. E. (2016). Bioconversion of rice straw waste by black soldier fly larvae (*Hermetia illucens* L.): Optimal feed rate for biomass production. *Journal of Entomology and Zoology Studies*, *4*, 1036–1041.
47. Erickson, M. C., Islam, M., Sheppard, C., Liao, J., Doyle, M. P. (2004). Reduction of *Escherichia coli* O157: H7 and *Salmonella enterica* serovar Enteritidis in chicken manure by larvae of the black soldier fly. *Journal of Food Protection*, *67*, 685–690.
48. Caligiani, A., Marseglia, A., Leni, G., Baldassarre, S., Maistrello, L., Dossena, A., Sforza, S. (2018). Composition of black soldier fly prepupae and systematic approaches for extraction and fractionation of proteins, lipids and chitin. *Food Research International*, *105*, 812–820.
49. Kyerematen, R. A., Boateng, B. A., Haruna, M., Eziah, V. Y. (2013). Decomposition and insect succession pattern of exposed domestic pig (*Sus scrofa* L.) carrion. *ARPN Journal of Agricultural and Biological Science*, *8*, 756–765.
50. Manning, P. (2020). Life after death: How insects rise from the dead and transform corpses into skeleton. https://phys.org/news/2020-10-life-death-insects-dead-corpses.html (accessed Dec 30, 2023).

CHAPTER 2

Biofertilizers: Key Contributors to Sustainable Agroforestry

VIPULA VYAS,[1*] SUNIL CHOUDHARY,[2] TANMAYA KUMAR BHOI,[1] KULDEEP SHARMA,[1] and SANGEETA SINGH[1]

[1]*Forest Protection Division, ICFRE-Arid Forest Research Institute (ICFRE-AFRI), Jodhpur, Rajasthan, India*

[2]*S.B.R.M. Government College, Nagaur, Rajasthan, India*

*Corresponding author

ABSTRACT

Agroforestry is an innovative land-use system that integrates agricultural crops with tree species to maximize land productivity while supporting ecological balance. With the rising global population, the demand for higher agricultural output has often been met through the excessive use of chemical fertilizers. However, this reliance has degraded soil health, depleted essential nutrients, and disrupted natural agronomic and ecological processes. Such practices impair soil structure, reduce nutrient retention, and weaken the biological systems vital for sustainable farming. To overcome these challenges, there is a pressing need to adopt eco-friendly alternatives that restore soil fertility and ensure long-term productivity. Organic farming, particularly the use of microbial biofertilizers, offers a viable solution. Microbial biofertilizers—comprising beneficial bacteria, fungi, and other inoculants—improve nutrient availability, enhance plant growth, suppress diseases, and increase crop yields sustainably. Their integration into agroforestry systems not only boosts agricultural productivity but also supports biodiversity and environmental conservation. By combining tree-based systems with microbial inputs, farmers can achieve

Multitrophic Interactions for Sustainable Agroforestry: Harnessing the Power of Beneficial Insects and Ecological Dynamics. Hanuman Singh Jatav, Vishnu D. Rajput, Ipsita Samal, and Tanmaya Kumar Bhoi (Eds).

DOI: 10.1201/9781779641250-2

sustainable yields, improve soil health, and contribute to climate resilience. This chapter highlights the synergy of agroforestry and microbial biofertilizers as a sustainable approach to modern agriculture.

2.1 INTRODUCTION

Approximately, 30% of the world's land area is affected by land degradation, which presents hitherto unseen social, economic, and environmental issues. Nearly three billion individuals live in areas where there is some degree of land deterioration [1–3]. Many of these are small-scale farmers who are also at the forefront of the problems caused by land degradation. A projected 2.2 billion hectares of degraded land are potentially available for restoration worldwide, and in order to address this issue, strategic policies that prioritize sustainable agriculture have been presented [4, 5]. Agroforestry has been recognized as one of the most useful methods for reclaiming and mitigating degraded areas. Growing trees next to crops in the lower layer of the soil is known as agroforestry [6]. Economic interactions are promoted by this integration [7]. Smallholder farmers can improve their ability to withstand the effects of climate change by implementing agroforestry techniques, which can improve food security, revenue streams, public health, and environmental stability. By enabling farmers to participate in off-farm activities and preserve stability, the majority of agroforestry systems tangentially support food security [8]. Agroforestry also conserves biodiversity, enhances soil fertility [9], and improves the quality of agroecosystems [10]. According to recent research, agroforestry practices and principles could be widely adopted to produce food and fiber sustainably in the global agriculture industry, which would be advantageous for both farmers and society at large [11].

Unsustainable exploitation of Earth's biological diversity has resulted from population growth and rising per capita consumption, which have led to a demand for a large amount of food supplies. By 2050, there will be 9.5 billion people on the planet, which means that these trends are going to persist [12]. A gradual decline in soil fertility, encompassing various physicochemical parameters and the soil's overall natural health, has been the consequence of the overuse of chemical fertilizers to meet this increasing demand. Both the environment and various health issues among the populace are negatively impacted by chemical fertilizers.

Therefore, it is imperative to develop sustainable and environmentally friendly technologies that can reduce the need for chemical fertilizers. These chemical fertilizers may be replaced by biofertilizers in order to preserve environmental health and agricultural productivity. Additionally, when contrasted with synthetic fertilizers, biofertilizers offer a more cost-effective and user-friendly alternative. Biofertilizers are comprised of beneficial microorganisms that play a vital role in improving nutrient uptake, promoting the synthesis of growth hormones, and fostering the production of beneficial phytochemicals in crops. This, in turn, leads to an augmentation of both the quantity and quality of crop production [13]. This chapter offers thorough and succinct explanations of all the different elements of plant growth-promoting rhizobacteria (PGPR)-based biofertilizers.

2.2 NATURE'S BOUNTY "BIOFERTILIZERS"

A common definition of biofertilizer is a substance that contains living organisms and, when applied, stimulates plant growth by giving the plant the primary nutrients that are necessary for growth [14]. Numerous advantageous microorganisms have a well-established track record of use in agriculture as biofertilizers [15]. Biofertilizers comprise a diverse array of microorganisms crucial for improving soil health and fostering the growth of plants. They include microorganisms with the ability to fix nitrogen (N_2 fixing), solubilize phosphorus and potassium, as well as endo- and ectomycorrhizal fungi. Additionally, there are other beneficial microorganisms categorized as (PGPRs [16]. These microorganisms work in synergy to support plant development and contribute to sustainable agricultural practices. Depending on how they work in the rhizospheres of plants, biofertilizers come in a variety of forms. A biofertilizer may either comprise a solitary microbial strain, possessing some plant growth-promoting attributes or a consortium of microbes exhibiting a diverse array of plant growth promoting (PGP) traits [17]. However, the various kinds of biofertilizers typically work in concert with one another to provide a sustainable and efficient means of attaining food security with the least amount of negative environmental effects. Biofertilizers and PGPRs have been widely acknowledged as crucial elements in integrated soil nutrient management strategies aimed at fostering sustainable crop production.

2.2.1 TYPES OF BIOFERTILIZERS

2.2.1.1 N2-FIXERS

While nitrogen is abundant in the atmosphere, its challenge in being fixed and utilized by plants acts as a constraint on its availability as a crucial nutrient. Nonetheless, a significant amount of nitrogen can be fixed by some microorganisms, some of which can also ally themselves with plants in different ways [18]. This characteristic lessens losses through denitrification, leaching, and volatilization and enables plants to absorb fixed nitrogen efficiently. These nitrogen-fixing biofertilizers are made up of bacteria that increase the amount of N supplied to crops by the soil through biological N_2 fixation. Long-term studies on the legume N_2-fixing rhizobia have demonstrated that inoculated plants have a higher amount of fixed nitrogen than uninoculated ones [19, 20]. Commercial biofertilizers have been effectively formulated with certain effective N-fixing strains, including *Azotobacter* spp. and *Rhizobium* [21]. These N-fixing bacteria can be further separated into two groups: those that live in the rhizosphere as free-living, symbiotic, and associative microbes without endophytic symbioses.

2.2.1.2 PHOSPHATE-SOLUBILIZING BACTERIA (PSB)

Biofertilizers are integral for augmenting the solubility of essential plant nutrients within the soil, thereby enhancing their accessibility and absorption by crops [22, 23]. Recent observations indicate a growing inclination toward phosphatic biofertilizers, primarily due to their capacity to ameliorate soil phosphorus (P) levels and act as biocontrol agents against agricultural pathogens [24]. The effectiveness of phosphorus solubilization is dependent on the type of P present in the soil, whether inorganic or organic. In many cases, agricultural soils naturally lack phosphorus, making organisms with phosphorus-solubilizing capabilities potential candidates for P biofertilizers [25]. Considering the limited availability of economically mineable P deposits, the adoption of phosphatic biofertilizers holds promise for significantly increasing crop yields [26]. Employing biofertilizers that solubilize phosphorus (PSB) holds great promise for maintaining soil phosphorus levels and fostering widespread, sustainable crop growth [27]. This approach not only addresses the deficiency of phosphorus in many agricultural soils but also opens up new possibilities for supporting extensive and environmentally conscious agricultural practices.

2.2.1.3 VAM

These are phosphate absorbers or biofertilizers that mobilize phosphorus. With over 80% of terrestrial plants, mycorrhizal fungi constitute obligate or facultative working mutualistic symbioses in which the fungus depends on the host for energy and photosynthates while offering a wide range of benefits to the hosts [28]. The fungus mycelium grows from the root surfaces of the host plant into the soil, expanding the surface area available for the plant to absorb and use nutrients more effectively, particularly from sources of insoluble phosphorus and other minerals like calcium, zinc, and copper and more [29]. Moreover, research has shown that mycorrhizal fungi play a significant role in enhancing soil aeration, water dynamics, and the ability to withstand the harmful effects of heavy metals and drought. These fungi also contribute to reducing plants' vulnerability to herbivores and root pathogens [28]. Arbuscular mycorrhizal (AM) fungi, such as those belonging to the Glomus species, are versatile symbionts that establish intercellular relationships within plant roots, forming distinctive vesicles and arbuscules. They are obligate partners of plants, assisting in nutrient and moisture uptake by extending the root system into deeper soil layers. AM fungi demonstrate nonspecificity in their associations, providing plants with enhanced access to essential resources. Additionally, they enhance the availability and mobility of phosphorus, a crucial nutrient that supports the growth and development of host plants [30]. Despite these benefits, large-scale industrial agriculture faces challenges in incorporating AM fungi due to their obligate nature and the difficulty in cultivating them.

2.2.1.4 K-SYNTHETIC BACTERIA (KSB)

The crop production is also frequently limited by K-solubilizing biofertilizers, and these biofertilizers hold equal significance in agricultural practices. Different studies have shown the ability of PGPR to solubilize potassium from rocks through acidification; consequently, KSBs are pivotal in enhancing crop growth and boosting productivity, including tomatoes [31, 32], wheat [33], maize [34], and many other crops. The research findings demonstrate how, in both controlled and uncontrolled environments, these bacteria can greatly enhance germination, nutrient uptake, growth, and crop yields. Additionally, using KSB as bioinoculants in the field significantly reduces the need for chemical fertilizers and is an environmentally beneficial method of producing crops [35]. In particular, native KSB are becoming a promising technology to mitigate K deficits in agricultural soils.

2.3 MODE OF ACTION

Biofertilizers play a crucial role in promoting plant growth both directly and indirectly. They achieve this by generating stimulating chemicals, facilitating enhanced absorption of nutrients from the soil, and providing protection to plants against pathogen infections [36]. PGPR enhances plant growth by producing a range of substances that influence the entirety of the microbial community within the rhizosphere. The positive effects of PGPR on both plant growth and stress response stem from a variety of mechanisms [37, 38]. PGPR's mode of action encompasses a combination of direct and indirect effects on both plants and the soil environment. This includes the production of substances that directly stimulate plant growth and the alteration of the microbial community in the rhizosphere. These bacteria contribute to nutrient availability and uptake, thus enhancing the overall health and vigor of plants. Understanding the mechanisms through which PGPR operate is crucial for unlocking their full potential in promoting sustainable agriculture. By comprehending these mechanisms, we can harness the benefits of PGPR to foster a more resilient and productive agricultural ecosystem.

2.3.1 NUTRIENT SOLUBILIZATION AND ACQUISITION

Many PGPR possess the ability to solubilize nutrients that are otherwise inaccessible to plants, such as phosphorus, iron, and zinc. They release organic acids and enzymes, which have the capability to decompose complex compounds, thereby liberating essential nutrients in a form that is accessible to plants [39, 40]. This enhances nutrient uptake and utilization by plants, leading to improved growth and development.

2.3.2 NITROGEN FIXATION

Certain PGPRs demonstrate the ability to convert nitrogen present in the atmosphere into easily accessible ammonium ions beneficial for plants [41]. This nitrogen-fixing capability contributes to enhanced plant growth and nutrient availability. This ability to fix nitrogen provides an additional source of this essential nutrient, particularly in nitrogen-deficient soils. Plants associated with nitrogen-fixing PGPR can benefit from increased nitrogen availability, leading to improved growth and development [42].

2.3.3 PHYTOHORMONE PRODUCTION

PGPRs are known to synthesize phytohormones, including some of the essential hormones like auxins, cytokinins, and gibberellins [43, 44]. These phytohormones play crucial roles in overseeing various plant processes, such as the growth of roots and shoots, seed germination, and flowering. The synthesis of these phytohormones by PGPR serves to stimulate and enhance plant growth and development, leading to the production of larger and healthier plants.

2.3.4 INDUCED SYSTEMIC RESISTANCE (ISR)

These biofertilizers have the ability to trigger the plant's immune system, resulting in the stimulation of ISR. They trigger a defense response in plants, which includes the production of antimicrobial compounds, enzymes, and defense-related proteins [45]. This primed defense response helps plants to protect themselves against a wide range of pathogens, including bacteria, fungi, and viruses. ISR provides long-lasting protection to plants, enhancing their resistance to diseases even after the removal of the PGPR [46].

2.3.5 BIOLOGICAL CONTROL OF PLANT PATHOGENS

Some PGPRs demonstrate antagonistic effects against plant pathogens. According to [47], these rhizobacteria are capable of generating antimicrobial compounds such as antibiotics, enzymes, and volatile organic compounds. These substances play a role in hindering the growth and development of plant pathogens. PGPR can also compete for nutrients and space with pathogens, reducing their ability to colonize and infect plants. Through biocontrol mechanisms, PGPR can suppress diseases caused by fungi, bacteria, and other pathogens, leading to healthier plants and reduced reliance on chemical pesticides [48].

2.3.6 IMPROVED TOLERANCE TO ABIOTIC STRESS

These rhizobacteria play a crucial role in enhancing plant tolerance to various abiotic stresses such as drought, salinity, and heavy metal toxicity. They achieve this by producing osmoprotectants, including proline and glycine betaine, which assist plants in maintaining cellular water balance, particularly in drought conditions [49]. Additionally, PGPRs are involved in synthesizing antioxidant enzymes that effectively scavenge reactive oxygen species, mitigating oxidative stress induced by abiotic stressors. Furthermore, these beneficial bacteria contribute to improved nutrient uptake and regulation of ion transport, thereby enhancing plant tolerance to challenges like salinity and heavy metal toxicity. Overall, PGPRs offer multifaceted support to plants, contributing to their resilience against a range of environmental stresses.

2.3.7 IMPROVED SOIL HEALTH

PGPRs play a significant role in enhancing the general well-being and fertility of the soil. They can modify the soil environment by improving soil structure, increasing organic matter decomposition, and enhancing nutrient cycling [50]. These rhizobacteria (PGPR) not only play an important role in promoting plant growth but also in fostering the development of other beneficial microbes of the soil. One example is the positive interaction with mycorrhizal fungi, contributing to an enhanced nutrient uptake process and ultimately improving overall plant performance (Figure 2.1).

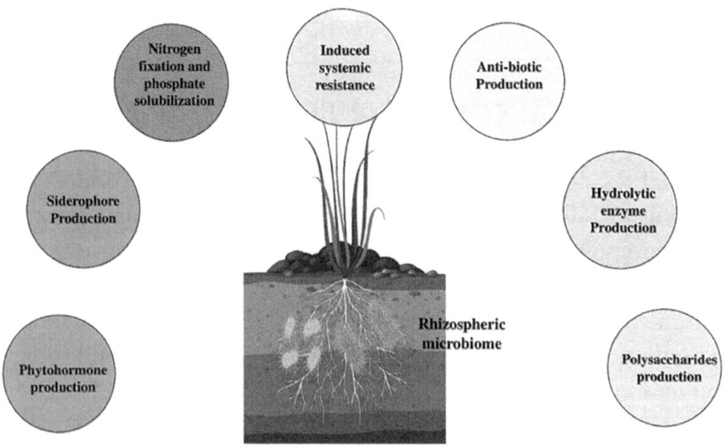

FIGURE 2.1 Action mechanism of biofertilizers.

2.4 BIOFERTILIZERS AND AGROFORESTRY

2.4.1 METHODS OF APPLICATION

Most of the PGPRs are commonly distributed as conventional inoculants based on carriers offering advantages in terms of affordability and straightforward manufacturing processes. The mass production process involves cultivating microorganisms, processing carrier materials, blending carrier material with the microbial culture, and packaging [51]. Ideally, carrier materials used in biofertilizer preparation should be economical, locally available, and easy to process [52]. The carrier materials utilized for high-quality biofertilizers should possess certain characteristics such as nontoxicity, organic composition, and biodegradability, along with a notable water-holding capacity [54]. These properties ensure that the carriers can harbor a substantial number of bacterial cells and sustain their viability over prolonged durations. Commonly employed carrier materials include charcoal, farmyard manure, neutralized vermiculite, peat soil/lignite, and press mud [55, 56]. Additionally, liquid formulations have emerged as an alternative for species like *Acetobacter*, *Azospirillum, Azotobacter*, and *Rhizobium*. Despite being comparatively costly, liquid formulations offer various advantages, including simplified production processes, higher cell counts, extended shelf life, minimal risk of contamination, stability during storage at temperatures up to 45°C, and enhanced soil efficiency [57]. Liquid formulations, despite being more expensive, have become increasingly popular thanks to their improved efficacy and performance. In terms of application, microbial biofertilizers are commonly applied through soil application, seedling, root dipping, and seed treatment. These practices contribute to the effective integration of biofertilizers into agricultural systems, promoting sustainable and environmentally friendly nutrient management.

1. **Direct application in the soil:** It involves directly adding treatment to the soil, either alone or in combination with other materials [58]. This method typically includes preparing a mixture of cow dung, microbial fertilizer, and rock phosphate, which is left in a shaded area overnight while maintaining its moisture content at 50%. Afterward, this prepared blend is applied onto the soil [62].

2. **Seedling Root Dipping:** During this method, the roots of seedlings are immersed in a water solution containing biofertilizer (such

as nitrogen-fixing *Azotobacter* or *Azospirillum* and phosphorus-solubilizing microbial biofertilizer) for an adequate duration [60, 61]. The duration of treatment varies depending on the specific crop being cultivated.

3. **Seed Treatment:** Seed treatment stands out as a highly efficient and cost-effective method widely utilized for all categories of inoculants. The procedure involves thoroughly mixing seeds with a slurry, ensuring a uniform coating, followed by shade-drying before being promptly sown within a 24-hour timeframe. In the case of liquid biofertilizers, the coating process varies depending on the seed quantity; smaller quantities may be coated in plastic bags, while larger quantities can be accommodated in buckets. Notably, seed treatment allows for the simultaneous application of two or more bacteria without any adverse antagonistic effects on each other, promoting a synergistic and effective approach to enhance seed and soil health [59].

2.4.2 ENVIRONMENTAL BENEFITS OF USING BIOFERTILIZERS

The environmental benefits of microbial biofertilizers are multifaceted. They have emerged as a sustainable and environmentally friendly alternative to traditional chemical fertilizers, offering a multitude of benefits that contribute to the preservation of the ecosystem. One of the key advantages lies in the significant reduction of chemicals in agricultural practices [63, 64]. Unlike synthetic fertilizers that often contain harmful chemicals, microbial biofertilizers harness the power of beneficial microorganisms to enhance nutrient availability in the soil. These microorganisms promote nutrient absorption and in turn reducing the need for chemical inputs. The utilization of microbial biofertilizers also plays a crucial role in controlling the loss of biodiversity in agricultural ecosystems [15, 65]. Synthetic fertilizers not only degrade soil quality over time but also pose a threat to the diverse array of organisms inhabiting the soil [66, 67]. Microbial biofertilizers, on the other hand, foster a balanced and harmonious environment where beneficial microorganisms collaborate with plants to maintain a healthy ecosystem. By promoting a symbiotic relationship between microorganisms and plants, microbial biofertilizers contribute to the conservation of soil biodiversity, which is essential for sustainable and resilient agricultural systems [68, 69]. Another significant environmental benefit of microbial biofertilizers is their role in climate change mitigation and carbon sequestration [70, 71]. The excessive use of chemical fertilizers releases greenhouse gases into the atmosphere, contributing to global warming and climate change. In contrast, microbial biofertilizers aid in carbon sequestration by enhancing the organic content of soil [72]. The microorganisms present in these biofertilizers facilitate the decomposition of organic residues, converting them into stable organic carbon. This process not only improves soil fertility but also helps in mitigating climate change by sequestering carbon in the soil, reducing the overall carbon footprint of agricultural activities [73, 74]. Microbial biofertilizers are essential for maintaining agricultural sustainability as they support soil health and structure. They aid in improving water retention, preventing erosion, and promoting nutrient cycling in the soil [75]. This results in diminishing the reliance on synthetic chemicals, and the long-term productivity of the soil is maintained. This sustainable approach ensures that agricultural practices do not compromise the well-being of the environment, addressing concerns related to soil degradation and loss of arable land (Figure 2.2).

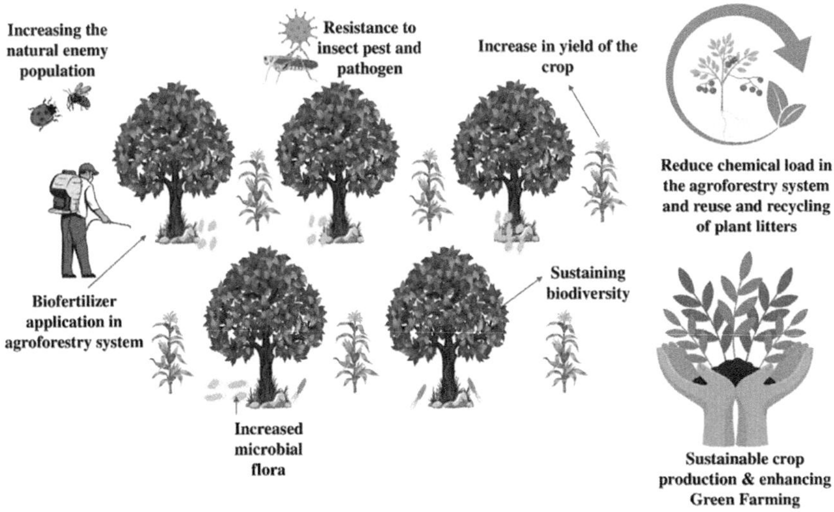

FIGURE 2.2 Various benefits of biofertilizers for sustainable agroforestry.

2.4.3 APPLICATION CHALLENGES FOR BIOFERTILIZER

Although biofertilizers are safe, economical, renewable, and environmentally friendly, there are certain application-related challenges. The adoption of the biofertilizer approach faces several challenges, such as slow behavior, complex formulations, vulnerability to moisture and temperature, absence of specific microbial strains, insufficient facilities for biofertilizer manufacturing, a shortage of skilled labor, seasonal demand influenced by microbial activity, regulatory constraints on manufacturing, and varying customer demands [76]. However, the efficacy of utilizing biofertilizers in forestry hinges upon critical elements such as the cultivation of microbial strains, preparation of inoculants, formulation processes, and the implementation of field-testing methodologies. Overcoming these challenges is crucial to realizing the full potential of biofertilizers in forestry.

2.4.3.1 COMPLICATIONS IN FORMULATION

The utilization of biofertilizers is gaining increased recognition, with a multitude of products now accessible on a global scale [77]. The effectiveness of biofertilizers containing live microbial cells heavily relies on the formulation used. Creating inoculants is a complex process with multiple stages. It involves selecting and utilizing one or more strains of microbe along with suitable additives as well as suitable carriers. The goal is to create a formulation that provides a safe environment for the microorganisms, protecting them from adverse conditions during storage, ensuring their survival, facilitating transportation, and aiding their establishment in soil after application [25]. The formulation's significance lies in its ability to effectively introduce microorganisms to the target area, enhancing their activity upon injection into the host. Additionally, cost-effectiveness is a crucial consideration in the selection of formulated products. In the realm of commercial biofertilizer production, the availability and affordability of diverse carriers and organic materials are vital factors that should be taken into account [78, 79].

2.4.3.2 CHANGES IN EFFICACY WITH THE CHANGE IN CLIMATE

The effectiveness of bioformulation technology is affected by the severe and erratic environmental conditions [53, 80]. During soil application, inoculants are exposed to both biotic environmental stresses (such as microflora and fauna) and abiotic stressors (such as soil pH, salinity, and the temperature). Introduced inoculums find it difficult to withstand harsh weather conditions in semiarid environments, such as droughts, high salinities, soil erosion, and even insufficient irrigation, which may cause the inoculum to rapidly deteriorate [81–83]. Nonetheless, this includes taking into account how the quickly shifting global climate and the conditions that are expected to exist in each region over the next several decades may impact the suitability and effectiveness of the product.

2.4.3.3 CONDITIONS OF STORAGE

Temperature, humidity, sunlight intensity, and carrier material properties all have an impact on the biological efficacy, microbial survival, and product storage life [84]. The carrier needs to be stable and an increased shelf life in order to preserve the microorganisms' viability while they are kept in the farmer's warehouse. The biofertilizer needs to have an adequate number of viable cells in storage for it to work well in the field [85]. Another major issue that can arise when using biofertilizers is their shelf life. Microbial cells employed in biofertilizers undergo degradation over time, exhibiting a restricted shelf life usually ranging from four to six months when kept and maintained between 20 and 25°C [86]. Special precautions are necessary for their storage and transportation. This added need for care and attention in handling significantly contributes to the overall cost of the product.

2.4.3.4 MARKET OF BIOFERTILIZERS

Due to growing health concerns, people are currently interested in organic farming. Due to this, the market for biofertilizers is growing in both developing and developed nations [87]. The global market for biofertilizers is projected to witness significant growth, with an anticipated increase from $1.88 billion in 2022 to $3.51 billion by 2027. This growth signifies a compound annual growth rate of 13.2% over the forecast period [88]. Presently, legumes and N_2-fixing inoculants are the market leaders for biofertilizers worldwide. The world market for biofertilizer is currently dominated by legumes and N_2-fixing inoculants. Phosphate solubilizers constitute approximately 15% of the global biofertilizer market, while other bioinoculants contribute to around 7%. The majority, comprising about 78% of the market, is represented by rhizobia-based inoculants [8].

2.4.3.5 LEGAL AFFAIRS, REGISTRATION, AND INTELLECTUAL PROPERTY

It is essential to secure registration permission prior to a product being on sale. It could be difficult to adapt really creative technologies into the current legal frameworks, though. In certain situations, regulatory bodies might have to hold discussions in order to create suitable frameworks [89]. Ensuring compliance with manufacturing procedures and supplying comprehensive evidence of safety and effectiveness, often substantiated by an impartial third party, are crucial components of this process. The expansion and production of microbe-based products face substantial impediments due to legal and regulatory challenges [90]. Countries and, in some cases, subnational jurisdictions have implemented risk assessment and testing policies to prevent the dissemination of harmful organisms, thereby creating significant barriers for the industry. The regulatory procedures governing biological or biobased products are intricate and extensive, displaying variations across different countries and subjurisdictions [91]. Registering such products involves intensive constraints, including prolonged timelines, intricate documentation, and high fees; also the firms aiming to introduce products globally face difficulties in meeting diverse regulatory requirements across regions. Usually, the process of product registration involves obtaining approval and certification from national regulatory authorities. This is followed by comprehensive inspections and assessments conducted by experts under the supervision of food safety and national commissions [92].

2.4.4 THE PRESENT SCENARIO OF USING BIOFERTILIZER IN AGROFORESTRY

Concerns about sustainable forest and agricultural management, particularly integrated nutrient management, have accelerated market momentum in many regions of the world, especially in emerging nations. As a result, there has been a growing emphasis on the adoption of PGPRs, including PSBs, seaweeds, *Azospirillum, Acetobacter, Rhizobium*, *Azotobacter*, and arbuscular mycorrhizal fungi (AMF) [93–95]. Surprisingly, there is an increasing demand for biofertilizers not only in Argentina, Europe (particularly in Germany, Italy, and Spain), Canada, China, but also in India, US, and Africa [40, 82, 96].

Due to their increasing need for organic products and recognition of the significant advantages these fertilizers offer, these countries are making significant efforts to promote the usage and development of biofertilizers [97].

Based on Research Institute of Organic Agriculture (Forschungsinstitut für biologischen Landbau in German) data [98], the area dedicated to organic agriculture worldwide has expectedly expanded from 69.4 million hectares in 2017 to 74.9 million hectares in 2020. The use of more organic farming practices has led to an increase in the utilization of organic products, such as biofertilizers, which have improved soil fertility. However, as they grow and participate in the nutrient cycle, plants gain from the application of biofertilizers as seeds or soil inoculants. Biofertilizers are gaining popularity in forestry and agriculture because they are less expensive than chemical fertilizers, enhance soil health, and boost productivity by 10–25% without damaging the environment or the soil [99]. Utilizing biofertilizers is a desirable alternative for forestry and agriculture since it can lessen soil erosion and increase water retention [100]. In an effort to increase agricultural yields, research has recently concentrated on using biofertilizers to improve soil fertility through phytoremediation of metal-contaminated soils [101], lowering soil salinization [102], and raising the soil's capacity to withstand harsh events [103]. A portion of the research has moved from the lab to the field for field testing. The goal of research has been to keep improving technology in order to lower costs, preserve quality, and make biofertilizers more appealing to farmers. Advanced, affordable, and ecologically friendly fertilizers have been made possible by new technologies like nanotechnology, which has been introduced as nutrients, nanobiofertilizers, nanofertilizers, and nano-pesticides [104, 105]. High-throughput sequencing is suggested as a valuable method to uncover the "missing components" within soil phosphorus cycling. These components, which may include microbes impacted by unsustainable intensive farming methods, play a vital role in restoring healthy and functional microbial communities in today's agriculture [106, 107] (Table 2.1).

2.5 FUTURE PROSPECTS/DIRECTION

Significant advancements have been made globally in utilizing PGP microbes as biofertilizers and biopesticides, leading to positive outcomes

TABLE 2.1 Use of Biofertilizers in Different Agroforestry Systems and Their Effect on Crop

Sl. No.	Agroforestry System	Name of the Biofertilizer Used	Resultant Effect	Country	Reference
1	Coffee (*Coffea* spp.)-based agroforestry	MYKORICH® biofertilizer (MRICH),	Inoculation with either MRICH or MRICH+BB (Biochar) treatments also resulted in positive effects on soil physicochemical characteristics, including organic matter, nitrogen (N), phosphorus (P), potassium (K), and cation exchange capacity.	Philippines	[108]
2	*Leucaena leucocephala*-based agroforestry	*Azotobactor chroococcum*+ Phosphate-Solubilizing Bacteria+Vermicompost	Increased growth yield of maize crop	India	[109]
3	Agri-silviculture of the plum (*Prunus domestica* L. Cv. Opal),	*Azotobacter chroococcum* (Az), *Thiobacillus* spp. (Th)., *Rhizophagus intraradices* (AMF)	The combined use of AMF and Az improved smooth vetch forage quantity and quality	Iran	[110]
4	Rubber-based ecosystem	Azotobacter and phosphate-solubilizing bacteria (PSB)	FYM, Azotobacter and phosphate-solubilizing bacteria (PSB) registered highest in all growth parameters of bird's eye chilli	India	[111]
5	*Calliandra houstoniana* tree-based agroforestry	Azospirillum + Rhizophagus	Increase the production of organic Hibiscus	Mexico	[112]
6	Albizia procera-based agroforestry	Rhizobium and the mycorrhiza *Glomus etunicatum*	Dual inoculated seedlings required a lower dosage of N fertilizer than those inoculated only with mycorrhizas.	India	[113]
7	Albizia procera-based agroforestry	*Brady rhizobium japonicum* and phosphate-solubilizing bacteria (PSB)	The available nitrogen was found to be maximum in treatment combinations of Brady rhizobium japonicum and Karanj cake (T7), i.e., 371.2 $kgha^{-1}$.	India	[114]
8	Nagpur mandarin (*Citrus reticulata* Blanco.)-based agroforestry	Azotobacter and PSB	The maximum seed yield, biological yield, and oil yield were obtained in mustard crop	India	[115]

for plant growth and soil health through direct interactions among different microbial types [57]. Recently, the concept of microbial consortiums for PGP has been proposed as a viable strategy to enhance the activity and viability of plant growth-promoting microbes [116]. However, widespread adoption of PGP microbes poses several challenges. First, transitioning from laboratory and greenhouse settings to field trials necessitates the development of innovative approaches for growing, storing, and handling these microbes. This includes addressing issues related to shipping, formulation, and application on a larger scale [25]. Second, public awareness and education are crucial due to prevailing misconceptions associating microbes with diseases [117]. Overcoming these misconceptions is essential before the large-scale implementation of these beneficial microbes in agriculture. Additionally, conventional practices involving chemical pesticides to combat plant diseases contribute to environmental and health concerns [105, 106, 118–123]. Hence, there is an urgent need for ecofriendly alternatives, with biofertilizers emerging as potential substitutes that not only address feeding the growing population but also enhance productivity and support plant growth under stress conditions [13]. Recognizing the significance of biofertilizers and incorporating them into modern agriculture is crucial for sustainable practices [124–136]. Shifting from merely growing plants to cultivating plant–microbial communities is proposed for sustainable agriculture. This approach aims to achieve high productivity with minimal energy and chemical inputs, while minimizing environmental impact [4]. However, this transformative process is in its early stages, requiring continued efforts and collaboration among experts in plant and microbial genetics, molecular biology, and ecology to successfully develop sustainable microbial-based agro technologies.

KEYWORDS

- **microbes**
- **biofertilizers**
- **agroforestry**
- **microbial inoculants**

REFERENCES

1. Wassie, S. B. (2020). Natural resource degradation tendencies in Ethiopia: A review. *Environmental Systems Research*, *9*(1), 1–29.
2. Hossain, A., Krupnik, T. J., Timsina, J., Mahboob, M. G., Chaki, A. K., Farooq, M., Bhatt, R., Sh, F., Hasanuzzaman, M. (2020). Agricultural land degradation: Processes and problems undermining future food security. In: *Environment, Climate, Plant and Vegetation Growth* (pp. 17–61). Springer International Publishing: Cham.
3. Eswaran, H., Lal, R., Reich, P. F. (2019). Land degradation: An overview. In: *Response to Land Degradation* (pp. 20–35).
4. Edrisi, S. A., Sarkar, P., Son, J., Prakash, N. T., Baral, H. (2022). Assessing the realization of global land restoration: A meta-analysis. *Anthropocene Science*, *1*(1), 179–194.
5. Gupta, S. R., Dagar, J. C., Teketay, D. (2020). Agroforestry for rehabilitation of degraded landscapes: Achieving livelihood and environmental security. *Agroforestry for Degraded Landscapes: Recent Advances and Emerging Challenges*, *1*, 23–68.
6. Dagar, J. C., Gupta, S. R., Teketay, D. (Eds.), (2020). *Agroforestry for Degraded Landscapes*. Springer.
7. Santiago-Freijanes, J. J., Mosquera-Losada, M., Rois-Dıaz, M., Ferreiro-Dominguez, N., Pantera, A., Aldrey, J. A., Rigueiro-Rodriguez A (2021) Global and European policies to foster agricultural sustainability: Agroforestry. *Agroforestry Systems*. doi: 10.1007/s10457-018-0215-9
8. Duffy, C., Toth, G. G., Hagan, R. P. O. et al. (2021). Agroforestry contributions to smallholder farmer food security in Indonesia. *Agroforestry Systems*, *95*, 1109–1124. doi: 10.1007/s10457-021-00632-8
9. Fahad, S., Chavan, S. B., Chichaghare, A. R., Uthappa, A. R., Kumar, M., Kakade, V., Pradhan, A., Jinger, D., Rawale, G., Yadav, D. K., Kumar, V., Farooq, T. H., Ali, B., Sawant, A. V., Saud, S., Chen, S., Poczai, P. (2022). Agroforestry systems for soil health improvement and maintenance. *Sustainability*, *14*(22), 14877.
10. Udawatta, R. P., Rankoth, L. M., Jose, S. (2021). Agroforestry for biodiversity conservation. In: *Agroforestry and Ecosystem Services* (pp. 245–274).
11. Pantera, A., Mosquera-Losada, M. R., Herzog, F. et al. (2021). Agroforestry and the environment. *Agroforestry Systems*, *95*, 767–774. doi: 10.1007/s10457-021-00640-8
12. Kopittke, P. M., Menzies, N. W., Wang, P., McKenna, B. A., Lombi, E. (2019). Soil and the intensification of agriculture for global food security. *Environment International*, *132*, 10.
13. Singh, S., Bhoi, T. K., Vyas, V. (2023). Interceding microbial biofertilizers in agroforestry system for enhancing productivity. In: *Plant Growth Promoting Microorganisms of Arid Region* (pp. 161–183). Springer Nature Singapore: Singapore.
14. Thomas, L., Singh, I. (2019). Microbial biofertilizers: Types and applications. In: *Biofertilizers for Sustainable Agriculture and Environment* (pp. 1–19).
15. Chakraborty, T., Akhtar, N. (2021). Biofertilizers: Prospects and challenges for future. In: *Biofertilizers: Study and Impact* (pp. 575–590).
16. Pankaj, U. (2020). Bio-fertilizers for management of soil, crop, and human health. In: *Microbes in Agriculture and Environmental Development* (pp. 71–86). CRC Press.

17. El-Deen, S. R. O., El-Azeem, A., Samy, A., Abd Elwahab, A. F., Mabrouk, S. S. (2020). Effects of phosphate solubilizing microorganisms on wheat yield and phosphatase activity. *Egyptian Journal of Medical Microbiology, 55*, 71–86. doi: 10.21608/ejm.2020.20675.113718.
18. Sanyal, D., Osorno, J. M., Chatterjee, A. (2020). Influence of rhizobium inoculation on dry bean yield and symbiotic nitrogen fixation potential. *Journal of Plant Nutrition, 43*, 798–810. doi: 10.1080/01904167.2020.1711946
19. Gedamu, S. A., Tsegaye, E. A., Beyene, T. F. (2021). Effect of rhizobial inoculants on yield and yield components of faba bean [Vicia fabae L.] on vertisol of Wereillu District, South Wollo. Ethiopa, *CABI Agriculture and Bioscience, 2*, 1–10. doi: 10.1186/ s43170-021-00025-y
20. Ketema, P., Tefera, T. (2022). Effectiveness of rhizobium strains on faba bean [Vicia fabae L.] at Gumer district, highland area of Southern Ethiopia. *Ukrainian Journal of Ecology, 12*, 13–18.
21. Adeleke, R. A., Raimi, A. R., Roopnarain, A., Mokubedi, S. M. (2019). Status and prospects of bacterial inoculants for sustainable Management of Agroecosystems. In: Giri, B., Prasad, R., Wu, Q. S., Varma, A. (Eds.), *Biofertilizers for Sustainable Agriculture and Environment* (pp. 137–172). Springer International Publishing: Cham.
22. Ahmad, M., Adil, Z., Hussain, A., Mumtaz, M. Z., Nafees, M., Ahmad, I., et al. (2019). Potential of phosphate solubilizing bacillus strains for improving growth and nutrient uptake in mungbean and maize crops. *Pakistan Journal of Agricultural Sciences, 56*, 283–289. doi: 10.21162/PAKJAS/19.7455
23. Abdelmoteleb, A., Gonzalez-Mendoza, D. (2020). Isolation and identification of phosphate solubilizing bacillus spp. from Tamarix ramosissima rhizosphere and their effect on growth of Phaseolus vulgaris under salinity stress. *Geomicrobiology Journal, 37*, 901–908. doi: 10.1080/01490451.2020.1795321
24. Soumare, A., Boubekri, K., Lyamlouli, K., Hafidi, M., Ouhdouch, Y., Kouisni, L. (2020). From isolation of phosphate solubilizing microbes to their formulation and use as biofertilizers: Status and needs. *Frontiers in Bioengineering and Biotechnology, 7*, 425. doi: 10.3389/fbioe.2019.00425
25. Vassilev, N., Vassileva, M., Martos, V., Garcia del Moral, L. F., Kowalska, J., Tylkowski, B., et al. (2020). Formulation of microbial inoculants by encapsulation in natural polysaccharides: Focus on beneficial properties of carrier additives and derivatives. *Frontiers in Plant Science, 11*, 270.
26. Kalayu, G. (2019). Phosphate solubilizing microorganisms: Promising approach as biofertilizers. *International Journal of Agronomy, 2019*, 1–7.
27. Fatima, F., Ahmad, M. M., Verma, S. R., Pathak, N. (2022). Relevance of phosphate solubilizing microbes in sustainable crop production: A review. *International Journal of Environmental Science and Technology, 19*(9), 9283–9296.
28. Rizvi, A., Ahmed, B., Khan, M. S., Umar, S., Lee, J. (2021a). Sorghumphosphate solubilizers interactions: Crop nutrition, biotic stress alleviation, and yield optimization. *Frontiers in Plant Science, 12*, 746780. doi: 10.3389/fpls.2021.746780
29. Thakur, P., Singh I (2018) Biocontrol of soilborne root pathogens: An overview. In: *Root Biology, Soil Biology* (pp 181–220). Springer. doi: 10.1007/978-3-319-75910-4_7 5078

30. Singh, I., Giri B (2017) Arbuscular mycorrhiza mediated control of plant pathogens. In: *Mycorrhiza—Nutrient Uptake, Biocontrol, Ecorestoration* (pp. 131–160). Springer: Cham.
31. Kalyanasundaram, G. T., Syed, N., Subburamu, K. (2021). Recent developments in plant growth-promoting rhizobacteria [PGPR] for sustainable agriculture. *Recent Developments in Applied Microbiology and Biochemistry, 2*, 181–192.
32. Reyes-Castillo, A., Gerding, M., Oyarzúa, P., Zagal, E., Gerding, J., Fischer, S. (2019). Plant growth-promoting rhizobacteria able to improve NPK availability: Selection, identification and effects on tomato growth. *Chilean Journal of Agricultural Research, 79*, 473–485. doi: 10.4067/S0718-58392019000300473
33. Raji, M., Thangavelu, M. (2021). Isolation and screening of potassium solubilizing bacteria from saxicolous habitat and their impact on tomato growth in different soil types. *Archives of Microbiology, 203*, 3147–3161. doi: 10.1007/ s00203-021-02284-9.
34. Laxita, L., Shruti, S. (2020). Isolation and characterization of potassium solubilizing microorganisms from South Gujarat region and their effects on wheat plant. *Mukta Shabad, 9*, 7483–7496.
35. Akintokun, A. K., Ezaka, E., Akintokun, P. O., Shittu, O. B., Taiwo, L. B. (2019). Isolation, screening and response of maize to plant growth promoting. *Scientia Agriculturae Bohemica, 50*, 181–190. doi: 10.2478/sab-2019-0025
36. Imran, M., Shahzad, S. M., Arif, M. S., Yasmeen, T., Ali, B., Tanveer, A. (2020). Inoculation of potassium solubilizing bacteria with different potassium fertilization sources mediates maize growth and productivity. *Pakistan Journal of Agricultural Sciences, 57*, 1045–1055
37. del Carmen Orozco-Mosqueda, M., del Carmen Rocha-Granados, M., Glick, B. R., Santoyo, G. (2018). Microbiome engineering to improve biocontrol and plant growth-promoting mechanisms. *Microbiological Research, 208*, 25–31.
38. Backer, R., Rokem, J. S., Ilangumaran, G., Lamont, J., Praslickova, D., Ricci, E., Subramanian, S., Smith, D. L. (2018). Plant growth-promoting rhizobacteria: Context, mechanisms of action, and roadmap to commercialization of biostimulants for sustainable agriculture. *Frontiers in Plant Science, 9*, 1473.
39. Bajracharya, A. M. (2019). Plant growth promoting rhizobacteria [PGPR]: Biofertliser and Biocontrol agent—Review article. *Journal of Balkumari College, 8*, 42–45.
40. Daniel, A. I., Fadaka, A. O., Gokul, A., Bakare, O. O., Aina, O., Fisher, S., Burt, A. F., Mavumengwana, V., Keyster, M., Klein, A. (2022). Biofertilizer: The future of food security and food safety. *Microorganisms, 10*(6), 1220.
41. Timofeeva, A., Galyamova, M., Sedykh, S. (2022). Prospects for using phosphate-solubilizing microorganisms as natural fertilizers in agriculture. *Plants, 11*, 2119.
42. Santoyo, G., Urtis-Flores, C. A., Loeza-Lara, P. D., Orozco-Mosqueda, M. D. C., Glick, B. R. (2021). Rhizosphere colonization determinants by plant growth-promoting rhizobacteria [PGPR]. *Biology, 10*(6), 475.
43. 42. Aasfar, A., Bargaz, A., Yaakoubi, K., Hilali, A., Bennis, I., Zeroual, Y., Meftah Kadmiri, I. (2021). Nitrogen fixing Azotobacter species as potential soil biological enhancers for crop nutrition and yield stability. *Frontiers in Microbiology, 12*, 628379.
44. EL Sabagh, A., Islam, M. S., Hossain, A., Iqbal, M. A., Mubeen, M., Waleed, M., Reginato, M., Battaglia, M., Ahmed, S., Rehman, A., Arif, M., Athar, H.-U.-R.,

Ratnasekera, D., Danish, S., Ali Raza, M., Rajendran, K., Mushtaq, M., Skalicky, M., Brestic, M., Soufan, W., Fahad, S., Pandey, S., Kamran, M., Datta, R., Abdelhamid, M. T. (2022). Phytohormones as growth regulators during abiotic stress tolerance in plants. *Frontiers in Agronomy*, *4*, 765068.

45. Rai, P. K., Singh, M., Anand, K., Saurabh, S., Kaur, T., Kour, D., Yadav, A. N., Kumar, M. (2020). Role and potential applications of plant growth-promoting rhizobacteria for sustainable agriculture. In: *New and Future Developments in Microbial Biotechnology and Bioengineering* (pp. 49–60). Elsevier.
46. Zehra, A., Raytekar, N. A., Meena, M., Swapnil, P. (2021). Efficiency of microbial bio-agents as elicitors in plant defense mechanism under biotic stress: A review. *Current Research in Microbial Sciences*, *2*, 100054.
47. Singh, S., Bhoi, T. K., Majhi, P. K., Vyas, V., Singh, I., Khan I et al. Emergent tools and techniques in diagnosis of soil-borne phytopathogens. In: *Detection. Diagnosis and Management of Soil-borne Phytopathogens*. Springer Nature Singapore: Singapore.
48. Shameer, S., Prasad, T. N. V. K. V. (2018). Plant growth promoting rhizobacteria for sustainable agricultural practices with special reference to biotic and abiotic stresses. *Plant Growth Regulation*, *84*, 603–615.
49. Verma, P. P., Shelake, R. M., Das, S., Sharma, P., Kim, J. Y. (2019). Plant growth-promoting rhizobacteria [PGPR] and fungi [PGPF]: Potential biological control agents of diseases and pests. In: *Microbial Interventions in Agriculture and Environment: Volume 1: Research Trends, Priorities and Prospects* (pp. 281–311).
50. Gowtham, H. G., Singh, S. B., Shilpa, N., Aiyaz, M., Nataraj, K., Udayashankar, A. C., Amruthesh, K. N., Murali, M., Poczai, P., Gafur, A., Almalki, W. H., Sayyed, R. Z. (2022). Insight into recent progress and perspectives in improvement of antioxidant machinery upon PGPR augmentation in plants under drought stress: A review. *Antioxidants*, *11*(9), 1763.
51. Moore, J. A., Abraham, P. E., Michener, J. K., Muchero, W., Cregger, M. A. (2022). Ecosystem consequences of introducing plant growth promoting rhizobacteria to managed systems and potential legacy effects. *New Phytologist*, *234*(6), 1914–1918.
52. Maheshwari, P., Gayathry, G., Sankar, P. M., Sangeetha, P., Anandaraj, P. (2022). Mass Multiplication and Production Cost Analysis of Phosphate Solubilizing Microorganisms. In: *Industrial Microbiology Based Entrepreneurship: Making Money from Microbes* (pp. 287–302). Springer Nature Singapore: Singapore.
53. Aloo, B. N., Mbega, E. R., Makumba, B. A., Tumuhairwe, J. B. (2022). Effects of carrier materials and storage temperatures on the viability and stability of three biofertilizer inoculants obtained from potato [*Solanum tuberosum* L.] rhizosphere. *Agriculture*, *12*(2), 140.
54. Saif, S., Abid, Z., Ashiq, M. F., Altaf, M., Ashraf, R. S. (2021). Biofertilizer formulations. In: *Biofertilizers: Study and Impact* (pp. 211–256).
55. Tripti, Kumar, A., Kumar, V., Anshumali, Bruno, L. B., Rajkumar, M. (2022). Synergism of industrial and agricultural waste as a suitable carrier material for developing potential biofertilizer for sustainable agricultural production of eggplant. *Horticulturae*, *8*(5), 444.
56. Asha, K., Krishna, H., Raju, K. K., Guruprasad, V. (2023). Role of bio-fertilizers towards sustainable agricultural development: A review. *Journal of Advanced Applied Scientific Research*, *5*(5).

57. Behera, B., Paramasivan B. (2022). Research trends and market opportunities of microalgal biorefinery technologies from circular bioeconomy perspectives. *Bioresource Technology, 351*, 127038.
58. Singh S., Bhoi T. K., Vyas V., Sharma K., Singh I. and Nirwan B. (2022). *Variations in Sturdiness Quotient [SQ] of Khejri Seedlings*. The Indian Forester.
59. Ibeh, C. U., Chime, E. U., Nweke, I. A., Nsoanya, L. N. (2019). The effect of two biofertilizers under two crop combination on microbial population and early plant growth in sandy loam soil. *British Journal of Environmental Sciences*, *7*(4), 31–39
60. Singh, P., Singh, J., Ray, S., Rajput, R. S., Vaishnav, A., Singh, R. K., Singh, H. B. (2020). Seed biopriming with antagonistic microbes and ascorbic acid induce resistance in tomato against Fusarium wilt. *Microbiological Research*, *237*, 126482.
61. Amenaghawon, A. N., Anyalewechi, C. L., Kusuma, H. S. (2021). Fabrication approaches for biofertilizers. In: *Biofertilizers: Study and Impact* (pp. 491–515).
62. De, L. C., De, T., Biswas, S. S., Kalaivanan, N. S. (2021). Organic plant nutrient, protection and production management. In: *Advances in Organic Farming* (pp. 115–131). Woodhead Publishing.
63. Raj, A., Jhariya, M. K., Khan, N., Banerjee, A., Meena, R. S. (2021). Ecological intensification for sustainable development. In: *Ecological Intensification of Natural Resources for Sustainable Agriculture* (pp. 137–170).
64. Qaim, M. (2020). Role of new plant breeding technologies for food security and sustainable agricultural development. *Applied Economic Perspectives and Policy*, *42*(2), 129–150.\
65. Giller, K. E., Hijbeek, R., Andersson, J. A., Sumberg, J. (2021). Regenerative agriculture: An agronomic perspective. *Outlook on Agriculture*, *50*(1), 13–25.
66. Nishad, R., Ahmed, T., Rahman, V. J., Kareem, A. (2020). Modulation of plant defense system in response to microbial interactions. *Frontiers in Microbiology*, *11*, 1298.
67. Baweja, P., Kumar, S., Kumar, G. (2020). Fertilizers and pesticides: Their impact on soil health and environment. *Soil Health*, 265–285.
68. Tripathi, S., Srivastava, P., Devi, R. S., Bhadouria, R. (2020). Influence of synthetic fertilizers and pesticides on soil health and soil microbiology. In: *Agrochemicals Detection, Treatment and Remediation* (pp. 25–54). Butterworth-Heinemann.
69. Fasusi, O. A., Cruz, C., Babalola, O. O. (2021). Agricultural sustainability: Microbial biofertilizers in rhizosphere management. *Agriculture*, *11*(2), 163.
70. Khatoon, Z., Huang, S., Rafique, M., Fakhar, A., Kamran, M. A., Santoyo, G. (2020). Unlocking the potential of plant growth-promoting rhizobacteria on soil health and the sustainability of agricultural systems. *Journal of Environmental Management*, *273*, 111118.
71. Singh, S., Bhoi, T. K., Khan, I., Vyas, V., Athulya, R., Rathi, A., Samal, I. (2023). Climate change drivers and soil microbe-plant interactions. In: *Climate Change and Microbiome Dynamics: Carbon Cycle Feedbacks* (pp. 157–176). Springer International Publishing: Cham.
72. Kumar, S., Sindhu, S. S., Kumar, R. (2022). Biofertilizers: An ecofriendly technology for nutrient recycling and environmental sustainability. *Current Research in Microbial Sciences*, *3*, 100094.

73. Mason, A. R. G., Salomon, M. J., Lowe, A. J., Cavagnaro, T. R. (2023). Microbial solutions to soil carbon sequestration. *Journal of Cleaner Production, 417*, 137993.
74. Holka, M., Kowalska, J., Jakubowska, M. (2022). Reducing carbon footprint of agriculture—Can organic farming help to mitigate climate change? *Agriculture, 12*(9), 1383.
75. Ozlu, E., Arriaga, F. J., Bilen, S., Gozukara, G., Babur, E. (2022). Carbon footprint management by agricultural practices. *Biology, 11*(10), 1453.
76. Singh, T. B., Ali, A., Prasad, M., Yadav, A., Shrivastav, P., Goyal, D., Dantu, P. K. (2020). Role of organic fertilizers in improving soil fertility. In: *Contaminants in Agriculture: Sources, Impacts and Management* (pp. 61–77).
77. Nasir, J., Ashfaq, M., Baig, I., Khair, S. M., Mangan, T., Allan, C., et al. (2021). *Representative Agricultural Pathways and Socioeconomic Benefits of Groundwater Management Interventions in Punjab, Sindh and Balochistan Provinces, Pakistan.* Institute for Land, Water and Society: Albury, NSW.
78. Kumawat, K. C., Nagpal, S., Sharma, P. (2021). Present scenario of bio-fertilizer production and marketing around the globe. In: *Biofertilizers* (pp. 389–413). Woodhead Publishing.
79. Raimi, A., Roopnarain, A., Adeleke, R. (2021). Biofertilizer production in Africa: Current status, factors impeding adoption and strategies for success. *Scientific African, 11*, e00694.
80. Khan, A., Panthari, D., Sharma, R. S., Punetha, A., Singh, A. V., Upadhayay, V. K. (2023). Biofertilizers: A microbial-assisted strategy to improve plant growth and soil health. In: *Advanced Microbial Techniques in Agriculture, Environment, and Health Management* (pp. 97–118). Academic Press.
81. Ijaz, M., Ali, Q., Ashraf, S., Kamran, M., Rehman, A. (2019). Development of future bioformulations for sustainable agriculture. In: *Microbiome in Plant Health and Disease: Challenges and Opportunities* (pp. 421–446).
82. Saha, S., Paul, D., Poudel, T. R., Basunia, N. M., Hasan, T., Hasan, M., Li, B., Reza, R., Haque, A. R.,Hanif, A.,Sarker, M.,Roberts, N. J., Khoso, M. A., Wu, H., Shen, H. L. (2023). Biofertilizer science and practice for agriculture and forestry: A review. *Journal of Applied Biology and Biotechnology, 11*(Issue), 31–44.
83. Fahad, S., Sonmez, O., Saud, S., Wang, D., Wu, C., Adnan, M., Turan, V. (Eds.), (2021). *Sustainable Soil and Land Management and Climate Change*. CRC Press.
84. Afiukwa, C. A., Igwe, D. O., Ubi, B. E. (2021). Biotechnology role in climate change adaptation and mitigation for sustainable crop production. In: *Handbook of Climate Change Management: Research, Leadership, Transformation* (pp. 1–27). Springer International Publishing: Cham.
85. Rai, P. K., Rai, A., Sharma, N. K., Singh, T., Kumar, Y. (2023). Limitations of biofertilizers and their revitalization through nanotechnology. *Journal of Cleaner Production, 418*, 138194.
86. Gautam, K., Sirohi, C., Singh, N. R., Thakur, Y., Jatav, S. S., Rana, K., Chitara, M., Meena, R. P., Singh, A. K., Parihar, M. (2021). Microbial biofertilizer: Types, applications, and current challenges for sustainable agricultural production. In: *Biofertilizers* (pp. 3–19). Woodhead Publishing.

87. Stella, M., Theeba, M., Illani, Z. I. (2019). Organic fertilizer amended with immobilized bacterial cells for extended shelf-life. *Biocatalysis and Agricultural Biotechnology*, *20*, 101248.
88. Saritha, M., Tollamadugu, N. P. (2019). The status of research and application of biofertilizers and biopesticides: Global scenario. In: *Recent Developments in Applied Microbiology and Biochemistry* (pp. 195–207). Elsevier: the Netherlands.
89. FAQs. (2023). Biofertilizers global market report 2023-by product [Nitrogen Fixing, Phosphate Solubilizing, Other Products], By crop type [Cereals and Grains, Oilseeds and Pulses, Fruits and Vegetables, Other Crop Types], By form [Liquid, Carrier based], By Application [Seed Treatment, Soil Treatment]-Market Size, Trends, And Global Forecast 2023–2032.
90. Tabassum, B., Khan, A., Tariq, M., Ramzan, M., Khan, M. S. I., Shahid, N., et al. (2017). Bottlenecks in commercialisation and future prospects of PGPR. *Applied Soil Ecology*, *121*, 102–117. doi: 10.1016/j.apsoil.2017.09.030
91. Mawar, R., Manjunatha, B. L., Kumar, S. (2021). Commercialization, diffusion and adoption of bioformulations for sustainable disease management in Indian arid agriculture: Prospects and challenges. *Circular Economy and Sustainability*, *1*, 1367–1385.
92. Basu, A., Prasad, P., Das, S. N., Kalam, S., Sayyed, R. Z., Reddy, M. S., et al. (2021). Plant growth promoting rhizobacteria [PGPR] as green bioinoculants: Recent developments, constraints, and prospects. *Sustainability*, *13*, 140. doi: 10.3390/su13031140
93. Balasubramani, N., Vincent, A. (Eds.), (2021). Certificate Course on Integrated Nutrient Management [CCINM] for Fertiliser Dealers [E-book]. Hyderabad: National Institute of Agricultural Extension Management [MANAGE].
94. Maharana, R., Dobriyal, M. J., Behera, L., Sukhadiya M. (2018). Enhancement of seedling vigour through biofertilizers application in gamhar [Gmelinaarborea roxb.]. *International Journal of Chemical Studies*,*6*, 54–60.
95. Mącik, M., Gryta, A., Frąc, M. (2020). Biofertilizers in agriculture: An overview on concepts, strategies and effects on soil microorganisms. *Advances in Agronomy*, *162*, 31–87.
96. Koovalamkadu Velayudhan, P., Singh, A., Korekallu Srinivasa, A. (2021). Exploring the global research trends in biofertilizers: A bibliometric approach. *3 Biotech*, *11*(6), 304.
97. Joshi, S. K., Gauraha, A. K. (2022). Global biofertilizer market: Emerging trends and opportunities. *Trends of Applied Microbiology for Sustainable Economy*, 689–697.
98. FiBL. (2016). Global biofertilizers market; 2022. Retractive from https:// www.mordorintelligence.com/industry-reports/global-biofertilizersmarket-industry?
99. Rashid, A., Mir, M. R., Hakeem, K. R. (2016). Biofertilizer use for sustainable agricultural production. In: *Plant, Soil and Microbes* (pp. 163–180). Springer: Berlin.
100. Asoegwu, C. R., Awuchi, C. G., Nelson, K. C., Orji, C. G., Ulunma, N. O., Egbufor, U. C., et al. (2020). A review on the role of biofertilizers in reducing soil pollution and increasing soil nutrients. *Himalayan Journal of Agriculture*, *1*, 34–38.
101. Liu, A., Wang, W., Zheng, X., Chen, X., Fu, W., Wang, G., et al. (2022). Improvement of the Cd and Zn phytoremediation efficiency of rice [Oryza sativa] through the inoculation of a metal-resistant PGPR strain. *Chemosphere*, *302*, 134900

102. Abbas, R., Rasul, S., Aslam, K., Baber, M., Shahid, M., Mubeen, F., et al. (2019). Halotolerant PGPR: A hope for cultivation of saline soils. *Journal of King Saud University, 31*, 1195–1201.
103. Hossain, A., Ali, M. E., Maitra, S., Bhadra, P., Rahman, M. M., Ali, S., et al. (2022). The role of soil microorganisms in plant adaptation to abiotic stresses: Current scenario and future perspectives. In: *Plant Perspectives to Global Climate Changes* (pp. 233–278). Elsevier: the Netherlands.
104. Thirugnanasambandan T. (2018). *Advances and Trends in NanoBiofertilizers*. Available from: SSRN 3306998
105. Bhatnagar, S., Mahanta, D. K., Vyas, V., Samal, I., Komal, J., Bhoi, T. K. (2023). Storage pest management with nanopesticides incorporating silicon nanoparticles: A novel approach for sustainable crop preservation and food security. *Silicon, 16*, 471–483.
106. Mitter, E. K., Tosi, M., Obregón, D., Dunfield, K. E., Germida, J. J. (2021). Rethinking crop nutrition in times of modern microbiology. Innovative biofertilizer technologies. *Frontiers in Sustainable Food Systems, 5*, 606815.
107. Kong, Z., Hart, M., Liu H. (2018). Paving the way from the lab to the field: Using synthetic microbial consortia to produce high-quality crops. *Frontiers in Plant Science, 9*, 1467.
108. Victoria, K. S., Gonzales, K. K., Gestiada, E. C., Aggangan, N. S. (2023 Feb 1). Influence of biochar and biofertilizer mixture on Coffea spp. planted in an agroforestry ecosystem. In: *IOP Conference Series: Earth and Environmental Science*, (Vol. 1145, No. 1, p. 012008). IOP Publishing.
109. Yadav, M., Umrao, R., Rout S. (2019). Effects of organic, inorganic and bio-fertilizers on the growth of maize under Subabul (*Leucaena leucocephala*) based agroforestry system. *Journal of Plant Development Sciences, 11*(4), 189–199.
110. Heydarzadeh, S., Jalilian, J., Pirzad, A., Jamei, R., Petrussa E. (2022). Fodder value and physiological aspects of rainfed smooth vetch affected by biofertilizers and supplementary irrigation in an agri-silviculture system. *Agroforestry Systems, 96*, 221–232.
111. Kannur, S., Patil, S. J., Inamati, S. S. (2020). Studies on nutrient management of bird's eye chilli under rubber based agroforestry system. *Journal of Tree Sciences, 39*(1), 35–41.
112. Silva-Galicia, A., Larsen, J., Álvarez-Espino, R., Ceccon E. (2023). Agroecological and agroforestry strategies to improve organic hibiscus productivity in an Indigenous non-governmental organization from Mexico. *Renewable Agriculture and Food Systems, 38*, e3.
113. Muhadiono, I., Supiandi, S., Mansjoer, I. K., Garcia, M. U. Agroforestry technology: Rhizobium and endomycorrhizal infection in the roots of Albizia procera (Roxb.) Benth. as biofertilizer for the future. *Agroforestry Technology: Rhizobium and Endomycorrhizal Infection in the Roots of Albizia procera (Roxb.) Benth. as Biofertilizer for the Future, 1990*(39), 107–114.
114. Tiwari, M., Toppo, P., Sahu, M., Toppo, S., Singh L. Effect of nutrients on growth and yield of ginger under bamboo-based agroforestry system. *The Pharma Innovation Journal, SP-11*(9), 2924–2926.

115. Singh, B., Pandey, S. B. S., Jayant, A. K. Growth and yield of mustard (Brassica juncea L.) in Nagpur mandarin-based agroforestry system in South-Eastern Rajasthan. *Annals of Agricultural Research, 44*(3), 325–332.
116. Tabacchioni, S., Passato, S., Ambrosino, P., Huang, L., Caldara, M., Cantale, C., Hett, J., Del Fiore, A., Fiore, A., Schlüter, A., Sczyrba, A., Maestri, E., Marmiroli, N., Neuhoff, D., Nesme, J., Sørensen, S. J., Aprea, G., Nobili, C., Presenti, O., Bevivino, A. (2021). Identification of beneficial microbial consortia and bioactive compounds with potential as plant biostimulants for a sustainable agriculture. *Microorganisms, 9*(2), 426.
117. Kour, D., Rana, K. L., Yadav, A. N., Yadav, N., Kumar, M., Kumar, V., Vyas, P., Dhaliwal, H. S., Saxena, A. K. (2020). Microbial biofertilizers: Bioresources and eco-friendly technologies for agricultural and environmental sustainability. *Biocatalysis and Agricultural Biotechnology, 23*, 101487.
118. Toju, H., Peay, K. G., Yamamichi, M., Narisawa, K., Hiruma, K., Naito, K., et al. (2018). Core microbiomes for sustainable agroecosystems. *Nature Plants,4*, 247–57.
119. Vyas, V., Bhoi, T. K., Samal, I., Singh, S., Mahanta, D. K. (2024). Revitalizing degraded soils with agroforestry interventions: Opportunities, challenges, and future direction. In: *Agroforestry to Combat Global Challenges: Current Prospects and Future Challenges* (pp. 529–549).
120. Giri, V., Bhoi, T. K., Samal, I., Komal, J., Majhi, P. K. (2024). Exploring the agroforestry systems for ecosystem services: A synthesis of current knowledge and future research directions. In: *Agroforestry to Combat Global Challenges: Current Prospects and Future Challenges* (pp. 503–528).
121. Bhoi, T. K., Samal, I., Majhi, P. K., Komal, J., Mahanta, D. K., Pradhan, A. K., Saini, V., Nikhil Raj, M., Ahmad, M. A., Behera, P. P., Ashwini, M. (2022). Insight into aphid mediated Potato Virus Y transmission: A molecular to bioinformatics prospective. *Frontiers in Microbiology, 13*, 1001454.
122. Mahanta, D. K., Jangra, S., Ghosh, A., Sharma, P. K., Iquebal, M. A., Jaiswal, S., Baranwal, V. K., Kalia, V. Chander, S. (2022). Groundnut bud necrosis virus modulates the expression of innate immune, endocytosis, and cuticle development-associated genes to circulate and propagate in its vector, Thrips palmi. *Frontiers in Microbiology, 13*, 773238.
123. Mahanta, D. K., Komal, J., Samal, I., Bhoi, T. K., Dubey, V. K., Pradhan, K., Nekkanti, A., Gouda, M. N. R., Saini, V., Negi, N., Bhateja, S., Jat, H. K., Jeengar, D. (2023a). Nutritional aspects and dietary benefits of "Silkworms": Current scenario and future outlook. *Frontiers in Nutrition, 10*, 1121508.
124. Samal, I., Bhoi, T. K., Vyas, V., Majhi, P. K., Mahanta, D. K., Komal, J., Singh, S., Kumar, P. V. D., Acharya, L. K. (2023a). Resistance to fungicides in entomopathogenic fungi: Underlying mechanisms, consequences, and opportunities for progress. *Tropical Plant Pathology, 49*, 5–17.
125. Mahanta, D. K., Bhoi, T. K., Komal, J., Samal, I., Nikhil, R. M., Paschapur, A. U., Singh, G., Kumar, P. V. D., Desai, H. R., Ahmad, M. A., Singh, P. P., Majhi, P. K., Mukherjee, U., Singh, P., Saini, V., Shahanaz, Srinivasa, N., Yele, Y. (2023b). Insect-pathogen crosstalk and the cellular-molecular mechanisms of insect immunity: Uncovering the underlying signaling pathways and immune regulatory function of non-coding RNAs. *Frontiers in Immunology, 14*, 1169152.

126. Samal, I., Bhoi, T. K., Majhi, P. K., Murmu, S., Pradhan, A. K., Kumar, D., Saini, V., Paschapur, A. U., Raj, M. N., Ankur, Manik, S., Behera, P. P., Mahanta, D. K., Komal, J., Alam, P., Balawi, T. A. (2023b). Combatting insects mediated biotic stress through plant associated endophytic entomopathogenic fungi in horticultural crops. *Frontiers in Plant Science*, *13*, 1098673.
127. Komal, J., Desai, H. R., Samal, I., Mastinu, A., Patel, R. D., Kumar, P. D., Majhi, P. K., Mahanta, D. K., Bhoi, T. K. (2023). Unveiling the genetic symphony: Harnessing CRISPR-Cas genome editing for effective insect pest management. *Plants*, *12*(23), 3961.
128. Samal, I., Bhoi, T. K., Mahanta, D. K., Komal, J. (2023c). Establishing the role of silicon (Si) in plant resistance to insects: A bibliometric approach. *Silicon*, *16*, 2119–2128.
129. Majhi, P. K., Samal, I., Bhoi, T. K., Pattnaik, P., Pradhan, C., Gupta, A., Mahanta, D. K., Senapati, S. K. (2023). Climate-smart agriculture: An integrated approach to address climate change and food security. In: *Climate Change and Insect Biodiversity* (pp. 221–245). CRC Press.
130. Mahanta, D. K., Samal, I., Komal, J., Bhoi, T. K., Majhi, P. K., Ahmad, M. A. (2023c). Understanding anthropogenic climate change, its consequences on insect pests, and techniques in forecasting and monitoring pest dynamics: A contemporary scenario. In: *Climate Change and Insect Biodiversity* (pp. 44–67). CRC Press.
131. Bhoi, T. K., Samal, I., Mahanta, D. K., Komal, J., Jinger, D., Sahoo, M. R., Achary, G. C., Nayak, P., Sunani, S. K., Saini, V., Raghuraman, M., Singh, S. (2023). Understanding how silicon fertilization impacts chemical ecology and multitrophic interactions among plants, insects and beneficial arthropods. *Silicon*, *15*(6), 2529–2549.
132. Bhoi, T. K., Samal, I., Mahanta, D. K., Komal, J., Majhi, P. K., Ankur. (2024). Exploring Melatonin's potential as an alternative strategy for protecting plants from biotic stresses. In: *Melatonin in Plants: A Pleiotropic Molecule for Abiotic Stresses and Pathogen Infection* (pp. 223–242). Springer Nature Singapore: Singapore.
133. Samal, I., Bhoi, T. K., Pradhan, A. K., Mahanta, D. K. (2023d). Plantix app: A success story of artificial intelligence in plant protection. *Van Sangyan*, *24*.
134. Mahanta, D. K., Bhoi, T. K., Komal, J., Samal, I., Mastinu, A. (2024). Spatial, spectral and temporal insights: Harnessing high-resolution satellite remote sensing and artificial intelligence for early monitoring of wood boring pests in forests. *Plant Stress*, *11*, 100381.
135. Samal, I., Mahanta, D. K., Bhoi, T. K., Komal, J., Jatav, H. S., Jatav, S. S., Sathyanarayana, E. (2024). *Approaches of Biochar in Ecosystem Management: Current Scenario and Future Perspectives*. IntechOpen
136. Samal, I., Bhoi, T. K., Mahanta, D. K., Komal, J., Singh, S. (2024). Biorational pest management: Potentials, unintended consequences, and future concerns. In: *Biorationals and Biopesticides: Pest Management* (pp. 47–75).

CHAPTER 3

Conservatory Measures for Natural Enemies in Pest Control in Agroforestry: Sustainable Multitrophic Interactions

SATYABRATA SARANGI,[1*] SUMAN SAMILITA DASH,[1] and ARADHANA PANDA[2]

[1]*Department of Entomology, Odisha University of Agriculture and Technology, Bhubaneswar, Odisha, India*

[2]*Department of Entomology, School of Agriculture, DRIEMS University, Tangi, Cuttack, Odisha, India*

*Corresponding author

ABSTRACT

Natural enemies such as ants, beetles, spiders, and wasps play a crucial role in regulating insect pest populations within agroforestry systems. These systems, which integrate trees with agricultural crops, not only enhance biodiversity and pollination but also promote natural pest control. However, the outcomes of such interactions are often unpredictable, depending on ecological complexity and landscape dynamics. Parasitoids and predators require alternative hosts or prey to complete their life cycles, particularly when their primary resources are scarce. Therefore, incorporating native perennial flowering plants that provide nectar and pollen becomes central to habitat management. Such plantings ensure a continuous supply of food resources, improving the survival and efficiency of beneficial insects. Factors like habitat structure, patchiness,

Multitrophic Interactions for Sustainable Agroforestry: Harnessing the Power of Beneficial Insects and Ecological Dynamics. Hanuman Singh Jatav, Vishnu D. Rajput, Ipsita Samal, and Tanmaya Kumar Bhoi (Eds).

DOI: 10.1201/9781779641250-3

quality, and the dispersal ability of organisms strongly influence the success of biological control. Conservation strategies, such as silvo-arable agroforestry systems, create a "win–win" scenario by reducing pest pressure while boosting pollinators and natural enemies. Beyond ecological benefits, conservation plantings also offer aesthetic value and provide educational opportunities for communities, particularly in urbanizing landscapes. For effective implementation, careful planning regarding the size, composition, and distribution of conservation areas is essential. Such agroecological strategies ultimately optimize yields and enhance sustainability through natural pest regulation.

3.1 INTRODUCTION

Globally, agriculture must respond to the expanding demands brought about by market competitiveness and the mounting obligation to uphold environmental stewardship. Farmers used to depend mainly on chemical pesticides, which led to several ecological imbalances. It is necessary to support innovative methods to ensure both sustainable agriculture and the environment in the long run. One such strategy is the integration of forestry and agriculture to produce crops and an effective land management system. Agroforestry combines agricultural land with plants from woody forests [56, 115].

In agroforestry systems, the tree canopy structure affects the microclimate, which provides an environment that is advantageous to natural enemies. The connection promotes different ecosystem services (ESs), including biomass production, biodiversity enhancement, soil and water conservation, and pollination services [107]. Rather than emphasizing the preservation of the ecosystem's natural services and processes to increase agricultural production, the primary goal of sustainable intensification is to cut off foreign inputs [112]. According to reports, pollination services have increased, coupled with a significant increase and decrease in the biocontrol agent and pest population, respectively [116].

Research on how agroforestry affects functional biodiversity is desperately needed [115]. Resources are arranged in this agroecological phenomenon such that all the elements coexist and the dynamics of the food cycle are preserved. In these systems, a range of food sources, hosts, prey, and appropriate habitats are available for the natural enemies [61]. Additionally, pest damage is limited to a point where plants can express their defense mechanisms and thereby recover from the losses they have caused. Using natural resources carefully for ecosystem management requires an understanding of the numerous trophic interactions and techniques as well as the integration of these systems.

Harmony in nature can be achieved by manipulating the tritrophic relationship, which involves the herbivore, host, and their natural enemies. Natural enemies always have access to food and shelter from forest trees [77]. A component of biological control in conservation is habitat management, whereby agricultural manipulation has increased biodiversity by offering a mosaic of resources and niches for various species. Predators and parasites can flourish in this diverse habitat, which in turn supports a wide variety of natural enemies. This balance allows the ecosystem to function better by giving the predators and parasites access to hatching grounds, refuge cover, migratory routes, and overwintering locations. For example, *Eucalyptus torelliana* [101] trees planted in windbreaks employ their hairy leaves to gather grass pollen, giving predators another place to hide [28].

Certain microhabitats are preferred by different species of natural enemies for feeding, nesting, or procreation. More floristic diversity in agroforestry produces additional food, which supports the persistence and activity of the beneficial faunal population [53]. Among them are pollinators, which support both the wild population and higher crop yields [117]. Since 90% of blooming plants are entomophilous, pollination is a valuable function [57, 58]. Agroforestry offers the vertical structure that birds and spiders require. The nectar and pollen found in forest vegetation are highly valuable for parasitic ichneumonids and syrphids. Since syrphids are aphids' predators [63], they are widely distributed in regions with an abundance of great floral diversity [88].

Ladybird beetles (Coccinellidae) have been employed for aphids feeding on goldenrods as backup prey when their preferred prey is not there [4]. Because there is a suitable minute prey on juniper (*Juniperus* spp.), the predator *Amara* spp. (Coleoptera: Carabidae) is known to be abundant in juniper grasslands, in New Mexico compared to any other locations [109]. Supporting the populations of endemic and native natural enemies and promoting their conservation would significantly reduce the pest population, maintaining it below economic threshold level. Agroforest habitats have been observed to include an enormous diversity of bio-control agents not only are these natural enemies essential for pest management, but they also lessen the need for chemical pesticides, which lowers pollution levels in the environment, minimizes harm to nontarget organisms, and supports sustainable farming methods. This is consistent with the integrated pest management (IPM) tenets, which highlight the application of biocontrol agents in the management of pests. Natural enemies and other species need this connectedness to move about since it aids in their migratory, genetic exchange, and long-term survival.

An environmentally friendly method of controlling pests in agroforestry systems is the preservation of natural enemies. Agroforestry promotes the existence and activity of natural enemies by fostering a diverse and habitat-rich environment, which helps to maintain resilient and ecologically balanced agricultural landscapes. To allow the crops to act as a breeding habitat for beneficial insects, some strips of weeds should be left inside or surrounding the forest trees [104].

3.2 DIFFERENT TYPES OF NATURAL ENEMIES RECORDED UNDER THE AGROFORESTRY SYSTEM

To manage insect pests and maintain the general health and sustainability of the agroecosystem, agroforestry systems provide a vast range of natural enemies that cooperate. Predators, parasitoids, and pathogens are the three main groupings of these natural enemies that can be roughly divided depending on their life cycles and eating preferences.

3.2.1 PREDATORS (FIGURE 3.1)

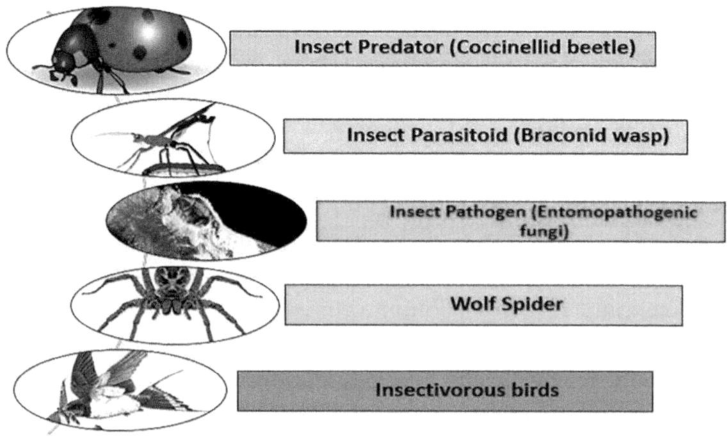

FIGURE 3.1 Different natural enemies observed in agroforestry ecosystem for pest control.

Ladybird (Coccinellidae): Aphids and other soft-bodied insects are widely consumed by ladybugs, sometimes referred to as lady beetles. They are prevalent in the systems of agroforestry and are essential for managing insect populations. According to some researchers, coccinellid species [18, 49, 86] that prey on insect pests of field and orchard crops overwinter at field edges in grass and tree wastes, providing habitat for the survival of birds and small mammals feeding on them.

Ground Beetles (Carabidae): Numerous kinds of ground beetles are helpful predators that consume a range of insect pests, such as caterpillars and beetle grubs. They are frequently discovered in agroforestry systems' soil and leaf litter [45].

Rove Beetles (Staphylinidae): Polyphagous insect predator that consumes both larvae and a variety of soft-bodied insect species.

Lacewings (Chrysopidae): Aphids, mites, and other tiny insects provide the food source for lacewing larvae, which are predatory insects [124]. Adult lacewings are useful pollinators in agroforestry systems because they also eat nectar and pollen.

Hoverflies (Syrphidae): Hoverflies have two distinct functions. Aphids, thrips, and other soft-bodied pests are preyed upon by the larvae, which also help with pollination when they are adults.

Hoverflies are drawn to agroforestry systems by their varied vegetation, which also creates ideal conditions for their reproduction [75]. Including a border of *Phacelia tanacetifolia* has promoted the syrphid population in cabbage *Brassica oleracea* L., leading to a reduction in aphid populations [124]. It is recommended that growers mow alternate grass strips between tree rows to maximize advantage because crop plants can also serve as sources of nectar and pollen. The impact of adult syphids on wheat aphids when they are cultivated next to one another was investigated [21]. After dissecting the adult syrphids, it was discovered that while their larvae feed on the aphid *Rhopalosiphum padi* (L.), their densities are decreased because they feed on the pollens of the flowering canola (*Brassica napus* L.).

Predatory Ants: Ants (Hymenoptera: Formicidae) proved to be a predator of several insects. In Africa, intercropping of cocoa with coconut palm trees and olive trees reported an increase in the abundance of *Oecophylla longinoda*, which was found to be a predator of a hemipteran pest of cocoa and a coreid pest of coconuts [121].

Predatory Bugs (Hemiptera): Predatory insects such as damsel bugs and assassin bugs consume a broad range of pests such as aphids, mites, and caterpillars. They are advantageous natural enemies in environments with agroforestry.

Mites: Predatory mites are tiny bio-control agents that are abundant in trees and crops. Because of the neighboring *Eucalyptus torelliana* F. windbreaks, the predatory mite *Amblyseius victoriensis* re-established itself in citrus plantations following a chemical disruption [101]. In a scenario comparable to this, numbers of the predatory mite *Amblyseius tutsi*, Pritchard & Baker were found to be high nine months after they were released at a density of one per leaf in certain citrus-growing regions of South Africa. Thus, it was determined that this tree is advantageous for controlling thrips and phytophagous mites [40]. According to research, growers of orange *Citrus aurantium* L. that maintain a ground cover of Rhodes grass (*Chloris gayana* Kunth) encourage the growth of *Amblyseius victoriensis* (Womersley), a predatory mite that feeds on the air-blown pollen of the Rhodes grass, which in turn provides control of a phytophagous eriophyid, *Tegolophus australis* Keifer [97].

3.2.2 PARASITOIDS

Braconid Wasps (Braconidae): Braconid wasps are parasitoids that deposit their eggs within or on top of other insects' larvae. Once the host is consumed, the wasp larvae eventually kill it. They work well to get rid of troublesome caterpillars [109].

***Trichogramma* Wasps (Trichogrammatidae)**: These microscopic wasps parasitize the eggs of several pests, including butterflies and moths. By burying their eggs inside of them, which the larvae feed on to stop them from hatching and finishing their life cycle, they are useful for managing the early phases (egg) of pest development, for example, wasp reporting controlling spruce budworm eggs [80].

Ichneumonid wasps: Ichneumonid wasps are parasitoids that prey on a variety of insect pests, particularly caterpillars, just like braconid wasps. They play a vital role in controlling the number of pests, lepidopterans [87].

3.2.3 PATHOGENS

Entomopathogenic fungi: Natural insect pathogens include fungi like *Beauveria bassiana* and *Metarhizium anisopliae*. They are compatible with agroforestry systems that sustain a variety of microhabitats, and they infect and kill pests, especially those in the soil. Shade trees in coffee favor the entomopathogenic fungus *Beauveria bassiana* (Bals.-Criv.) Vuill., which controls the coffee berry borer [77].

Bacillus thuringiensis (Bt.): Bacteria known as Bt generate proteins that are poisonous to some insect pests. In agroforestry and other agricultural systems, it is applied as a biological insecticide to selectively manage pests such as beetles, caterpillars, and others.

3.2.4 SPIDERS

Insects are naturally preyed upon by a variety of spider species [68]. By creating webs and going on hunts, they are essential in the management of many pests [69]. Spiders can find appropriate homes in agroforestry because of the greenery there. In China, a substantial increase in spider population was observed in rice fields when grown as an intercrop with cypress trees, causing a significant reduction in brown plant hopper, *Nilaparvata lugens* [89]. Also, it is an important predator of the Douglas-fir tussock moth (*Orgyia pscudotsugata* (McDunnough), western spruce budworm (*Choristoneura occidentalis* Freeman), and other defoliators [80] (Figure 3.2).

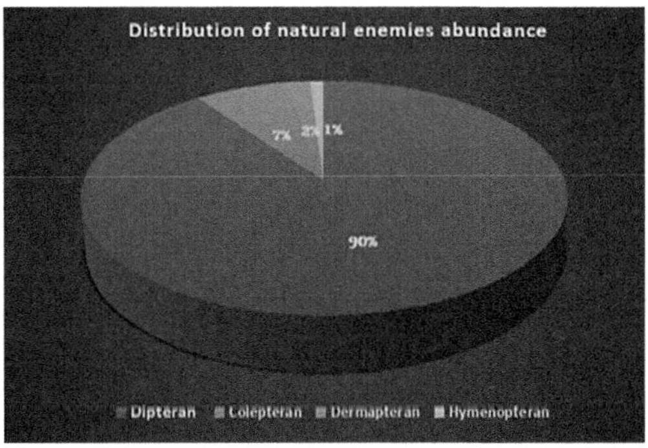

FIGURE 3.2 Abundance of natural enemies collected in areas of Agroforestry systems, Federal district, Brazil [104].

3.2.5 PARASITIC NEMATODES

Some nematodes parasitize the larvae of insects. Once within the host's body, they discharge bacteria that cause the insect to die. These nematodes can be incorporated into agroforestry systems and are efficient against pests that live in the soil.

3.2.6 BIRDS

Because they consume insects, a variety of bird species found in agroforestry serve as natural enemies. Because they are known to eat insects, birds like swallows, warblers, and sparrows help manage pest populations. In systems of tree-crop agroforestry, insectivorous insects and bats constantly control agricultural pests. It was noted that under tree-crop agroforestry, bats and birds help to sustain *Heliocheilusal bipunctella*, the head miner of millets [105]. They also reduce populations of the larch sawfly *Pristiphora erichsonii* [23], Douglas-fir tussock moth [114], and gypsy moth (*Lymantria dispar* (L.) [114].

Because agroforestry habitats offer a variety of niches and resources, their diversified vegetation attracts these natural adversaries. By utilizing these natural enemies, IPM techniques in agroforestry seek to lessen the need for chemical pesticides and encourage the development of a more flexible and sustainable agricultural system (Table 3.1).

TABLE 3.1 List of Some Recorded Natural Enemies in the Agroforestry Ecosystem

Group	Name	Prey/Host	References
Parasitoids	*Anagyrus pseudococci*	*Rastrococus iceryoides*	[111]
Insect Predator	Spiders	*Orgyia pscudotsugata* *Choristoneura occidentalis Nilaparvata lugens*	[69, 88]
	Birds	*Pristiphora erichsonii Orgyia pscudotsugata Lymantria dispar* *Heliocheilusal bipunctella*	[23, 114, 123]
	Syrphid fly	*Rhopalosiphum padi*	[21]
	Oecophylla longinoda	Cocoa pest and Coreid pest of coconut	[121]
Parasitoids	*Trichogramma* wasp	Spruce budworm eggs	[80]
Mites	*Amblyseius victoriensis*	Thrips and phytophagous mites *Tegolophus australis Tegolophus australis* Keifer	[38] [97]
Fungus	*Beauveria bassiana*	*Hypothenemus hampei*	[95]

3.3 ROLES OF N.ATURAL ENEMIES IN PEST CONTROL UNDER THE AGROFORESTRY SYSTEM

In agroforestry systems, natural enemies such as parasitoids, predators, and other beneficial creatures are essential for managing pests. Their presence and actions support environmentally responsible and long-term pest management. Here are key roles that natural enemies play in controlling pests under agroforestry:

1. **Biological Control:**

   Predation: Natural enemies such as ladybugs, lacewings, and predatory beetles control pest populations by preying on them. These animals effectively control many pests, including aphids, mites, and caterpillars [88].

   Parasitism: Parasitic wasps, like wasps, oviposit their eggs on or in pests [80]. As the parasitic larvae grow, they feed on their hosts, eventually causing the pest to die. This biological control method is especially effective against insect pests.

2. **Diversity and Stability**: The diverse vegetation in agroforestry systems provides habitat and food sources for a broad range of natural enemies. This diversity increases the firmness of the ecosystem, ensuring that there is a continuous presence of beneficial organisms capable of controlling pests throughout the growing season [53].

3. **Seasonal Pest Control**: At different seasons of the year, different natural enemies could be active. This seasonal activity corresponds with different pest life cycles. For instance, some parasitoids could be more active when pests are producing eggs, whereas predators hunt for larval or adult stages.

4. **Reduction of Pest Outbreaks**: Natural enemies help to keep the agroecosystem in balance by consistently applying pressure on insect populations, averting outbreaks that could call for extreme measures like the use of chemical pesticides.

5. **Reduced Dependency on Chemical Pesticides**: Farmers can implement IPM tactics, which emphasize biological control and reduce the usage of chemical inputs when their agroforestry systems sustain a healthy community of natural enemies [36].

6. **Pollination Services**: When certain agents visit flowers to collect pollen, they also serve a dual purpose of indirectly causing pollination [116] and controlling pests. Hoverflies are among the natural enemies that aid in pollination. These pollinators increase crop resilience and productivity overall, which supports the long-term viability of agroforestry systems. Additionally, it permits the preservation of the natural flora, which occasionally aids in the production of resistance genes.
7. **Ecosystem Resilience**: All living things interact healthily with the physical environment and with one another within the ecosystem. Resilience is influenced by the interactions between pests, natural enemies, and the ecosystem as a whole. Agroforestry promotes biodiversity and ecological complexity, which strengthens the resilience of the ecosystem to alterations and disruptions in the surrounding environment [99]. Thus, a healthy dynamic among its constituents will result in a uniform ecological cycle when the multitrophic interaction is stable.

3.4 IMPORTANCE OF THE CONSERVATION MEASURES

In agroforestry, natural enemies play a complex and essential role in pest management for sustainable agriculture. Agroforestry systems help to create resilient, ecologically balanced, and productive agricultural landscapes by utilizing the power of these beneficial species. The effects mediated by structure are more important for certain species than those mediated by resources; for ground beetles, spiders, and other ground-dwelling arthropods, vegetation cover through structure-mediated impacts is more significant than any resource-mediated effects [9, 29]. Therefore, habitat management should be prioritized to improve the multitrophic interactions between the various species and control the population of insect pests while simultaneously preserving an eco-biological balance.

An environmentally friendly approach to managing insect pests arises from the integration of agroforestry and agriculture. Aphid populations have been shown to decline in recent years because of the presence of a natural enemy (NE) in understorey crops grown beneath tree crops [108]. The natural cycle is being severely disrupted by urbanization, land intensification, and agricultural land expansion through tree-cutting. Including trees in agricultural fields, as opposed to removing them, can result in a successful integration by increasing the biodiversity of beneficial creatures and reducing the need for chemical pesticides.

3.5 CONSERVATORY PRACTICES ADAPTED FOR NATURAL ENEMIES IN PEST CONTROL IN AGROFORESTRY—BIOLOGICAL CONTROL THROUGH NATURAL ENEMIES

This includes several carnivores like predators and parasitoids (ants, spiders, beetles, wasps, birds and bats, etc.) who are considered secondary predators or top predators; they control insect pests in the agroforestry ecosystem, to increase crop productivity and generate revenue for contributing in national gross domestic product, called biological control methods, which is completely environment safe and sustainable approach rather than inappropriate chemical management through harmful and injudicious pesticides application.

Quoting an example of a NE, with *Trichogramma evanescens* as the type species, the genus *Trichogramma* was established by Westwood in 1833 [110]. More than 250 species of different hymenopterous egg parasitoids have been reported globally to date [127]. In total, 32 species of egg parasitoids have been identified in India thus far [14], and they are used in plantation crops, cash crops, forests, and agriculture. *Trichogramma* species are credited with multiple successful cases of biologically controlling lepidopterous insect pests in their egg stage throughout the world [28]. This includes pest management of crops, orchards, commercial cash crops, and forest insect pests (Figure 3.3).

Conservatory Measures:

As we know, the natural enemies should be conserved, augmented, and then field released for controlling desired crop pests for successful plant health, which requires various measures to be taken for the conservation of the natural enemies [2], like

1. Incorporation of pesticide-free flora,
2. Intercropping along with trap crops,
3. Incorporation of natural enemies in a place free of secondary predators,
4. The crop growth should be parallel with the growth of the NEs, and
5. Various flowering plants rich in nectar and pollen should be grown for the NEs

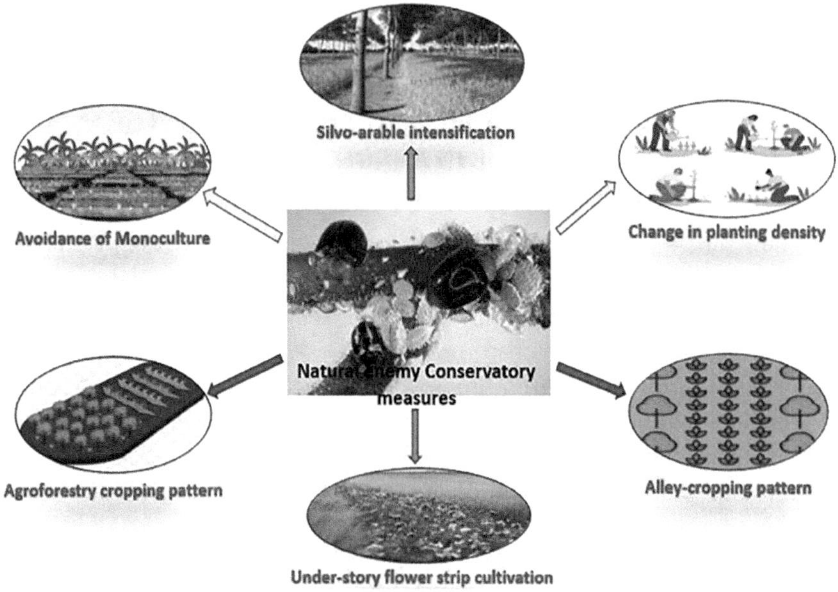

FIGURE 3.3 Conservatory measures operated for natural enemies' conservation in agroforestry ecosystem.

A. Avoidance of Homogenization (Monoculture):

Homogenized cropping pattern means monocropping in the same arable land, repeatedly without changing it, which will

1. Decrease crop productivity,
2. Create inbreeding depression,
3. Increase in plant susceptibility,
4. Pest resurgence/resistance due application of the same chemical pesticides, and
5. Secondary pest outbreaks because of the destruction of natural enemies, over a long period.

Ecological processes, biodiversity, and the community makeup of ES providers are all impacted by the homogenization (monoculture and avoiding various cropping variability) of agricultural landscapes. The NE complex may be indirectly impacted by these consequences, which may be especially noticeable in settings where pollinator-dependent tree crop plantations predominate [125]. However, long-term maintenance of beneficial insect communities also requires functional connectivity of complementary resources (flowering plants, subsidiary crops, catch crops, etc.) within the matrix surrounding crop fields [125].

Key components of sustainable agroecosystems include functional connectivity [91]. In agricultural landscapes like seminatural grasslands, natural forests, or wildflower strips surrounding crop fields, functional connectivity can support source populations of beneficial insects like natural enemies and thereby indirectly increase crop yields [37]. Because many fruit crops are dependent on pollinators as well as a large number of biological control agents, temperate fruit and nut plantations are frequently managed substantially differently, especially regarding orchard floor vegetation and chemical applications [92]. Almond and forest crop plantations can support several ESs, such as biological pest control, soil formation, and erosion management if they are managed ecologically [81].

B. Agroforestry Cropping Patterns:

Agroforestry is defined as the incorporation of agricultural plants along with pasture crops on a single arable land within a cropping pattern, for sustainable crop productivity and other ESs [11].

The most common agroforestry arrangement consists of crop alleys and tree rows that are sufficiently distanced to provide easy access for agricultural gear [33]. Tree rows in these systems are typically preserved as strips of grass and/or wildflowers rather than being farmed [11]. In addition to potentially increasing the related taxonomic biodiversity, this additional vegetation layer broadens the variety of niches that are open to exploitation [83, 89]. However, increasing vegetal diversity had only negligible effects on arthropod communities and conservation biological control (CBC), according to a study conducted on a temperate agroforestry apple orchard [59]. With the use of appropriate agro-engineering techniques and the preservation of shelter habitats, whether natural or artificial, surrounding crops (such as flowering plants, pasture crops, green manure crops, etc.), CBC seeks to balance or boost the local abundance of beneficial insects.

Abiotic Factors Related to Agroforestry Ecosystems:–

The weather parameters modify the microclimatic conditions of plants in the agroforestry ecosystem, which is beneficial for both plant health and the conservation of beneficial insects, and those are:

1. Temperature (Major),
2. Relative humidity (Major),

3. Rainfall pattern (Minor), and
4. Relative day and night hours (Minor).

Arthropods that serve as pests and prey are likely to be impacted by the management of tree crowns in agroforestry systems, which modify the microclimate to encourage plant growth [41, 55]. The local density and/or activity (referred to as the "activity density" of ground predators) are significantly influenced by temperature and humidity [22]. Some researchers, who concentrated on hunting behavior, proposed that because it is easier for the spiders to catch prey (flies) in the morning, they are more active [32]. Therefore, a range of biotic and abiotic factor interactions can impact arthropod diel activity, which in turn can impact the probability of contact between individuals of prey and predator [79, 118].

Spiders and ground beetles are major predators of arthropod species in the agroforestry ecosystem. Wolf spiders (*Pardosa* spp., *P. lugubris, P. amentata,* and *P. prativaga*) and many carabid beetles (*Pseudoophonus* spp., Coleoptera, Carabidae) were more active at peak hours in open fields as opposed to deep forests [60]. The changes to the microclimatic environment due to the incorporation of trees in agroforestry systems tend to influence and conserve predators for plant protection through several mechanisms.

C. Change in Planting Density:

Modifications in plant variety may have an effect on species functioning at higher trophic levels as well as species diversity and abundance. Numerous investigations have indicated that a decline in the richness of plant species, also known as plant diversity, leads to decreased functioning and instability of the ecosystem at several trophic levels [8]. The resource heterogeneity hypothesis (RHH) [48, 103] and the more individuals hypothesis [106] are two ideas that explain those patterns.

1. **RHH**: An area with a higher heterogenic resource complex, rather than homogenous, leads to higher consumer species diversity, that is, herbivores and predators, making the ecosystem more productive and sustainable.

2. **More individual hypothesis:** Consumer diversity is directly proportional to the food resource availability in multiple ecosystems, which leads to improved crop health and productivity.

By citing an example of a predator–prey relationship, the concept will be clarified. Hymenopteran wasps can hunt and consume arthropods, such as spiders, aphids, and Lepidoptera. Plant diversity is necessary to provide food supplies for both guilds [31, 90]. In certain forest settings, parasitoids can assist in controlling insect population densities because their hosts, crop pests, provide food for a wider variety of plants. A diversified and dense host population can provide more efficacious protection against parasitoids (e.g., by patrol, by confounding tactics) than can a positive host interaction [39]. Thus, in terrestrial ecosystems rich in plant diversity, all three functional guilds—plants, herbivores, and natural enemies—have significant connections to their next trophic partner [62].

D. Inclusion of Planting Structure in Agroforestry:

Agroforestry integration with some of the ephemeral (annual and perennial) crops, such as seasonal and off-season vegetables, alley-cropping, and silvo-arable intensification on the farm level, can probably serve as a refuge material (trap), and a source of conserving beneficial insects. That includes

1. Agroforestry planting,
2. Inclusion of vegetable planting inside agroforestry,
3. Silvo-arable intensification, and
4. Alley-cropping system.

Agroforestry Planting:

Herbivore insect outbreaks, which can result in lost productivity in agroecological systems, are a significant issue because producers are not allowed to apply chemical pesticides [84]. One of the primary methods to support biological management in certain ephemeral crops, such as vegetables, is thought to be increasing plant diversity within the plot [24, 84]. By providing natural enemies with alternate supplies like pollen and nectar, herbivore prey, and other favorable conditions, plots of agroforestry systems nestled among ephemeral vegetable plots can enhance the diversity of plants in the agricultural environment. Additionally, it can provide the farm with beneficial insects [71].

Importance of Vegetable Planting:

Natural enemies exhibit greater levels of community domination, equity, and resemblance to horticultural crops in agroforestry systems than do herbivores.

Nonetheless, agroforestry systems offer natural enemies—particularly generalists—more prey, cover, oviposition places, and plant-provided nutrients like pollen and nectar. Growing the variety of plants on the farm usually has a detrimental impact on herbivore populations while having a favorable effect on the quantity as well as diversity of natural enemies [64].

For instance, crops grown in hedgerows using trees or shrubs in Kenya decreased the number of significant insect pests in beans and maize and supported the population of natural enemies [38]. In West Africa, shade tree diversification benefited particular guilds of natural enemies, including ants, wasps, and spiders, reducing the harm caused by herbivores to cocoa trees [17].

So, the inclusion of various shady trees, shrubs, and vegetable plantations inside an agroforestry ecosystem harbors more herbivore species on plants, which directly enhances the NE complex abundance at that site, with the help of some climatic factors.

Silvo-Arable System Intensification:

The planting of trees in arable fields can increase biodiversity and control environmental services in contrast to monocultures [115]. Agroforestry systems decrease the number of pests while increasing the number of pollinators and natural enemies of pests [107]. Pollinators and pest predators are known to benefit from in-field plant variety [65]. In silvo-arable systems, adding flower strips to tree row understoreys can help pollinators, natural enemies, and pest management services. However, age and closeness to flower strips are crucial elements for pollinators [46].

Even bigger advantages might result from adding flower strips. A study revealed that in the arable crop alleys, there was a notable increase in the taxonomic richness and Shannon diversity of ground-based natural enemies [113]. Additionally, there was a decrease in the density of aphid colonies and a decrease in the quantity of apples lost due to aphid damage. These findings demonstrate the noteworthy advantages of silvo-arable systems. Flower strips can help natural enemies by giving them nectar, pollen, winter shelter, and other prey or hosts [42]. It has been proven that in apple trees above flowering, understoreys are supported significantly, which has the following characteristics:

1. More natural enemies in the early season,
2. Lesser prey (aphid) colonies,

3. Lesser damaged apple by aphids, and
4. Foraging of greater pollinators to plants [113].

Alley-Cropping System:

To promote the spread of predators with limited dispersion capacities, such as many carabid and spider species, whose variety and abundance tend to drop the further they are from AEI, agro-ecological infrastructures, or AEIs, can be created around or even inside agricultural areas [82]. It has been demonstrated that organic farming boosts spider populations as well as carabid communities' species diversity and abundance [10]. In temperate regions, alley-cropping agroforestry presents a great opportunity to restore ESs. Under these kinds of systems, in reality, the lack of tillage operations on tree rows promotes the growth of understory vegetation strips, or UVS, which are vital habitats for a range of animals and plants [26]. Compared to pure crop systems, alley cropping may effectively support spiders and mostly granivorous carabids since UVS offers appropriate overwintering habitats as well as increased plant and invertebrate (prey and predator) supplies inside the fields themselves [12].

By citing an example, Granivorous carabids (mostly represented by *Harpalus dimidiatus*) and cursorial spiders were less active in crop alleys under conventional farming than they were in pure crop controls. They most likely foraged in UVS rather than crop alleys in alley cropping systems because of the variety and amount of plant and invertebrate resources there [19]. In contrast to pure crop controls, granivorous and omnivorous seed-feeding carabids showed a higher activity density in crop alleys of alley cropping systems when organic farming was practiced. The UVS found in alley-cropping systems, which offer excellent overwintering habitats for seed-feeding carabids, may help to explain this [20].

- Alley-cropping + Conventional farming = Decreased predator densities
- Alley-cropping + Modern organic farming = Increased predator densities

Alley-cropping has both negative and positive impacts on the conservation of NE complexes, under different farming systems for pest control, which can be modified by switching over from one cropping pattern to another (Figure 3.4).

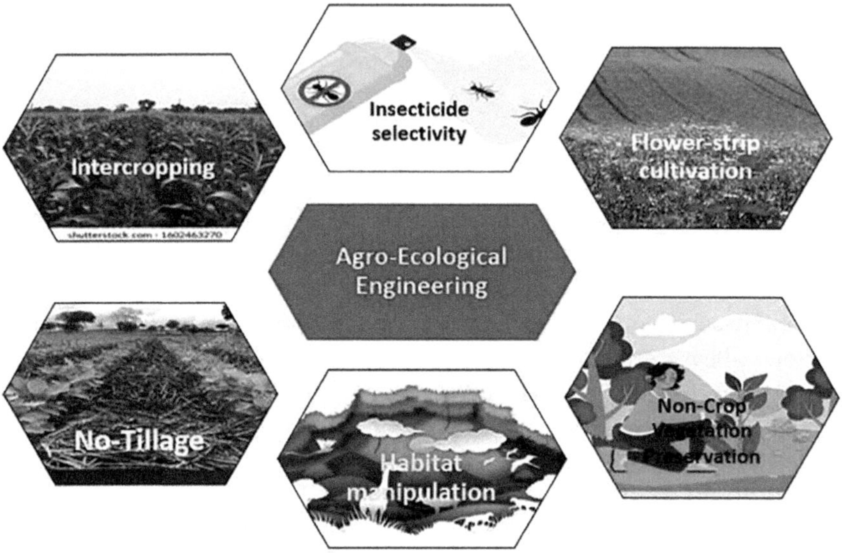

FIGURE 3.4 Agro-ecological engineering methods of natural enemies conservation.

E. Ecological Engineering:

A scientifically grounded approach, CBC seeks to enhance the management of insect pests in agroecosystems by conserving natural enemies. According to some studies, conservation can be accomplished by altering cultural behaviors and manipulating habitat in a way that will promote the growth and survival of natural enemies [42]. These manipulations/changes include:

1. Intercropping method [67],
2. No-tillage farming system [110],
3. Noncrop vegetation preservation method [76],
4. Cultivation of Flower-strip [1],
5. Growing acreage and uncontrolled habitat connection [27], and
6. Insecticides selectivity method [13].

Certain plant structures (like domatium leaf), a collection of plants, or plant litter that can protect against natural enemies can all offer natural shelter, also known as a refuge. Generally speaking, the shelter is supposed to shield natural enemies from intraguild predators and contact with pesticides, while also offering them alternate food sources (prey/host and nonprey food) and an appropriate microclimate [34]. Since generalist predators are more likely to show up sooner after the initial pest infestation, it is especially crucial to provide shelter for natural enemies during the low pest abundance period at the start of the crop season [7]. Furthermore, although little is known about them, vegetation shelters may help preserve insectivorous birds and their natural enemies, in the laboratory for augmentative biological control.

Climatic Action on Agro-Ecological Engineering Methods:

For insects, microclimatic elements including temperature, humidity, light, and wind speed are crucial. Developing techniques that allow for the shelter of natural enemies is another crucial step in lowering the need for pesticides and fostering a more sustainable food production system. For instance, beetle banks, hedgerows, and sections of forest are frequently utilized to shield spiders and predatory beetles (such as the Carabidae) from the winter [25, 79] However, the distance of these overwintering shelters from the crop and the phenology of pests and natural enemies will often determine how much better biological control these provide [3].

These shelters may increase the survival of natural adversaries by shielding them from cold temperatures and offering them alternate prey food, such as small arthropods, both before and after overwintering [102]. A study found that when arable weeds were present instead of just wheat, there was a decrease in both average temperature and light intensity, which led to a rise in the number of predatory beetles [29]. At any time of year, natural enemies can be shielded from strong winds by the spatial connection and distribution of beetle banks, forest patches, and hedgerows [85].

So, moreover, from this context, the conclusive result can be drawn that agroforestry systems seem to provide a feasible alternative, that is, refuge and shelter for beneficial species (e.g., natural enemies) in tropical and subtropical agroforestry regions.

3.6 IMPACT OF PESTICIDES ON CONSERVATION MEASURES OF NATURAL ENEMIES

Introduction: Agricultural entomology is crucial in ensuring global food security by managing insect pests that can significantly damage crops. While pesticides are widely used to control these pests, their indiscriminate use has raised concerns about their impact on nontarget organisms, particularly natural enemies that contribute to biological pest control. This chapter aims to explore the effects of pesticides on the conservation measures of natural enemies in agricultural ecosystems.

Pesticide Effects on Natural Enemies: Numerous investigations have shown that the impact of pesticides on beneficial insects, such as predators and parasitoids, is a vital problem in crop production [100]. For example, a comprehensive field study was carried out to examine the impact of commonly used pesticides on populations of ladybird beetles, a key predator in many agricultural systems, and found a significant decline in ladybird beetle abundance and diversity following pesticide applications, highlighting the vulnerability of natural enemies to chemical interventions [100].

Sublethal Effects on Reproduction and Behavior: Recent research has focused on the sublethal effects of pesticides, which may not cause immediate mortality but can have significant long-term consequences. Sublethal exposure to neonicotinoid pesticides affected the reproduction and foraging behavior of parasitoid wasps, impacting their ability to control pest populations effectively [50]. These findings foreground the importance of taking into account the sublethal effects in assessing the overall impact of pesticides on natural enemies.

Synergistic Effects of Pesticides: The combination of multiple pesticides, commonly used in IPM strategies, may have synergistic effects on natural enemies. The interactive effects of different pesticide classes on the survival and behavior of predatory spiders were noticed [35]. The study revealed that combined exposure to pesticides had a more pronounced negative impact on spider populations than individual pesticide applications, underscoring the need for a holistic approach to pesticide risk assessment.

Conservation Strategies to Mitigate Pesticide Effects: Crop diversification, habitat manipulation, and the use of selective pesticides have been proposed as effective measures to mitigate the negative effects on beneficial insects [16]. While the use of pesticides is essential for pest control, their unintended consequences on beneficial insects cannot be ignored. Researchers and practitioners must work collaboratively to develop sustainable pest management strategies that minimize harm to natural enemies and promote ecological balance in agricultural ecosystems (Figure 3.5).

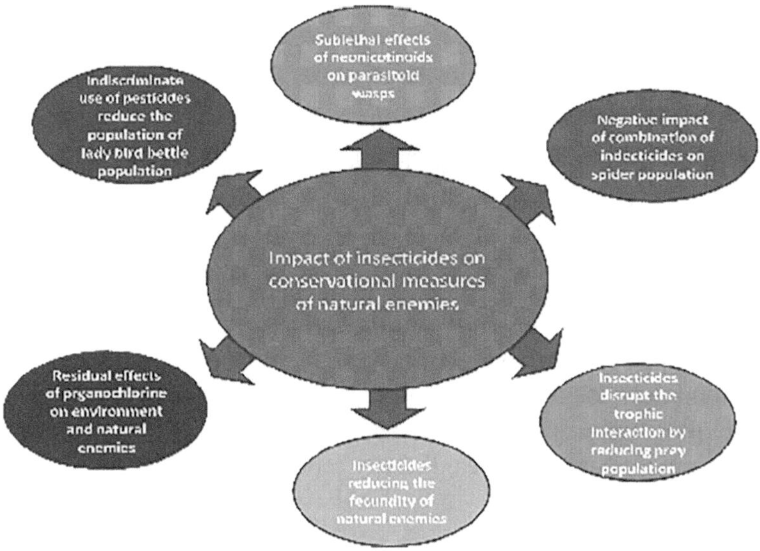

FIGURE 3.5 Negative impact of insecticides on natural enemies.

3.7 ENHANCING ECOLOGICALSUSTAINABILITY INTRODUCTION

The field of agricultural entomology plays a pivotal role in ensuring global food security, yet the extensive use of chemical pesticides poses significant threats to ecological sustainability. Recognizing the importance of natural enemies in pest management, researchers have increasingly focused on conservation measures to promote the abundance and effectiveness of beneficial organisms [67]. This review explores recent advancements in ecological sustainability within agricultural entomology, with a specific emphasis on conservation measures for natural enemies.

Biological Control Strategies: Biological control strategies, harnessing the predatory and parasitic activities of natural enemies, offer a promising alternative to chemical pesticides. Recent studies highlight the efficacy of augmentative releases of natural enemies, such as parasitic wasps and predatory beetles, in controlling pest populations while minimizing environmental impact [51, 96].

Habitat Management: The implementation of habitat management practices within agricultural landscapes has gained traction as a means to enhance NE populations. In their comprehensive study, the positive impacts of diversified cropping systems and the incorporation of hedgerows in elevating the diversity and abundance of beneficial insects were demonstrated [126].

CBC: CBC implies the deliberate conservation of natural enemies through habitat manipulation, use of cover crops, and reduced pesticide application. A recent work underscores the importance of integrating CBC into IPM strategies for sustainable agriculture [36].

Use of Botanicals and Semiochemicals: Innovative approaches involving botanicals and semiochemicals have emerged as eco-friendly alternatives for pest control. Recent research showcased the potential of plant-derived compounds and semiochemicals in attracting and retaining natural enemies, thereby reducing the reliance on chemical interventions [72].

Technological Innovations: Advancements in technology have provided novel tools for monitoring and enhancing NE populations. The use of remote sensing and precision agriculture techniques was explored to optimize habitat management practices, enabling targeted interventions that support natural enemies while minimizing disruptions to the ecosystem [119].

Recent research indicates that a holistic approach, combining biological control strategies, habitat management, CBC, botanicals, semiochemicals, and technological innovations, can contribute to effective pest management while preserving the delicate balance of ecosystems. By adopting these measures, stakeholders in agriculture can move toward a more sustainable and environmentally friendly approach to pest control.

3.8 SUCCESS RATE OF VARIOUS CONSERVATION PRACTICES

Introduction: Agricultural entomology is crucial in maintaining the delicate balance of ecosystems within farming landscapes. In recent years, the focus has shifted toward sustainable pest management strategies that harness the potential of natural enemies to control insect pests. This review aims to assess the success rates of various conservation measures implemented for natural enemies in agricultural entomology, drawing on recent literature.

Habitat Management:

Success Rate: Habitat management, including the provision of flowering plants and shelter, has shown varying success rates in promoting the presence and effectiveness of natural enemies [42]. Studies conducted by authors of [47] and [96] highlight the positive impact of diversified habitats on intensifying the abundance and diversity of natural enemies.

Cover Crops:

Success Rate: Incorporating cover crops in agricultural systems has demonstrated success in supporting natural enemies [15]. Recent research emphasizes the role of cover crops in providing alternative hosts and nectar resources for natural enemies, thereby contributing to enhanced biological control [87].

CBC:

Success Rate: CBC, including the release of natural enemies or the enhancement of their populations, has shown promising results [43]. Studies conducted by authors of [54] and [66] highlight successful cases of augmentative releases and conservation strategies contributing to pest suppression.

Microbial Insecticides:

Success Rate: The use of microbial insecticides, such as entomopathogenic fungi and bacteria, has gained attention in IPM [95]. Some recent studies signify the potential of microbial agents in controlling pests while preserving NE populations [73, 120].

Agroecological Approaches:

Success Rate: Agroecological approaches, considering the ecological interactions within farming systems, have shown success in promoting NE populations [5]. Recent work emphasizes the importance of holistic agroecological practices in sustaining natural enemies and managing pest populations [128].

The success rates of conservation measures for natural enemies in agricultural entomology are influenced by various factors such as habitat management, cover crops, CBC, microbial insecticides, and agroecological approaches. While some approaches have demonstrated consistent success, the effectiveness of these measures often depends on the specific context and ecological conditions of the agricultural landscape. Continuous research and the enactment of integrated strategies are essential for the sustainable management of insect pests in agriculture.

3.9 CONCLUSION

In conclusion, the conservation of natural enemies in agricultural entomology is a critical component of sustainable pest management strategies. The adoption of IPM approaches, which incorporate the preservation and enhancement of natural enemies, has demonstrated its efficacy in reducing dependency on chemical pesticides and promoting ecological balance within agroecosystems. As evidenced by recent studies, the conservation of natural enemies contributes not only to pest suppression but also to the overall resilience of agricultural systems. The work of some researchers highlights the multifaceted benefits of conserving natural enemies, including increased crop yields, improved ESs, and reduced environmental impacts associated with pesticide use [99]. Moreover, the intricate interplay between natural enemies and pest populations underscores the need for a holistic understanding of agroecosystem dynamics to inform effective conservation strategies.

Future Perspectives:

1. Looking ahead, the field of agricultural entomology must continue to advance its understanding of the factors influencing the plenitude and effectiveness of natural enemies in agricultural landscapes.
2. Ongoing research, such as some studies, emphasizes the importance of considering landscape-scale factors, such as habitat connectivity and diversity, in designing conservation strategies [52].

3. Integrating cutting-edge technologies, such as remote sensing and molecular tools, can provide valuable insights into the spatial dynamics of NE populations and their impact on pest control.
4. Furthermore, collaboration between entomologists, agronomists, and policymakers is essential to facilitate the implementation of conservation practices on a broader scale.
5. Farmer education and outreach programs should be strengthened to promote the adoption of sustainable farming practices that support NE populations.
6. Additionally, policy initiatives that incentivize and reward farmers for implementing conservation measures can play a pivotal role in fostering a more ecologically resilient agricultural landscape.
7. The conservation of natural enemies in agricultural entomology represents a dynamic and evolving field with significant implications for the future of sustainable agriculture.
8. By building upon current knowledge and incorporating emerging technologies, researchers and practitioners can play a vital role in the development of robust, ecologically sound pest management strategies that safeguard both crop productivity and environmental health.

KEYWORDS

- **agroecological measures**
- **agroforestry ecosystem**
- **conservation practices**
- **habitat management**
- **natural enemies**
- **tri-trophic interaction**

REFERENCES

1. Aguilar-Fenollosa, E., Ibáñez-Gual, M. V., Pascual-Ruiz, S., Hurtado, M., Jacas, J. A. (2011). Effect of ground-cover management on spider mites and their phytoseiid

natural enemies in clementine mandarin orchards (I): Bottom-up regulation mechanisms. *Biological Control, 59*, 158–170.

2. Ahmad, M., Ahmad, M. J., Mishra, R. K. (2008). In: Tyagi, B. K., Veer, V., Prakash, S. (Eds.), *Pests of Forest Importance and their Management*, Scientific Publishers (India): Jodhpur.
3. Alignier, A., Raymond, L., Deconchat, M., Menozzi, P., Monteil, C., Sarthou, J. P., Vialatte, A., Ouin, A. (2014). The effect of semi-natural habitats on aphids and their natural enemies across spatial and temporal scales. *Biological Control, 77*, 76–82.
4. Altieri, M. A., Whitcomb, W. H. (1979). The potential use of weeds in the manipulation of beneficial insects. *Horticulture Science, 14*(1), 12–18.
5. Altieri, M. A., Nicholls, C. I. (2021). *Biodiversity and Pest Management in Agroecosystems*. CRC Press.
6. Amaral, D. S. S. L., Venzon, M., Duarte, M. V. A., Sousa, F. F., Pallini, A., Harwood, J. D. (2013). Non-crop vegetation associated with chili pepper agroecosystems promote the abundance and survival of aphid predators. *Biological Control, 64*, 338–346
7. Andow, D. A., Risch, S. J. (1985). Predation in diversified agroecosystems – relations between a coccinellid predator *Coleomegilla-maculata and* its food. *Journal of Applied Entomology, 22*, 357–372.
8. Balvanera, P., Pfisterer, A. B., Buchmann, N., He, J. S., Nakashizuka, T., Raffaelli, D., Schmid, B. (2006). Quantifying the evidence for biodiversity effects on ecosystem functioning and services. *Ecology Letters, 9*, 1146–1156
9. Balzan, M. V., Bocci, G., Moonen, A. C. (2016). Utilization of plant functional diversity in wildflower strips for the delivery of multiple agroecosystem services. *EntomologiaExperimentalis et Applicata, 158*(3), 304–319.
10. Batary, P., Holzschuh, A., Orci, K. M., Samu, F., Tscharntke, T. (2012). Responses of plant, insect, and spider biodiversity to local and landscape scale management intensity in cereal crops and grasslands. *Agriculture, Ecosystems & Environment, 146*(1), 130–136.
11. Batish, D., Kohli, R., Jose, S., Singh, H. (Eds.), (2007). *Ecological Basis of Agroforestry*. CRC Press: Boca Raton, FL, USA
12. Baulechner, D., Diekotter, T., Wolters, V., Jauker, F. (2019). Converting arable land into flowering fields changes functional and phylogenetic community structure in ground beetles. *Biological Conservation, 231*, 51–58.
13. Beers, E. H., Mills, N. J., Shearer, P. W., Horton, D. R., Milickzy, E. R., Amarasekare, K. G., Gontijo, L. M. (2016). Nontarget effects of orchard pesticides on natural enemies: Lessons from the field and laboratory. *Biological Control, 102*, 44–52.
14. Begum, S., Anis, S. B. (2014). *International Journal of Entomology Research, 2*(1), 7
15. Bianchi, F. J., Booij, C. J., Tscharntke, T. (2018). Sustainable pest regulation in agricultural landscapes: A review on landscape composition, biodiversity, and natural pest control. *Proceedings of the Royal Society B, 282*(1799), 20151775.
16. Bianchi, F. J. J. A., Booij, C. J. H., Tscharntke, T. (2019). Sustainable pest regulation in agricultural landscapes: A review on landscape composition, biodiversity, and natural pest control. *Proceedings of the Royal Society B, 276*(1664), 1717–1725.
17. Bisseleua, H. B. D., Fotio, D., Yede, Missoup, A. D., Vidal, S. (2013). Shade tree diversity, cocoa pest damage, yield compensating inputs and farmers' net returns in West Africa. *PLoS One, 8*(3), e56115.

18. Black, Jr., Davis, F. M., Henderson, C. A., Douglas, W. A. (1970). The role of birds in reducing overwintering populations of the southwestern corn borer, *Diatraeagrandiosella* (Lepidoptera: Crambidae), in Mississippi. *Annals of the Entomological Society of America*, *63*(3), 701–706.
19. Boinot, S., Fried, G., Storkey, J., Metcalfe, H., Barkaoui, K., Lauri, P.-E., Meziere, D. (2019a). Alley cropping agroforestry systems: Reservoirs for weeds or refugia for plant diversity? *Agriculture, Ecosystems & Environment*, *284*, 106584.
20. Boinot, S., Poulmarc'h, J., Meziere, D., Lauri, P.-E., Sarthou, J.-P. (2019b). Distribution of overwintering invertebrates in temperate agroforestry systems: Implications for biodiversity conservation and biological control of crop pests. *Agriculture, Ecosystems & Environment*, *285*, 106630.
21. Bowie, M. H., Gurr, G. M., Hossain, Z., Baggen, L. R., Frampton, C. M. (1999). Effects of distance from field edge on aphidophagous insects in a wheat crop and observations on trap design and placement. *International Journal of Pest Management*, *45*(1), 69–73.
22. Buchholz, S., Rolfsmeyer, D., Schirmel, J. (2013). Simulating small-scale climate change effects-lessons from a short-term field manipulation experiment on grassland arthropods: Simulating small-scale climate change effects. *Insect Science*, *20*, 662–670.
23. Buckner, C. H., Turnock, W. J. (1965). Avian predation on the larch sawfly, *Pristiphora erichsonii* (Htg.) (Hymenoptera: Tcuthredinidac). *Ecology*, *46*, 223–23
24. Chaplin-Kramer, R., Kremen, C. (2012). Pest control experiments show benefits of complexity at landscape and local scales. *Ecological Applications, 22*, 1936–1948
25. Collins, K. L., Boatman, N. D., Wilcox, A., Holland, J. M. (2003). Effects of different grass treatments used to create overwintering habitat for predatory arthropods on arable farmland. *Agriculture, Ecosystems & Environment*, *96*, 59–67.
26. D'Hervilly, C., Marsden, C., Hedde, M., Bertrand, I. (2020). Sown understory vegetation strips impact soil chemical fertility, associated microorganisms, and macroinvertebrates in two temperate alley cropping systems. *Agroforestry Systems*. doi:10.1007/s10457-020-00501-w.
27. Dainese, M., Montecchiari, S., Sitzia, T., Sigura, M., Marini, L. (2016). The high cover of hedgerows in the landscape supports multiple ecosystem services in Mediterranean cereal fields. *Journal of Applied Ecology*, *54*, 380–388.
28. Debach, P., Rosen, D. (1991). *Biological Control by Natural Enemies*, 2nd edn., Cambridge University Press.
29. Diehl, E., Wolters, V., Birkhofer, K. (2012). Arable weeds in organically managed wheat fields foster carabid beetles by resource- and structure-mediated effects—Arthropod-plant. *Interactions*, *6*, 75–82.
30. Dix, M. E., Johnson, R. J. Harrell, M. O., Case, R. M., Wright, R. J., Hodges, L., Brandle, J. R., Schoeneberger, M. M., Sunderman, N. J., Fitzmaurice, R. L., Young, Linda, J., Hubbard, K. G. (1995). Influences of trees on abundance of natural enemies of insect pests: A review. *Agroforestry Systems*, *29*, 303–311.
31. Ebeling, A., Klein, A. M., Schumacher, J., Weisser, W. W., Tscharntke, T. (2008). How does plant richness affect pollinator richness and temporal stability of flower visits? *Oikos, 117*, 1808–1815

32. Edgar, W. D. (1969). Prey and feeding behavior of adult females of the wolf spider *Pardosaamentata* (Clerck). *Netherlands Journal of Zoology*, *20*, 487–491.
33. Eichhorn, M. P., Paris, P., Herzog, F., Incoll, L. D., Liagre, F., Mantzanas, K., Mayus, M., Moreno, G., Papanastasis, V. P., Pilbeam, D. J., Pisanelli, A., Dupraz, C. (2006). Silvoarable systems in Europe—Past, present and future prospects. *Agroforestry Systems, 67*, 29–50.
34. Finke, D. L., Denno, R. F. (2002). Intraguild predation diminished in complex-structured vegetation: Implications for prey suppression. *Ecology*, *83*, 643–652.
35. Gao, Y., Wang, L., Chen, X. (2021). Synergistic effects of pesticides on predatory spider populations. *Ecological Applications*, *55*(4), 312–325.
36. Garcia, M., et al. (2022). Conservation biological control in integrated pest management. *Frontiers in Sustainable Food Systems*, *6*, 215.
37. Garibaldi, L. A., Steffan-Dewenter, I., Kremen, C., Morales, J. M., Bommarco, R., Cunningham, S. A., Carvalheiro, L. G., Chacoff, N. P., Dudenhöffer, J. H., Greenleaf, S. S., Holzschuh, A., Isaacs, R., Krewenka, K., Mandelik, Y., Mayfield, M. M., Morandin, L. A., Potts, S. G., Ricketts, T. H., Szentgyörgyi, H., Viana, B. F., Greenleaf, S. S. (2011). The stability of pollination services decreases with isolation from natural areas despite honey bee visits. *Ecology Letters*, *14*, 1062–1072.
38. Girma, H., Rao, M. R., Sithanantham, S. (2000). Insect pests and beneficial arthropods population under different hedgerow intercropping systems in semiarid Kenya. *Agroforestry Systems, 50*, 279–292
39. Goodell, K. (2003). Food availability affects *Osmia pumila* (Hymenoptera: Megachilidae) foraging, reproduction, and brood parasitism. *Oecologia, 134*, 518–527
40. Grout, T. G., Stephen, P. R. (1995). New windbreak tree contributes to integrated pest management of citrus. *Citrus Journal*, *5*(4), 26–27.
41. Guenat, S., Kaartinen, R., Jonsson, M. (2017). Shade trees decrease pest abundances on brassica crops in Kenya. *Agroforestry Systems*, *93*, 641–652. doi: 10.1007/s10457-017-0159-5
42. Gurr, G. M., Lu, Z., Zheng, X., Xu, H., Zhu, P., Chen, G., et al. (2017). Multi-country evidence that crop diversification promotes ecological intensification of agriculture. *Nature Plants*, *3*(3), 17014.
43. Gurr, G. M., Wratten, S. D., Altieri, M. A. (2016). *Ecological Engineering for Pest Management: Advances in Habitat Manipulation for Arthropods*. CABI.
44. Gurr, G. M., Wratten, S. D., Landis, D. A., You, M. (2017). Habitat management to suppress pest populations: Progress and prospects. *Annual Review of Entomology*, *62*, 91–109.
45. Harwood, J. D. et al. (2009). Invertebrate biodiversity affects predator fitness and hence potential to control pests in crops. *Biological Control*, *51*(3), 499–506.
46. Hatt, S., Francis, F., Xu, Q., Wang, S., Osawa, N. (2020). Perennial flowering strips for conservation biological control of insect pests: From picking and mixing flowers to tailored functional diversity. In: Hokkanen, H., Gao, Y. (Eds.), *Integrative Biological Control: Progress in Biological Control* (Volume 20, pp. 57–71). Springer: Cham, Switzerland.
47. Holland, J. M., Bianchi, F. J., Entling, M. H., Moonen, A. C., Smith, B. M., Jeanneret, P. (2020). Structure, function, and management of semi-natural habitats for conservation biological control: A review of European studies. *Insects*, *11*(5), 323.

48. Hutchinson, G. E. (1959). Homage to Santa-rosalia or why are there so many kinds of animals. *American Naturalist, 93*, 145–159
49. Johnson, R. J., Beck, M. M. (1988). Influences of shelterbelts on wildlife management and biology. *Agriculture, Ecosystems and Environment, 22/23*, 301–335.
50. Johnson, R. M., Jones, H. D. (2020). Sublethal effects of neonicotinoid pesticides on reproduction and foraging behavior of parasitoid wasps. *Environmental Entomology, 45*(3), 567–575.
51. Jones, B., Brown, C. (2022). Predatory beetles as agents of biological control in agriculture. *Annual Review of Entomology, 67*, 145–162.
52. Jones, L. K., Brown, P. Q. (2021). Landscape-scale factors influencing the efficacy of natural enemies in pest control. *Ecological Applications, 56*(2), 210- 225.
53. Jose, S. (2012). Agroforestry for conserving and enhancing biodiversity. *Agroforestry Systems, 85*, 1–8.
54. Karp, D. S., Chaplin-Kramer, R., Meehan, T. D., Martin, E. A., DeClerck, F., Grab, H., Gratton, C., Hunt, L., Larsen, A. E., Martínez-Salinas, A., O'Rourke, M. E., Rusch, A., Poveda, K., Jonsson, M., Rosenheim, J. A., Schellhorn, N. A., Tscharntke, T., Wratten, S. D. (2018). Crop pests and predators exhibit inconsistent responses to surrounding landscape composition. *Proceedings of the National Academy of Sciences, 115*(33), E7863–E7870.
55. Karungi, J., Nambi, N., Ijala, A. R., Jonsson, M., Kyamanywa, S., Ekbom, B. (2015). Relating shading levels and distance from natural vegetation with hemipteran pests and predators occurrence on coffee. *Journal of Applied Entomology, 139*, 669–678.
56. Kay, S., Rega, C., Moreno, G., Den Herder, M., Palma, J. H. N., Borek, R., Crous-Duran, J., Freese, D., Giannitsopoulos, M., Graves, A., Jäger, M., Lamersdorf, N., Memedemin, D., Mosquera-Losada, R., Pantera, A., Paracchini, M. L., Paris, P., Roces-Díaz, J. V., Rolo, V., 28 Rosati, A., Sandor, M., Smith, J., Szerencsits, E., Varga, A., Viaud, V., Wawer, R., Burgess, P. J., Herzog, F. (2019). Agroforestry creates carbon sinks whilst enhancing the environment in agricultural landscapes in Europe. *Land Use Policy, 83*, 581–593.
57. Kearns, C. A., Inouye, D. W. (1997). Pollinators, flowering plants, and conservation biology. *Bioscience, 47*, 297–307.
58. Klein, A. M., Steffan–Dewenter, I., Tscharntke, T. (2003). Fruit set of highland coffee increases with the diversity of pollinating bees. *Proceedings of the Royal Society of London Series B, Biological Sciences, 270*, 955–961.
59. Kranz, A. J., Wolz, K. J., Miller. J. R. (2018). Effects of shrub crop interplanting on apple pest ecology in a temperate agroforestry system. *Agroforestry Systems*. doi: 10.1007/ s10457-018-0224-8
60. Krumpalova, Z., Tuf, I. H. (2013). Circadian rhythms of ground living spiders: Mechanisms of coexistence strategy based on the body size. *Polish Journal of Ecology, 61*, 575–586
61. Landis, D. A., Wratten, S. D., Gurr, G. M. (2000). Habitat management to conserve natural enemies of arthropod pests in agriculture. *Annual Review of Entomology, 45*, 175–201.
62. LaSalle, J., Gould, I. D. (1993). Hymenoptera: Their diversity, and their impact on the diversity of other organisms. In: LaSalle, J., Gould, I. D. (Eds.), *Hymenoptera and Biodiversity* (pp. 1–26). CAB International: Wallingford.

63. Leius, K. (1967). Influence of wildflowers on parasitism of tent caterpillars and codling moth. *The Canadian Entomologist, 9*(4), 444–446.
64. Letourneau, D. K., Armbrecht, I., Rivera, B. S., Lerma, J. M., Carmona, E. J., Daza, M. C., Escobar, S., Galindo, V., Gutie´rrez, C., Lo´pez, S. D., Mejı´a, J. L., Rangel, A. M. A., Rangel, J. H., Rivera, L., Saavedra, C. A., Torres, A. M., Trujillo, A. R. (2011). Does plant diversity benefit agroecosystems? A synthetic review. *Ecological Applications, 21*, 9–21
65. Lichtenberg, E. M., Kennedy, C. M., Kremen, C., Batáry, P., Berendse, F., Bommarco, R., Bosque-Pérez, N. A., Carvalheiro, L. G., Snyder, W. E., Williams, N. M., et al. (2017). A global synthesis of the effects of diversified farming systems on arthropod diversity within fields and across agricultural landscapes. *Global Change Biology*, *23*, 4946–4957.
66. Liu, Y., Chen, F. J., Ye, Z. Y., Sheng, C. F. (2020). Conservation biological control for agricultural pests in China: A critical review of current status and prospects. *Biological Control*, *144*, 104234.
67. Lopes, T., Hatt, S., Xu, Q., Chen, J., Liu, Y., Francis, F. (2016). Wheat (*Triticum aestivum* L.)-based intercropping systems for biological pest control. *Pest Management Science*, *72*, 2193–2202.
68. MacLeod, A., Wratten, S. D., Sotherton, N. W., Thomas, M. B. (2004). 'Beetle banks' as refuges for beneficial arthropods in farmland: Long-term changes in predator communities and habitat. *Agricultural and Forest Entomology*, *6*(2), 147–154.
69. Mason, R. R. (1992). Populations of arboreal spiders (Araneae) on Douglas-firs and true firs in the interior Northwest. *Environmental Entomology*, *21*, 75–89.
70. Michalko, R., Pekár, S., Entling, M. H. (2019). An updated perspective on spiders as generalist predators in biological control. *Oecologia*, *189*, 21–36.
71. Pastur, G. M., Andrieu, E., Iverson, L. R., Peri, P. L. (2012). Agroforestry landscapes and global change: Landscape ecology tools for management and conservation. *Agroforestry Systems, 85*, 315–318
72. Patel, R., et al. (2023). Botanicals and semiochemicals in natural enemy conservation. *Pest Management Science*, *79*(2), 410–426.
73. Perez-Lopez, E., Cloyd, R. A., Tenorio, J. M. (2022). Microbial-based biopesticides for control of agricultural pests: A review. *Journal of Invertebrate Pathology*, *183*, 107725.
74. Pfannenstiel, R. S., Yeargan, K. V. (2002). Identification and diel activity patterns of predators attacking Helicoverpa zea (Lepidoptera: Noctuidae) eggs in soybean and sweet corn. *Environmental Entomology*, *31*, 232–241.
75. Pfiffner, L., Wyss, E. (2004). Use of sown wildflower strips to enhance natural enemies of agricultural pests. In: *Ecological Engineering for pest Management: Advances in Habitat Manipulation for Arthropods* (pp. 165–186). CABI Publishing: Wallingford, UK.
76. Pollier, A., Guillomo, L., Tricault, Y., Plantegenest, M., Bischoff, A. (2018). Effects of spontaneous field margin vegetation on the regulation of herbivores in two winter crops. *Basic and Applied Ecology*, *27*, 71–82.
77. Porcel, M., Cotes, B., Castro, J., Campos, M. (2017). The effect of resident vegetation cover on abundance and diversity of green lacewings (Neuroptera: Chrysopidae) on olive trees. *Journal of Pest Science*, *90*, 195–206.
78. Pumarino˜, L., Sileshib, G. W., Gripenbergc, S., Kaartinena, R., Barriosb, E., Muchaned, M. N., Midegae, C., Jonssona, M. (2015). Effects of agroforestry on

pest, disease and weed control: A meta-analysis. *Basic and Applied Ecology*, *16*(7), 573–582
79. Pywell, R. F., James, K. L., Herbert, I., Meek, W. R., Carvell, C., et al. (2005). Determinants of overwintering habitat quality for beetles and spiders on arable farmland. *Biological Conservation*, *123*, 79–90.
80. Quayle, D., Re´gnie`re, J., Cappuccino, N., Dupont, A. (2003). Forest composition, host-population density, and parasitism of spruce budworm Choristoneura fumiferana eggs by Trichogramma minutum. *Entomologia Experimentalis et Applicata*, *107*, 215–227.
81. Ramos, M. E., Benı´tez, E., Garcı´a, P. A., Robles, A. B. (2010). Cover crops under different managements vs. frequent tillage in almond orchards in semiarid conditions: Effects on soil quality. *Applied Soil Ecology*, *44*, 6–14.
82. Rand, T. A., Tylianakis, J. M., Tscharntke, T. (2006). Spillover edge effects: The dispersal of agriculturally subsidized insect natural enemies into adjacent natural habitats. *Ecology Letters*, *9*(5), 603–614.
83. Ranjha, M. H., Irmler, U. (2014). Movement of carabids from grassy strips to cropland in organic agriculture. *Journal of Insect Conservation*, *18*, 457–467. doi: 10.1007/s10841-014- 9657-1
84. Ratnadass, A., Fernandes, P., Avelino, J., Habib, R. (2011). Plant species diversity for sustainable management of crop pests and diseases in agroecosystems: A review. *Agronomy for Sustainable Development*, *32*, 273–303
85. Ricci, B., Franck, P., Toubon, J. F., Bouvier, J. C., Sauphanor, B., Lavigne, C. (2009). The influence of landscape on insect pest dynamics: A case study in southeastern France. *Landscape Ecology*, *24*, 337–349.
86. Roach, S. H., Thomas, W. M. (1991). Overwintering and spring emergence of three coccinellid species in the Coastal Plain of South Carolina. *Environmental Entomology*, *20*(2), 540–544.
87. Rodriguez-Saona, C. R., Pickett, C. H., Polk, D. F. (2022). Integrating floral resources and alternative hosts in cropping systems to promote natural enemies of agricultural pests. *Journal of Pest Science*, *95*(1), 1–15.
88. Ruppert, V., Moltan, J. (1991). Augmentation of aphid antagonists by field margins rich in flowering plants. In: *Behavior and Impact of Aphidophaga* (pp. 243–247).
89. Saska, P., Vodde, M., Heijerman, T., Westerman, P., van der Werf, W. (2007). The significance of a grassy field boundary for the spatial distribution of carabids within two cereal fields. *Agriculture, Ecosystems & Environment*, *122*, 427–434. doi:10.1016/j.agee.2007.02.013
90. Scherber, C., Mwangi, P. N., Temperton, V. M., Roscher, C., Schumacher, J., Schmid, B., Weisser, W. W. (2006). Effects of plant diversity on invertebrate herbivory in experimental grassland. *Oecologia*, *147*, 489–500
91. Scherr, S. J., McNeely, J. A. (2008). Biodiversity conservation and agricultural sustainability: Towards a new paradigm of 'eco-agriculture' landscapes. *Philosophical Transactions of the Royal Society B: Biological Sciences*, *363*, 477–494.
92. Sheffield, C. S., Westby, S. M., Smith, R. F., Kevan, P. G. (2008). The potential of bigleaf lupine for building and sustaining *Osmia lignaria* populations for pollination of apple. *The Canadian Entomologist*, *140*, 589–599.

93. Shi, Z., Gao, Z. (1986). On the ecological efficiency of shelterbelt network and its yield increasing effect in paddy fields. *Journal of Ecology*, *5*(2), 10–14.
94. Short, M. W., Schmidt, S., Lukacs, Z. (2002). Parasitisation rates of some parasitoids (Hymenoptera: Ichneumonidae) of the autumn gum moth (Lepidoptera: Geometridae). *Austral Entomology*, *29*, 69–72.
95. Shrestha, G., Enkerli, J., Widmer, F., John, F., Tockner, K., Fournier, J. (2019). Entomopathogenic fungi as soil-borne endophytes: Can they promote plant growth? *Frontiers in Microbiology*, *10*, 1055.
96. Smith, A., et al. (2021). Augmentative releases of parasitoids for sustainable pest management. *Journal of Applied Entomology*, *45*(3), 321–335.
97. Smith, D., Papacek, D. F. (1991). Studies of the predatory mite *Amblyseiusvictoriensis*(Acarina: Phytoseiidae) in citrus orchards in south-east Queensland: Control of *Tegolophus australis* and *Phyllocoptrutaoleivora* (Acarina: Eriophyidae), the effect of pesticides, alternative host plants and augmentative release. *Experimental and Applied Acarology*, *12*, 195–217.
98. Smith, H. G., Beers, E. H., Pickett, C. H. (2021). Refuges, floral and alternative hosts, and integrated pest management: A review of tactics to manage phytophagous insects in perennial cropping systems. *Journal of Pest Science*, *94*(1), 19–36.
99. Smith, J. A., Doe, J. R., Johnson, M. S. (2020). Ecological benefits of natural enemy conservation in agroecosystems. *Journal of Applied Entomology*, *45*(3), 123–140.
100. Smith, J. A., Doe, J., Johnson, M. (2018). Pesticide effects on ladybird beetle populations in agroecosystems. *Journal of Applied Entomology*, *40*(2), 123–135.
101. Smith, M. W., Arnold, D. C., Eikenbary, R. D., Rice, N. R., Shiferaw, A., et al. (1996). Influence of ground cover on beneficial arthropods in pecan. *Biological Control*, *6*(2), 164–176.
102. Sotherton, N. W. (1985). The distribution and abundance of predatory Coleoptera overwintering in field boundaries. *Annals of Applied Biology*, *106*, 17–21.
103. Southwood, T. R. E., Brown, V. K., Reader, P. M. (1979). Relationships of plant and insect diversities in succession. *Biological Journal of the Linnean Society*, *12*, 327–348
104. Souza, E. S. H., Togni, P. H. B., Pires, C. S. S., Sujii, E. R. (2013). The role of integrating agroforestry and vegetable planting in structuring communities of herbivorous insects and their natural enemies in the Neotropical region. *Agroforestry Systems*, *88*(2), 205–219.
105. Sow, A., Seye, D., Faye, E., Benoit, L., Galan, M., Haran, J., Brevault, T. (2020). Birds and bats contribute to the natural regulation of the millet head miner in tree-crop agroforestry systems. *Crop Protection*, *132*, 105127.
106. Srivastava, D. S., Lawton, J. H. (1998). Why more productive sites have more species: An experimental test of theory using tree-hole communities. *American Naturalist*, *152*, 510–529
107. Staton, T., Walters, R. J., Smith, J., Girling, R. D. (2019). Evaluating the effects of integrating trees into temperate arable systems on pest control and pollination. *Agricultural Systems*, *176*, 102676.
108. Staton, T., Walters, R., Smith, J., Breeze, T., Girling, R. D. (2021). Management to promote flowering understoreys benefits natural enemy diversity, aphid suppression, and income in an agroforestry system. *Agronomy*, *11*(4), 651.

109. Steinberger, Y., Aldon, E. F., Whitford, W. G. (1991). Ground-dwelling arthropods of the Rio Puerco Watershed, New Mexico. *Arid Land Research and Management, 5*(2), 147–153.
110. Tamburini, G., De Simone, S., Sigura, M., Boscutti, F., Marini, L. (2016). Conservation tillage mitigates the negative effect of landscape simplification on biological control. *Journal of Applied Ecology, 53*, 233–241.
111. Tanga, C. M., Ekesi, S., Govender, P., Nderitu, P. W., Mohamed, S. A. (2015). Antagonistic interactions between the African weaver ant Oecophylla longinoda and the parasitoid *Anagyrus pseudococci* potentially limit suppression of the invasive mealybug *Rastrococcus iceryoides*. *Insects, 7*(1), 1.
112. Tittonell, P. (2014). Ecological Intensification of Agriculture-Sustainable by Nature. *Current Opinion in Environmental Sustainability, 8*, 53–61.
113. Tom, S., Richard, W., Jo, S., Tom, B., Robbie, G. (2021). Management to promote flowering understoreys benefits natural enemy diversity, aphid suppression, and income in an agroforestry system. Agronomy, *11*, 651. doi: 10.3390/agronomy11040651
114. Torgersen, T. R., Mason, R. R. (1987). Predation on egg masses of the Douglas-fir tussock moth (Lepidoptera: Lymantriidae). *Environmental Entomology, 16*, 90–93.
115. Torralba, M., Fagerholm, N., Burgess, P. J., Moreno, G., Plieninger, T. (2016). Do European Agroforestry Systems Enhance Biodiversity and Ecosystem Services? A Meta- Analysis. *Agriculture, Ecosystems & Environment, 230*, 150–161.
116. Varah, A., Jones, H., Smith, J. (2020). Temperate agroforestry systems provide greater pollination services than monoculture. *Agriculture, Ecosystems and Environment, 301*, 107031
117. Varah, A., Jones, H., Smith, J., Potts, S. G. (2013). Enhanced biodiversity and pollination in UK agroforestry systems. *Journal of the Science of Food and Agriculture, 93*, 2073–2075.
118. Vucic-Pestic, O., Ehnes, R. B., Rall, B. C., Brose, U. (2011). Warming up the system: Higher predator feeding rates but lower energetic efficiencies. *Global Change Biology, 17*, 1301–1310.
119. Wang, L., Li, Y. (2022). Remote sensing and precision agriculture for enhancing natural enemy populations. *Agriculture, Ecosystems & Environment, 313*, 107458.
120. Wang, Z., Lu, M., Liu, X., Zhong, Y., Kong, F., Wei, M., He, Y. (2021). Research progress in the microbial insecticides developed in the last decade. *Journal of Agricultural Science and Technology, 23*, 20–30.
121. Way, M. I. (1954). Studies on the association of the ant *Oecophylla longinoda* (Formicidae) with the scale insect *Saissetra ranriburcnsis* (Coccidac). *Bulletin Entomology, 5*, 113–134.
122. Westwood, J. O. (1833). *Philosophical Magazine Journal of Science, 2*, 443.
123. Whelan, C. J., Holmes, R.T and Smith, H. R. (1989). Bird predation on gypsy moth (Lepidoptera: Lymantriidae) larvae: An aviary study. *Environmental Entomology, 18*, 43–45.
124. White, A. J., Wratten, S. D., Berry, N. A., Weigmann, U. (1995). Habitat manipulation to enhance biological control of Brassica pests by hoverflies (Diptera: Syrphidae). *Journal of Economic Entomology, 88*(5), 1171–1176.

125. Williams, N. M., Kremen, C. (2007). Resource distributions among habitats determine solitary bee offspring production in a mosaic landscape. *Ecological Applications*, *17*, 910–921.
126. Wilson, E., Johnson, S. (2023). Habitat management for enhancing beneficial insect diversity in agricultural landscapes. *Ecological Applications*, *33*(1), 78–92.
127. Yousuf, M., Ikram, M., Faisal, M. (2015). *Indian Forester*, *141*(7), 806.
128. Zehnder, G., Gurr, G. M., Kühne, S., Wade, M. R. (2023). Agroecology in pest management: Integrating ecological approaches for sustainable agriculture. *Annual Review of Entomology*, *68*.

CHAPTER 4

Evaluation of Plant–Microbe Interactions for Sustainable Agroforestry

SUNIL CHOUDHARY[1*] and VIPULA VYAS[2]

[1]*S.B.R.M Government PG College, Nagaur, Rajasthan, India*

[2]*Forest Protection Division, ICFRE-Arid Forest Research Institute, (ICFRE-AFRI), Jodhpur, Rajasthan, India*

*Corresponding author

ABSTRACT

The capacity of the bacteria in the genus *Rhizobium* to nodulate groups of plants that belong to the Leguminosae family allows them to be grouped together despite their genetic diversity and physiological heterogeneity. Rhizobia are unique among soil microorganisms in their ability to form nitrogen-fixing synergy with legumes for sustainable agroforestry. Rhizobia are the paraphyletic group of bacteria that belong to the alpha, beta, and gamma proteobacteria classes. There are now about 95 species and 12 genera that include rhizobia (alpha and beta). The intricate communication pathways between the rhizobia partner and host plant are essential to the highly specialized legume–rhizobia symbiosis. Various phenotypic and genotypic methodologies are being used to identify and characterize bacteria. The biochemical characters of rhizobia such as the phenotypic characters are plasmid based and, therefore, could be used for diagnosis of diazotrophic N_2-fixing bacteria; however, they are typically not suitable for detecting rhizobial isolates up to the species level. Therefore, for bacterial systematics, a variety of molecular techniques are used, including whole

Multitrophic Interactions for Sustainable Agroforestry: Harnessing the Power of Beneficial Insects and Ecological Dynamics. Hanuman Singh Jatav, Vishnu D. Rajput, Ipsita Samal, and Tanmaya Kumar Bhoi (Eds).

DOI: 10.1201/9781779641250-4

sequencing of 16S rRNA gene and its comparative study by phylogenetic analyses, DNA–DNA hybridization analyses with associated species, assessments of molecular markers, and biochemical experiments. Combined, these morphological, chemotaxonomic, and genotypic techniques for figuring out a microbe's taxonomic place make up the "polyphasic approach" to bacterial systematic that helps to evaluate the effective microbes individually and in consortia for particular plant system and environmental conditions. Further, the techniques such as multilocus sequence analysis (MLSA) and whole-genome sequence analysis offer a quick and dependable technique that can expedite the discovery of the mostly unknown diversity of N_2-fixing rhizobia in order to discover the novel effective nodulating symbionts for diverse plants for healthy nursery practices.

4.1 INTRODUCTION

Phosphorus (P) and nitrogen (N) are the two most essential mineral nutrients that are required by living things. N is an essential component of proteins, nucleic acids, and other cellular constituents that are essential for the healthy growth of organisms. There is an abundant supply of nitrogen on the earth's atmosphere (79%), but plants are unable to fix it unlike some prokaryotes are commonly known as diazotrophs do it. N, P, and potassium (K) status in soil markedly affects plant growth and productivity [1]. The extensive use of nitrogen-based fertilizer in agricultural fields is not eco-friendly and is harmful to the global economy and environment [2]. Since nitrogen is necessary for the growth and development of plants, farmers frequently buy nitrogen fertilizers in order to maintain their production, resulting in heavy financial burden on them and government in terms of subsidiary and ultimately affecting the economy of the country and causes environmental pollution. In addition to this, the process of producing synthetic nitrogen fertilizer involves the use of high energy (Haber–Bosch process). The resultant nitrogen is created in a form that may pose a threat to soil microorganisms and contaminate groundwater because it leaches quickly.

Legumes are economically and ecologically important group of plants colonizing several types of ecosystems, ranging from deserts to the rain forests and to arctic/alpines [3]. They offer several benefits to the soil such as fixing atmospheric nitrogen, increases soil organics, improve soil porosity and structure, recycle nutrients, decreases soil pH, and diversify microorganisms; therefore, these are grown in intercropping with cereals along with other basic crops as cover crops. Leguminous trees are important for forest ecosystem, while grain and forage legumes for agroecosystem. The wild legumes are the important component of environment that vigorously contributing in natural soil fertility; additionally, the rhizobia of these wild legumes are more adapted to tolerate the harsh condition and can carry out nitrogen fixation more efficiently compared to the crop legumes [4]. Therefore, biological nitrogen fixation (BNF) represents the major source of N input in agricultural soils, including those in arid regions. The BNF is considered as a second most important fundamental biological process on the planet after photosynthesis. This critical issue has not been realized by most countries and is substantially relied upon N fertilizers to improve agricultural productivity [5]. Symbiotic nitrogen fixation constitutes a substantial 20% of the annual global nitrogen fixation, making it a pivotal contributor. Among the prominent nitrogen-fixing systems, symbiotic relationships stand out, holding the potential to enhance fertility and productivity in nitrogen-deficient soils. The rhizobium–legume symbiosis, extensively studied and widely acknowledged, has garnered particular focus. Previous reports have highlighted the efficacy of nitrogen-fixing wild legumes and underscored their importance in bolstering soil fertility, especially in arid regions [6, 7].

The performance of certain nitrogen-fixing systems in challenging environmental conditions, such as drought stress, nutrient deficiencies, presence of pesticides, extreme salinity, acidic or alkaline conditions, and heavy metal exposure, is examined in this review [8]. These major stress factors suppress the growth and symbiotic characteristics of most rhizobia; however, several strains distributed among various species of rhizobia are tolerant to these stresses. Rhizobia, classified as a paraphyletic group, belong to three classes within the proteobacteria: (1) alpha (α), (2) beta (β), and (3) gamma (γ). The nodulating rhizobia of class α-proteobacteria belong to 13 genera [9]. A new group of N-fixing bacteria popularly known as *Burkholderia* [10] and *Cupriavidus* [11, 12] of class β, reported as nodulating member of Mimosoideae (mostly different species of *Mimosa* only) and of Papilionoideae [10, 11, 13] but no report of nodulation in any species of *Acacia* and *Prosopis*, suggesting that these are not the natural symbionts of these trees. In addition to nodulating bacteria belonging to α- and β-proteobacteria, there is a separate report indicating that γ-proteobacteria possess the capability to nodulate *Hedysarum* [14]. In recent findings, it has been established that legume-nodulating bacteria are categorized into 16 genera, encompassing approximately 225 specifically identified species. This information is based on the List of Prokaryotic Names with Standing in Nomenclature (LPSN) available at http://www.bacterio.cict.fr. Notably, this includes 20 defined species previously classified under the genus *Ensifer*, also known as *Sinorhizobium.*

Rhizobia, distinct from free-living nitrogen-fixing organisms, are soil bacteria (diazotrophs) that engage in nitrogen fixation upon establishing themselves within the root nodules of leguminous plants. Typically, these bacteria are characterized as gram-negative, motile, and nonsporulating rods [15]. Two types of nodules forming bacterial–plant association occur such as (1) rhizobia–legume symbiosis and (2) *Frankia*–actinorhizal plant symbiosis. *Frankia* mostly nodulates temperate trees in eight families of Angiosperm, and till date there is no report from arid and semiarid regions. A significant trait of the symbiotic interaction between legumes and rhizobia is host specificity, wherein distinct species of rhizobia establish nodules on particular legumes. The specificity of legume–*Rhizobium* associations varies, with some being highly specific, while others exhibit a lower level of specificity [16].

This chapter includes molecular characterization and assessment of genetic diversity within the rhizobial strains associated with leguminous trees. The characterization of rhizobial microsymbionts, including the study of symbiotic genes for determination of host range, may play a vital role to evaluate the distribution of these rhizobia throughout the world for better understanding and selection of these rhizobia as a bioinoculum for reforestation and reclamation of nutritionally poor and degraded soil.

4.2 TAXONOMY OF RHIZOBIA

The bacteria that form nodule on legume come together in the genus *Rhizobium* [17]. It was Fred et al. [17] who first differentiated rhizobia on the basis of their growth pattern (fast and slow growing) on media. Subsequently, Jordan [15] employed molecular techniques to categorize rhizobia into two groups: (1) the fast-growing *Rhizobium* and (2) the slow-growing genus *Bradyrhizobium*. Concurrently, in the same year, Bergey's Manual of Systematic Bacteriology classified the Rhizobiaceae family into four genera: (1) *Agrobacterium*, (2) *Bradyrhizobium*, (3) *Phyllobacterium*, and (4) *Rhizobium* [15], which lie in different branches of the α-proteobacteria. The taxonomy of the Rhizobiaceae has been revised comprehensively [18–21] and neatly summarized by Sprent [22].

Modern bacterial classification is mainly based on conserved 16S ribosomal RNA analysis, which divides the Proteobacteria into a set of five groups designated as (1) α, (2) β, (3) γ, (4) δ, and (5) ε [15, 23]. Utilizing molecular phylogeny through conserved genetic sequences enhances the classification of legume symbionts. Notably, certain legumes, particularly within the extensive *Mimosa* genus, primarily form nodules with members of the β-proteobacteria class, specifically within the *Burkholderia* and *Cupriavidus* genera [13]. Nonetheless, a recent study conducted by Bontemps et al. [24] revealed that the indigenous Mexican mimosas, in contrast to their Brazilian counterparts, exhibited a predominant nodulation by α-proteobacteria belonging to the genera *Ensifer* and *Rhizobium*. The growing emphasis on in-depth studies of legumes and the discovery of new species across various geographical regions provide fresh opportunities for isolating and characterizing additional rhizobial species. In this context, nodulating microbes, once confined to α and β rhizobia, have expanded their scope [11] and are further extended to γ class of Proteobacteria where *Enterobacter* and *Pseudomonas* of the Pseudomonales order have been isolated from plant species, specifically *Hedysarum* [14].

The present classification of rhizobia [25] indicates the existence of approximately 11 genera within the α-proteobacteria class. These include *Azorhizobium*, *Bradyrhizobium*, *Devosia*, *Ensifer*, *Methylobacterium*, *Mesorhizobium*, *Microvirga*, *Ochrobactrum*, *Phyllobacterium*, *Rhizobium*, and *Shinella*. Additionally, recent discoveries within the α-proteobacteria class have identified a new genus, *Neorhizobium*, which represents a distinct clade comprising *Rhizobium galegae*, *Rhizobium vignae*, *Rhizobium huautlense*, and *Rhizobium alkalisoli* [26], and *Aminobacter* isolated from the root nodules of *Anthyllis vulneraria and the total nodulating genus* in class has been increased to 13 [9]. *Burkholderia* and *Cupriavidus* in the β-proteobacteria and some unclassified strains in the γ-proteobacteria [14] have also been recently described.

4.3 LEGUME–RHIZOBIA ASSOCIATION: SYMBIOSIS

The symbiotic relationship between legumes and rhizobia is defined by the development of nodules. During this process, communication occurs between the host plant and bacteria, facilitated by a variety of chemical signals [27] that leads to the curling of root hairs and colonization of root nodule bacteria

(RNB) in the curled portion; further, fusion of endomembranous system leads to the formation of infection thread through which the bacteria travels to the cortex and fixes atmospheric nitrogen in light of nitrogenase enzyme [28, 29]. Bacteria infiltrate host plants through two distinct mechanisms: (1) intercellular and (2) intracellular. In the latter process, bacteria dissolve the plant cell wall, a phenomenon prevalent in the majority of legume hosts, including well-studied models such as *Lotus japonicus* and *Medicago truncatula*. On the other hand, in the intercellular process observed in numerous tropical legumes, bacteria gain entry through cracks in the epidermis, which typically form at sites where lateral or adventitious roots emerge [30, 31]. Bacteria within the plant cell remain enclosed by a plant-derived membrane; within this structure, the rhizobial bacteria differentiate into bacteroides, and these coupled with the surrounding plant membrane are referred as "symbiosomes," the cellular structures that are capable of nitrogen fixation [32]. The nodules with pink-colored leghemoglobin in the center are considered effective, while colorless appearance in cross-sections indicates the ineffective nodules in terms of nitrogen fixation [22, 33, 34]. Whether it is intracellular or intercellular infection, nodulation depends on the expression of many genes of plant and bacterial origins, and one of the early requirements is the expression of nodulation gene that stimulates the plant genes for nodulation and results in the formation of nodules [35–37]. Within bacteria, the genes responsible for nodulation (*nod*) and nitrogen fixation (*nif* and *fix*) are located on the *sym* loci. In fast-growing rhizobia, these genes are encoded on plasmids, while in slow-growing rhizobia, they are found on chromosomal islands. The symbiosis between legumes and rhizobia exhibits significant specificity, a characteristic attributed to the NOD factor encoded by the *nod* gene. The NOD factor plays a crucial role in determining the host range of a specific rhizobial strain in the legume–rhizobia symbiotic relationship [38–40]. The gene, which governs the structure of this factor, is called host-specific nodulation (*hsn*) genes, and different strains have different complements of *hsn* genes [40, 41]. Precise compatibility between rhizobia and legumes is required especially in the soils, which lack native rhizobial population [42].

4.4 ISOLATION AND PURIFICATION OF ROOT NODULE BACTERIA

For isolation of RNB, the roots are first washed thoroughly to remove adherent soil particles. When detaching nodules from the roots, forceps are utilized on the root appendages to minimize the risk of damaging the nodules.

The separated nodules, kept intact and undamaged, undergo immersion in 95% ethanol or isopropanol for 5–10 seconds to break surface tension and eliminate air bubbles from the tissue. Subsequently, these nodules are placed in a 0.1% (w/v) mercuric chloride solution for 3–5 minutes, followed by thorough rinsing (five to six times) with sterile water using sterile forceps under aseptic conditions. After surface sterilization, the nodules are crushed with a pair of blunt-tipped forceps in a sizable drop (approximately 50–100 μL) of sterile water within a petri dish. A single loop full of the resulting nodule suspension is streaked onto a Yeast Extract Mannitol Agar (YEMA) plate containing Congo red (CR), as outlined by Somasegaran and Hoben [28]. The cultures are purified through repeated streaking and by culturing single pure colony (Figure 4.1). The bacterial strains are also observed regularly for their gram-negative character to check any contamination.

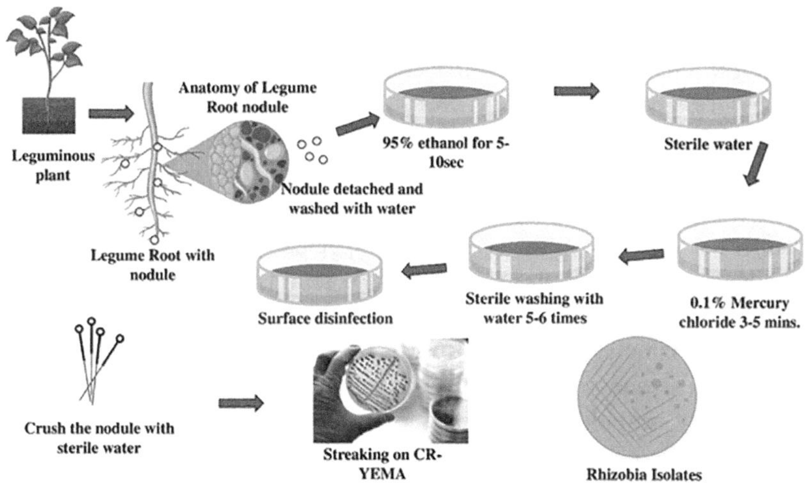

FIGURE 4.1 Methodology for isolation and purification of RNB.

4.5 CHARACTERIZATION OF RHIZOBIA

The conventional characterization and classification of microorganisms lack precision, prompting the adoption of advanced methods in bacterial systematics. Contemporary approaches include the utilization of complete 16S rRNA gene sequencing with comparative analysis through phylogenetic trees, DNA–DNA hybridization studies with closely related organisms, examination of molecular markers, biochemical assays, and evaluations based on physiological and morphological tests. Together, these methods—genotypic, chemotaxonomic, and phenotypic—constitute the "polyphasic approach" for bacterial systematic analysis [43, 44]. Additionally, techniques such as multilocus sequence analysis (MLSA) and whole-genome sequence analysis [45] offer rapid and reliable means for phylogenetic and taxonomic studies. These advanced methods accelerate the identification process, particularly beneficial in uncovering the still poorly understood diversity of nitrogen-fixing rhizobia in tropical regions. Recently, *in silico* DNA–DNA hybridization and average nucleotide identity (ANI) have been suggested to exclude wet DNA–DNA hybridization [46, 47]. Proteomics-based identification and characterization is also gaining importance as new techniques for characterizing bacteria [48].

4.5.1 PHENOTYPIC CHARACTERIZATION OF RHIZOBIA

Phenotypic characteristics for the characterization of RNB are considered less reliable compared to the higher serological techniques. This morphological aid provides a weaker base for the recognition of strains and provides a poor species diagnosis system [49]. Therefore, in general, it can be concluded that by adopting these phenotypic techniques, it is possible to identify strain within the species, but importance of the methods declines when the recognition of strains from different species or genera is required [50]. The diversity among the rhizobia based on phenotypic characters has been studied by a number of workers [51–53]. Although these phenotypic studies cannot be used alone for phylogenetic analysis, they provide the base for comprehensive polyphasic taxonomy [54]. A number of phenotypic tests such as NaCl tolerance, pH tolerance, acidification or alkalinizing response, and carbon utilization patterns (metabolic fingerprinting) can be performed for the initial grouping of RNB, as described by Choudhary et al. [55]. It had been concluded from many studies that the increasing salt concentrations decline the growth of rhizobia [56, 57]. Many studies showed a tolerance to pH 4–9 and growth at 513 mM salt concentration [58]. Utilization of multiple C source and different sensitivity to antibiotics is plasmid-based characters and may significantly affected by local environment and vary in strain to strain. Utilizing a greater number of carbohydrates may have an advantage for their survival in local environmental conditions [59]. In the contemporary context, the advent of automated applications such as BIOLOG and API has significantly simplified procedures, providing a more convenient method for conducting multiple sole carbon source tests within a single microtiter plate. Typically, *Rhizobium* species exhibit a broader capacity to utilize various carbohydrates compared to strains of *Bradyrhizobium* [18, 60]. These phenotypic characters are plasmid based; therefore, they are recurrently affected by horizontal gene transfer (HGT), and such transfer is also favored by environmental stresses [61]. HGT also creates phenotypic divergence within the strains of same species [62] and, thus, cannot be used as authentic aids to identify bacterial species.

4.5.2 BIOCHEMICAL CHARACTERIZATION OF RHIZOBIA

The biochemical characters of rhizobia include nitrate reductase, citrate utilization, indole production, and production of enzyme amylase; these characters such as the phenotypic characters are plasmid based and, therefore, could be used for the diagnosis of diazotrophic nitrogen-fixing bacteria but are usually not useful for diagnosing rhizobial strains to the species level. Reduction of nitrate can be assessed in YEMA medium containing 0.1% (w/v) KNO_3, as described by Lindström and Lehtomäki [63]. The catalase activity of rhizobial strains can be determined by the method used by Graham and Parker [64] and for gelatinase activity method described by El Idrissi et al. [65] can be utilized. Although the biochemical reactions such as urease, nitrate reductase, and oxidase are not very common in rhizobia, it has been concluded by many studies [52, 63, 65] who found that the production of these enzymes is not universal for all rhizobia, but most of the species of *Rhizobium* except *Rhizobium. lupini* had nitrate reductase activity. Sadowsky et al. [66] conducted a comparative analysis of biochemical characteristics between fast- and slow-growing soybean rhizobial strains. The examined rhizobia demonstrated positive traits for catalase, oxidase, urease, penicillinase, and nitrate reductase. However, none of the strains generated hydrogen sulfide or utilized citrate as a sole carbon source. Notably, a distinct difference in gelatinase activity was observed, with fast-growing soybean strains testing positive and slow growers exhibiting negative gelatinase activity.

4.5.3 PLANT GROWTH-PROMOTING ACTIVITIES OF RHIZOBIA

The plant growth-promoting rhizobacteria (PGPR) perform their activities by releasing a variety of substance in the rhizosphere in order to alter the whole microbial community in plant rhizosphere [67]. In general, PGPR contribute to the direct enhancement of plant growth by aiding in the acquisition of resources such as nitrogen, phosphorus [68], and essential minerals. Additionally, they regulate plant hormone levels [69]. PGPR also exert an indirect positive impact by mitigating the inhibitory effects imposed by diverse pathogens on plant growth and development [70]. This is achieved through acting as biocontrol agents and inducing resistance in host plants against both plant pathogens and abiotic stresses [71]. The rhizobia in addition to their N-fixing abilities have also been explored for their PGP activities such as phosphate solubilization [72], indole-3-acetic acid (IAA) production, ethylene reduction, siderophore production, and lytic enzyme and biocontrol activities against deleterious plant pathogens [73]. Further classification of these PGPR into extracellular and intracellular was carried out by Vessey [74] according to which the genera *Bacillus, Pseudomonas, Erwinia, Caulobacter, Serratia, Arthrobacter, Micrococcus, Flavobacterium, Chromobacterium, Agrobacterium,* and *Hyphomicrobium* come into the category of ePGPR, while the genera *Rhizobium, Bradyrhizobium, Ensifer, Azorhizobium, Mesorhizobium*, and *Allorhizobium referred as iPGPR.* Phosphorus (P) stands as the second crucial nutrient-limiting plant growth, following nitrogen. Similar to abundant atmospheric nitrogen, phosphorus exists as rock phosphate in the soil and is not readily available in a form accessible to plants. The absorbed forms of P for plants are monobasic ($H_2PO_4^-$) and the dibasic (HPO_4^{2-}) ions [75], and the concentrations of these elements in the soil typically range from 0.01 to 3.0 mg/L, representing a minor portion of the total phosphorus essential for plant growth. A media, Pikovskaya's, has been defined for screening phosphate-solubilizing bacteria (PSB) is widely used [72]. Among the diverse microorganisms possessing the capability to solubilize phosphorus, PSB stand out as potent bioagents. These microorganisms enhance the accessibility of phosphorus to plants by releasing various phosphatases that catalyze the hydrolysis of phosphoric esters [71]. This enzymatic activity plays a crucial role in making phosphorus available to plants in a form that might otherwise remain inaccessible. Notably, bacterial genera such as *Serratia, Rhizobium, Pseudomonas, Mycobacterium, Flavobacterium, Erwinia, Enterobacter, Burkholderia, Beijerinckia,*

Bacillus, and *Azotobacter* are identified as particularly significant PSB in this context [75]. In addition to P solubilization, another significance of PSB is their role in stimulating the efficiency of BNF [76]. Within RNB, the bacteria of genus *Rhizobium*, *Bradyrhizobium*, and *Mesorhizobium* are the potential P solubilizer [77, 78] that discharges certain low-molecular weight organic acids to dissolve the inorganic phosphates. The plant hormone IAA, commonly known as phytohormone, plays a pivotal role in fostering plant growth and development. It contributes to the enhancement of both root and shoot growth, elevates seedling vigor, and actively participates in the differentiation of vascular bundles. Furthermore, IAA is indispensable for the formation of nodules. A significant number of rhizospheric bacteria present in the crop rhizosphere possess the capability to produce and release auxins as secondary metabolites [79]. The auxins released by rhizobacteria influence the plant's endogenous pool of IAA [71], and reciprocally the plant IAA modulates gene expression of many rhizobacteria, thus playing a significant role in plant–microbe interactions. As far as the RNB is concern, it has been reported that many genera of Rhizobiaceae, including *Azorhizobium caulinodans, Bradyrhizobium japonicum, Bradyrhizobium elkanii, Mesorhizobium loti, Rhizobium japonicum, Rhizobium leguminosarum, Rhizobium lupine, Rhizobium meliloti, Rhizobium phaseoli, Rhizobium trifolii*, and *Ensifer* spp. [73, 78, 80, 81], were also showed IAA production and prove their potential to be used as efficient PGPR. *Rhizobia* in the form of biocontrol agents limit the growth of undesired organisms or pathogens, and, thus, the application of these bacteria to control soilborne pathogens is a potential eco-friendly procedure. There are varieties of processes through which RNB carry out their biocontrol activities; likewise, many of them compete with pathogens for nutrients and starve pathogens by displacing those [82] and some of the rhizobial strains, including *R. leguminosarum* bv. *trifolii, R. leguminosarum* bv. *viciae, R. meliloti, R. trifolii, Ensifer meliloti*, and *B. japonicum,* have been reported to secrete antibiotics and cell wall degrading enzymes that can inhibit the phytopathogens [83–85].

4.5.4 MOLECULAR CHARACTERIZATION OF RHIZOBIA

The characterization of rhizobia through phenotypic and biochemical methods provides a basic and important part in classification, but these techniques are less reliable and authentic; therefore, the significance of more potential methods in the form of molecular detection has been increased in the past decades [86]. Due to the considerable taxonomic diversity among rhizobia [87], there is a need for a robust classification system to characterize and identify effective nitrogen-fixing strains [88]. The advent of molecular genetic techniques, coupled with the availability of highly sensitive and accurate polymerase chain reaction (PCR)-based fingerprinting methods, has facilitated the development of precise characteristic patterns. These patterns, when visualized on agarose gels, offer distinct separation at the strain level [89]. This advancement has significantly streamlined microbial characterization, allowing for the differentiation of genera, species, and even closely related strains in a facile and rapid manner [90, 91]. Wide arrays of molecular methods have come into view nowadays for classification and identification of rhizobia that include (1) restriction fragment length polymorphisms (RFLPs), (2) random amplified polymorphic DNA (RAPD) using random primers, (3) DNA–DNA hybridization, (4) molecular analysis of symbiotic genes (*nod* and *nif*), (5) polyacrylamide gel electrophoresis (PAGE) of total proteins, and (6) analysis of conserved 16S ribosomal genes [92, 93]. The basis of shift from non-DNA-based microbial characterization to the DNA-based approaches is that these molecular techniques are very simple in application, less time-consuming, robust, and can distinguish between closely related species and also reflects phylogenetic relationships. In addition, these techniques do not require culturing the rhizobia from root nodules and having potential to identify the rhizobia directly from soils [93]. The choice of a method for molecular characterization is predominantly determined by the laboratory's requirements, desired level of resolution, and available resources.

The studies discussed above such as DNA–DNA hybridization and amplified fragment length polymorphisms (AFLPs) do not reveal exact species boundaries [94], as they do not consider HGT. Moreover, the implication of whole-genome analyses methods offers a clear version of diversity within and among species by taking account the phenomena of HGT, which is a quite important parameter for understanding the symbiotic efficiency and host range of rhizobia [40]. While investigations focused on particular genetic loci may be less labor intensive, they contribute to the understanding of HGT. Consequently, many studies adopt a polyphasic approach, utilizing genetic information from various specific loci and incorporating a diverse range of data types [95–97].

4.5.4.1 MULTILOCUS FINGERPRINTING ANALYSIS

The random amplification of polymorphic DNA [98] with the use of arbitrary primers that generate PCR-based genomic fingerprints for a specific organism is a way to explore the genetic diversity [99]. The applications of this method are widespread, including authentication of inoculant strains (for comparison of reisolate with parental strain), assessment of genetic diversity, and genetic relatedness among field populations of rhizobia [92, 100]. The diversity among the *Bradyrhizobium* strains by using GC-rich random primers has been assessed by many workers [100].

It was Richardson et al. [92], who constructed a *nifH*-directed 20 nucleotides primer (RPO-I) that corresponds to conserved *nifHDK* promoter elements of *R. leguminosarum* bv. *trifolii*, for classifying diverse collection of *R. meliloti*, *R. leguminosarum* biovars, and a wide range of *Rhizobium* and *Bradyrhizobium* strains [92]. The primer has its widespread utilities for identification of strains of rhizobia [92, 93]. In addition to its unique potential to evaluate the genetic diversity among the rhizobial strains, another attribute that makes the primer appropriate for genetic studies is capability to generate the amplicons at annealing temperatures as high as 65°C, which otherwise not possible with the contemporary primers [92]. The likely reason for the reproducibility of RAPD-PCR lies in the utilization of a primer that is specific to the sequence and the application of a high annealing temperature [101].

The rep-PCR, or repetitive element sequence-based PCR, is an innovative typing technique designed to distinguish microorganisms. This method employs primers that match interspersed repetitive consensus sequences, facilitating the amplification of DNA fragments of varying sizes. These fragments encompass sequences situated between the repetitive elements, allowing for effective microbial differentiation [102]. The enterobacterial repetitive intergenic consensus (ERIC) sequences of 126 bp with highly conserved central inverted repeats are located in extragenic regions of the bacterial genome that have been found to be very constructive for fingerprinting bacteria using PCR [103]. The operational mechanism of both ERIC and RAPD is comparable, with the former utilizing primers designed to match naturally existing genomic DNA sequences (Table 4.1) [93]. The importance of these methods for discrimination abilities is although found to be variable among different groups of organisms [93, 97, 104] but valuable for identification of strains of a single species [97]. Collectively, these methods provide useful tool for strain discrimination and also include noncoding genes [93, 97, 105].

The limitation or drawbacks of these PCR-based methods (ERIC, rep-PCR, RPO-I) is the lack of reproducibility that is largely due to the formation of heteroduplex between the homologous sequences and availability of multiple sites for amplification that leads to inconsistence genotyping [106].

TABLE 4.1 List of Primers for the Molecular Characterization of RNB

Primer	Sequence (5'→3')	References
	RAPD	
RPO-I	AATTTTCAAGCGTCGTGCCA	[151]
	16S ribosomal RNA	
FGPS6	GGAGAGTTAGATCTTGGCTCA	[152]
FGPS159	AAGGAGGGGATCCAGCCGCA	[152]
18F	AGAGTTTGATCCTGGCTCAG	[153]
1492R	CTACGGCTACCTTGTTACG	[153]
800F	GTAGTCCACGCCGTAAACGA	[154]
820R	CATCGTTTACGGCGTGGACT	[154]
1100F	AAGTCCCGCAACGAGCGCAA	[154]
	Sym genes	
nodA1	TGCRGTGGAARNTRNNCTGGGAAA	[155]
nodA2	GGNCCGTCRTCRAAWGTCARGTA	[155]
nodA3	TCATAGCTCYGRACCGTTCCG	[156]
nifHF	TACGGNAARGGSGGNATCGGCAA	[157]
nifHI	AGCATGTCYTCSAGYTCNTCCA	[157]
	Housekeeping genes	
recA6F	CGKCTSGTAGAGGAYAAATCGGTGGA	[137]
recA555R	CGRATCTGGTTGATGAAGATCACCAT	[137]
glnII-1	AACGCAGATCAAGGAATTCG	[135]
glnII-2	ATGCCCGAGCCGTTCCAGTC	[135]
TSdnaK2	GTACATGGCCTCGCCGAGCTTCA	[150]
TSdnaK3	AAGGAGCAGCAGATCCGCATCCA	[150]

Note. N = A, C, G, or T; R = A, G; W = A, T; Y = T, C; and S = G, C

4.5.4.2 RESTRICTION FRAGMENT LENGTH POLYMORPHISM

The RFLP is a non-PCR-based molecular marker and has the ability to show variation under both homozygous and heterozygous conditions and, thus, regarded as codominant marker [107]. The method involves the extraction of total genomic DNA followed by digestion with specific restriction endonucleases, which generates diverse pattern of band revealed on agarose gel electrophoresis [108]. The technique has been widely adopted for the analysis of bacterial genetic diversity and provides discrimination information at the species, subspecies, and strain levels [109]. Additionally, utilizing PCR-based locus-specific RFLP linked to ribosomal genes could prove advantageous for phylogenetic investigations. These methods typically exhibit high reproducibility and demonstrate strong consistency with both partial and complete gene sequencing, as evidenced in previous studies [110].

The PCR-RFLP of *16S* or *23S rRNA* or parts of both genes with or without the intergenic spacer (IGS) region is called amplified ribosomal DNA restriction analysis (ARDRA) [111]. The commonly employed method for studying microbial diversity relies on DNA polymorphism [112]. This involves amplifying specific loci within conserved regions using either universal- or genus-specific primer sets. The resulting amplified products are then subjected to digestion by tetracutting restriction endonucleases [113], followed by the separation of resulting fragments through high-density agarose gel electrophoresis. The resultant profiles are utilized either for clustering the microbial community into genotypic groups or for strain typing [108]. This method has found extensive application in assessing diversity among legume symbionts, particularly screening up to the genus level [114, 115]. In a study by Terefework et al. [116], PCR-RFLP using nine restriction enzymes was applied to the *16S* and *23S rRNA* genes of 42 rhizobial and agrobacterial strains. The aim was to determine the phylogenetic position of *R. galegae* and enhance understanding of the evolution of ribosomal operons. Furthermore, several studies have indicated that the IGS contains conserved portions of the genome, allowing for a higher level of discrimination, especially between closely related species. IGS has been effectively utilized for the identification of various rhizobial species [94].

4.5.4.3 RIBOSOMAL GENE (16S RRNA) ANALYSIS METHODS

Current bacterial phylogeny heavily depends on the analysis of conserved sequences within the *16S rRNA* gene, and to a certain extent, the *23S rRNA* gene. These genes serve as valuable tools for identifying new species, particularly due to their capacity to reveal limited variability among strains within bacterial species [117]. As these ribosomal sequences are functionally constant and also contributes in translation processes, therefore, many studies with the aim to differentiate the species of rhizobia comprehensively opted the sequence comparisons of *16S rRNA* genes to compare organisms and to infer phylogenies [118, 119]. In addition, the sequences of this conserved loci have been used extensively to characterize rhizobial strains at the species or higher level [120]. Partial or complete sequence analysis of this conserved loci (*16S rRNA*) has been used comprehensively as a rapid tool for characterization of rhizobial strains since Peter et al. [19] proposed the significance of full-length sequences in rhizobial phylogeny, while the partial sequences (by using single primer) contribute in rapid screening of bacterial strains [116]. Further, these sequences incorporate both conserved as well as variable regions, which add up in their significance to distinguish distantly and closely related organisms, respectively [121]. A correlation has been made between the DNA–DNA hybridization and sequence similarities of *16S rRNA* gene [122] and concluded that the strains, which share 70% genomic similarity in hybridization, show 97% sequence similarities for *16S rRNA* gene. For several decades, the most favored method to delineate new species involved a combination of *16S rRNA* gene phylogeny and DNA–DNA hybridization. In this approach, bacterial strains exhibiting a 70% or greater resemblance in their DNA–DNA hybridization, coupled with less than a 5% difference in the melting temperature, were regarded as belonging to the same species [45]. It is noteworthy that hybridization values of 50% or less lack significant informativeness, making DNA–DNA hybridizations unsuitable for estimating genetic distances between distantly related species [123]. The analysis of the 16S rRNA gene, considered a universal marker gene, has been instrumental in identifying various novel nitrogen-fixing symbionts within genera such as *Burkholderia*, *Methylobacterium*, *Ochrobactrum*, *Devosia*, *Ralstonia (Cupriavidus)*, and *Phyllobacterium* [124–126].

As discussed above, this technique has capability for resolving degree of relatedness between organisms above species level; nevertheless, it has limitation to explore diversity at interspecies level [94, 97, 122]. Moreover, evidences are also there for the incidence of chimeric *16S rRNA* sequences as well as multiple copy of different *16S rRNA* in certain rhizobia and for transfer of complete 16S sequences between unrelated rhizobia [127, 128] making the approach less effective to conclude novelty of sequences. The lone use of this gene sequence as marker for phylogenetic studies is also not suitable due to unequal evolutionary rates and phylogenetic tree-building artifacts [129]; therefore, it is suggested to consider the sequences of multiple loci for authentic evaluation of phylogeny [96, 97, 129].

4.5.4.4 MULTILOCUS SEQUENCE ANALYSIS METHODS

Besides ribosomal genes, various protein-coding housekeeping genes play a crucial role in the taxonomic analysis of rhizobia [130–132]. MLSA is often considered an alternative approach to traditional methods such as 16S rRNA sequence analysis and DNA–DNA hybridization [45]. This shift is primarily due to the limitations of more conserved rRNA gene sequences, which may not always provide sufficient discrimination between species. Consequently, MLSA is now recognized as a superior method compared to both DNA–DNA hybridization and 16S rRNA gene sequence analysis [133]. MLSA involves the use of highly divergent sequences from housekeeping genes (coding for proteins), allowing for the formation of sequence clusters at various taxonomic levels [134]. This higher degree of sequence divergence in housekeeping genes enhances the resolution and accuracy of species differentiation, making MLSA a valuable tool in the field of rhizobial taxonomy.

The routinely utilized housekeeping genes include *dnaK* [96]; *glnA* and *glnB* [135, 136]; *atpD* and *recA* [137]; and *dnaK*, *gyrB*, and *gltA* [138] for the construction of concatenated trees as they have profound usefulness in inferring phylogenetic relationships [93, 97]. The incongruent phylogenetic events in one or more lineages through the horizontal transfer of genes could be revealed with the use of MLSA [139]; additionally, they can be utilized to infer phylogenetic position and relationships among species of *Bradyrhizobium* [140], *Rhizobium* [141], *Ensifer* [105], and *Mesorhizobium* [142].

4.5.4.5 SEQUENCE ANALYSIS OF SYMBIOTIC (SYM) LOCI

Some auxiliary or accessory genes located on *sym* loci are involved in the legume symbiosis and commonly included in species description of new rhizobia. The genetic components encompassing nodulation genes (*nod*, *nol*, and *noe*) and nitrogen-fixation genes (*nif*, *fix*) play a crucial role in affirming the bacterium's capability to engage in nodulation and nitrogen fixation in symbiotic relationships with legume hosts, as evidenced by previous studies [95]. The bacterial phylogeny does not reflect its symbiotic characteristics and host range. Broader the host range, a greater number of plants can be nodulated as *Ensifer fredii* NGR234 carried out infection in roots of plants belonging to 112 genera of family Leguminosae, and additionally the genus *Parasponia* of family Cannabaceae showing a vast host range [143]. The rhizobia with narrow host range nodulate only few plants or may carry out nodulation in their parental host only; for example, the *R. leguminosarum* bv. *trifolii* form nodules only on plants of the genus *Trifolium* [144]. The sequence analysis of these symbiotic genes although utilizes for identification purposes to some extent, but their major role is to exploit geographic or host range relationships among bacterial strains [145]. Comparative studies of these symbiotic genes with housekeeping and core genes (16S) reflect the phenomena of HGT in the evolution of the *sym* loci [21, 40]. Accordingly, the phylogenetic trees constructed with these *sym* genes often show incongruence with the phylogenetic trees of protein-coding housekeeping and ribosomal genes again limiting their use as a tool to reveal taxonomic status [95–97]. Conclusively, the utilization of sequences encoding *nif* and *nod* genes has provided definitive evidence, indicating the effective isolation of diverse regional rhizobial populations over evolutionary time. This isolation has led to the emergence of locally adapted ecotypes or biotypes, as demonstrated in several studies [21, 95–97]. At present, rhizobia belonging to the same species are categorized into biovars (symbiovars) based on their specificity to symbiotic hosts. This classification primarily relies on the analysis of the *nodC* gene's phylogenetics [9]. The term "biovar" was first used to differentiate three clusters of bacteria *R. leguminosarum*: (1) *R. leguminosarum* bv. *trifolii,* (2) bv. *viciae* and (3) bv. *phaseoli* that showed nodulation in *Trifolium* sp., *Vicia* sp., and *Phaseolus* sp., respectively [15]. The notion of symbiovars appears to be applicable and effective for plants that exhibit a more constrained nodulation pattern, specifically those forming associations with specific rhizobia. However, in the context of legumes that are promiscuous in their nodulation, interacting with various rhizobia strains, there is an ongoing discussion regarding this matter [146]. The phylogenetic studies are, therefore, the determination of biovars based on the *nodC* [140, 147], *nodA* [148], and even *nifH* gene sequences [135, 137, 149–175]. To enhance the comprehension of *Rhizobium*–fabacean symbiosis, it is advisable to incorporate the characterization and phylogenetic analysis of symbiotic genes into the minimum standards for describing new fabacean symbionts (Figure 4.2).

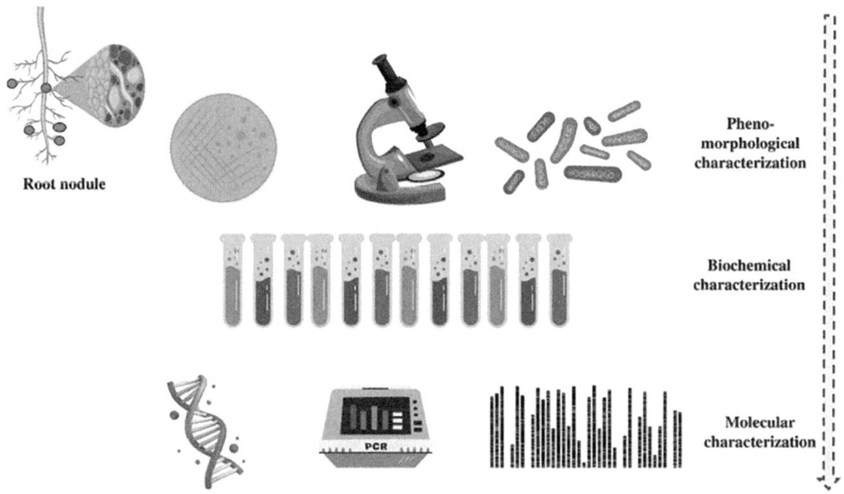

FIGURE 4.2 Characterization of RNB.

4.6 FUTURE DIRECTIONS

Future developments in bioinformatics and molecular approaches will further transform our knowledge and application of rhizobia in agroforestry and sustainable agriculture. The development and application of genomic technologies such as whole-genome sequencing and MLSA provide a potential avenue. By using these techniques, researchers may quickly and accurately identify rhizobial strains and explore the wide range of unexplored nitrogen-fixing symbionts. We can better comprehend these bacteria' symbiotic potential and adjust them to certain plant species and environmental situations by analyzing their genetic composition. This will improve soil health and agricultural output. Besides, thorough evaluations of the microbial communities in the rhizosphere will be possible because developments in metagenomics and high-throughput sequencing, which will provide insight into the intricate relationships that exist between rhizobia, plants, and other soil microbes. More promisingly, the development of synthetic biology techniques may enable the production of rhizobial strains with increased nitrogen-fixing efficiency and a wider host range, opening the door to more environmentally friendly farming methods. All things considered, the identification, and use of novel rhizobial symbionts will be fueled by ongoing advancements in microbial ecology, genetics, and biotechnology. This will enable the creation of specialized responses to a range of agricultural problems in the years to come.

KEYWORDS

- **biological nitrogen fixation**
- **rhizobia**
- **symbiosis**
- **diazotrophs**
- ***16S rRNA***
- **MLSA**

REFERENCES

1. Schlesinger, W. H., Lichter, J. (2001). Limited carbon storage in soil and litter of experimental forest plots under increased atmospheric CO_2. *Nature, 411*(6836), 466–469. doi: 10.1038/35078060
2. Smil, V. (1997). Global population and the nitrogen cycle. *Scientific American, 277*(1), 76–81. doi: 10.1038/scientificamerican0797-76
3. Morel, M. A., Braña, V., Castro-Sowinski, S. (2012). Legume crops, importance and use of bacterial inoculation to increase production. *Crop Plant, 12*, 217–240.
4. Zahran, H. H. (2001). Rhizobia from wild legumes: Diversity, taxonomy, ecology, nitrogen fixation and biotechnology. *Journal of Biotechnology, 91*(2–3), 143–153. doi: 10.1016/s0168-1656(01)00342-x
5. Giller, K. E. (2001). *Nitrogen Fixation in Tropical Cropping Systems*. CAB International.
6. Jha, P. K., Nair, S., Gopinathan, M. C., Babu, C. R. (1995). Suitability of rhizobia-inoculated wild legumes *Argyrolobium flaccidum, Astragalus graveolens, Indigofera gangetica* and *Lespedeza stenocarpa* in providing a vegetational cover in an unreclaimed limestone quarry. *Plant Soil, 177*(2), 139–149. doi: 10.1007/BF00010120
7. Zahran, H. H. (1998). Structure of root nodules and nitrogen fixation in egyptian wild herb legumes. *Biologia Plantarum, 41*(4), 575–585. doi: 10.1023/A:1001848617980
8. Zahran, H. H. (1999). *Rhizobium*-legume symbiosis and nitrogen fixation under severe conditions and in an arid climate. *Microbiology and Molecular Biology Reviews, 63*(4), 968–989. doi: 10.1128/MMBR.63.4.968-989.1999
9. Peix, A., Ramírez-Bahena, M. H., Velázquez, E., Bedmar, E. J. (2015). Bacterial associations with legumes. *Critical Reviews in Plant Sciences, 34*(1–3), 17–42. doi: 10.1080/07352689.2014.897899
10. Moulin, L., Munive, A., Dreyfus, B., Boivin-Masson, C. (2001). Nodulation of legumes by members of the beta-subclass of proteobacteria. *Nature, 411*(6840), 948–950. doi: 10.1038/35082070

11. Chen, W. M., Laevens, S., Lee, T. M., Coenye, T., De Vos, P., Mergeay, M., Vandamme, P. (2001). *Ralstonia taiwanensis* sp. nov., isolated from root nodules of *Mimosa* species and Sputum of a cystic fibrosis patient. *International Journal of Systematic and Evolutionary Microbiology*, *51*(5), 1729–1735. doi: 10.1099/00207713-51-5-1729
12. Vandamme, P., Coenye, T. (2004). Taxonomy of the genus *Cupriavidus*: A tale of lost and found. *International Journal of Systematic and Evolutionary Microbiology*, *54*(6), 2285–2289. doi: 10.1099/ijs.0.63247-0
13. Gyaneshwar, P., Hirsch, A. M., Moulin, L., Chen, W. M., Elliott, G. N., Bontemps, C., Estrada-de los Santos, P., Gross, E., dos Reis Jr., F. B., Sprent, J. I., Young, J. P. W., James, E. K. (2011). Legume-nodulating betaproteobacteria: Diversity, host range, and future prospects. *Molecular Plant-Microbe Interactions*, *24*(11), 1276–1288. doi: 10.1094/MPMI-06-11-0172
14. Benhizia, Y., Benhizia, H., Benguedouar, A., Muresu, R., Giacomini, A., Squartini, A. (2004). Gamma proteobacteria can nodulate legumes of the genus *Hedysarum*. *Systematic and Applied Microbiology*, *27*(4), 462–468. doi: 10.1078/0723202041438527
15. Jordan, D. C. (1984). Family III. In: *Bergey's Manual of Systematic Bacteriology* (pp. 234–244). Rhizobiaceae Conn 1938; Williams & Wilkins.
16. Dommergues, Y. R., Diem, H. G., Gauthier, D. L., Dreyfus, B. L., Cornet, F. (1984). Nitrogen-fixing trees in the tropics: Potentialities and limitations. In: *Advances in Nitrogen Fixation Research* (pp. 7–13). Springer: Netherlands.
17. Fred, E. B., Baldwin, I. L., McCoy, E. *Root Nodule Bacteria and Leguminous Plants (No. 5). UW-Madison Libraries Parallel Press*, 2002.
18. Young, J. P. W. (1996). Phylogeny and taxonomy of rhizobia. *Plant Soil*, *186*(1), 45–52. doi: 10.1007/BF00035054
19. Peter, J., Young, W., Haukka, K. E. (1996). Diversity and phylogeny of rhizobia. *New Phytologist*, *133*(1), 87–94. doi: 10.1111/j.1469-8137.1996.tb04344.x
20. Young, J. M., Kuykendall, L. D., Martínez-Romero, E., Kerr, A., Sawada, H. (2001). A Revision of *Rhizobium* Frank 1889, with an emended description of the genus, and the inclusion of all species of *Agrobacterium* Conn 1942 and *Allorhizobium undicola* de Lajudie et al. 1998 as new combinations: *Rhizobium radiobacter, R. Rhizogenes, R. Rubi, R. undicola* and *R. vitis*. *International Journal of Systematic and Evolutionary Microbiology*, *51*(1), 89–103. doi: 10.1099/00207713-51-1-89
21. Sawada, H., Kuykendall, L. D., Young, J. M. (2003). Changing concepts in the systematics of bacterial nitrogen-fixing legume symbionts. *Journal of General and Applied Microbiology*, *49*(3), 155–179. doi: 10.2323/jgam.49.155
22. Sprent, J. I. (2001). How and why did nodulation evolve in certain legumes? In: *Legumes down Under* (pp 86–87). Australian National University, Division of Botany and Zoology.
23. Zakhia, F., De Lajudie, P. (2001). Taxonomy of rhizobia. *Agronomie*, *21*(6–7), 569–576. doi: 10.1051/agro:2001146
24. Bontemps, C., Rogel, M. A., Wiechmann, A., Mussabekova, A., Moody, S., Simon, M. F., Moulin, L., Elliott, G. N., Lacercat-Didier, L., Dasilva, C., Grether, R., Camargo-Ricalde, S. L., Chen, W., Sprent, J. I., Martínez-Romero, E., Young, J. P., James, E. K. (2016). Endemic *Mimosa* species from Mexico prefer alphaproteobacterial rhizobial symbionts. *New Phytologist*, *209*(1), 319–333. doi: 10.1111/nph.13573

25. Berrada, H., Fikri-Benbrahim, K. (2014). Taxonomy of the rhizobia: Current perspectives. *Microbiology Research Journal, 4*(6), 616–639. doi: 10.9734/BMRJ/2014/5635
26. Mousavi, S. A., Österman, J., Wahlberg, N., Nesme, X., Lavire, C., Vial, L., Paulin, L., De Lajudie, P., Lindström, K. (2014). Phylogeny of the *Rhizobium–Allorhizobium–Agrobacterium* clade supports the delineation of *Neorhizobium* gen. nov *Systematic and Applied Microbiology, 37*(3), 208–215. doi: 10.1016/j.syapm.2013.12.007
27. Hungria, M., Stacey, G. (1997). Molecular signals exchanged between host plants and rhizobia: Basic aspects and potential application in agriculture. *Soil Biology and Biochemistry, 29*(5–6), 819–830. doi: 10.1016/S0038-0717(96)00239-8
28. Somasegaran, P., Hoben, H. J. (2012). *Handbook for Rhizobia: Methods in Legume-Rhizobium Technology*. Springer Science+Business Media.
29. Deans, J. D., Ali, O. M., Lindley, D. K., Nour, H. O. A. (1993). Rhizobial nodulation of *Acacia* tree species in Sudan: Soil inoculum potential and effects of peat. *Journal of Tropical Forest Science, 6*(1), 56–64.
30. Chandler, M. R., Date, R. A., Roughley, R. J. (1982). Infection and Root-Nodule Development in *Stylosanthes* species by *Rhizobium*. *Journal of Experimental Botany, 33*(1), 47–57.
31. Goormachtig, S., Capoen, W., James, E. K., Holsters, M. (2004). Switch from intracellular to intercellular invasion during water stress-tolerant legume nodulation. *Proceedings of the National Academy of Sciences of the United States of America, 101*(16), 6303–6308. doi: 10.1073/pnas.0401540101
32. Newcomb, W. (1981). Nodule morphogenesis and differentiation. *International Review of Cytology*, *13,* 247–298.
33. Corby, H. D. L., Polhill, R. M., Sprent, J. I. (1983). Taxonomy. *Nitrogen Fixation, 3*, 1–35.
34. Heichel, G. H., Vance, C. P. (1983). *Physiology and Morphology of Perennial Legumes*. ARS Reprints Collection.
35. Cullimore, J. V., Ranjeva, R., Bono, J. J. (2001). Perception of lipo-chitooligosaccharidic nod factors in legumes. *Trends in Plant Science, 6*(1), 24–30. doi: 10.1016/s1360-1385(00)01810-0
36. D'Haeze, W., Mergaert, P., Promé, J. C., Holsters, M. (2000). Nod factor requirements for efficient stem and root nodulation of the tropical legume *Sesbania rostrata*. *Journal of Biological Chemistry, 275*(21), 15676–15684. doi: 10.1074/jbc.275.21.15676
37. Oldroyd, G. E. (2001). Dissecting symbiosis: developments in nod factor signal transduction. *Annals of Botany, 87*(6), 709–718. doi: 10.1006/anbo.2001.1410
38. Spaink, H. P. (2000). Root nodulation and infection factors produced by rhizobial bacteria. *Annual Review of Microbiology, 54*(1), 257–288. doi: 10.1146/annurev.micro.54.1.257
39. Perret, X., Staehelin, C., Broughton, W. J. (2000). Molecular basis of symbiotic promiscuity. *Microbiology and Molecular Biology Reviews, 64*(1), 180–201. doi: 10.1128/MMBR.64.1.180-201.2000
40. Moulin, L., Béna, G., Boivin-Masson, C., Stępkowski, T. (2004). Phylogenetic analyses of symbiotic nodulation genes support vertical and lateral gene co-transfer within the *Bradyrhizobium* genus. *Molecular Phylogenetics and Evolution, 30*(3), 720–732. doi: 10.1016/S1055-7903(03)00255-0

41. Carlson, R. W., Sanjuan, J., Bhat, U. R., Glushka, J., Spaink, H. P., Wijfjes, A. H., van Brussel, A. A., Stokkermans, T. J., Peters, N. K., Stacey, G. (1993). The structures and biological activities of the lipo-oligosaccharide nodulation signals produced by type I and II strains of bradyrhizobium japonicum. *Journal of Biological Chemistry, 268*(24), 18372–18381. doi: 10.1016/S0021-9258(17)46854-2
42. Brockwell, J., Bottomley, P. J., Thies, J. E. (1995). Manipulation of rhizobia microflora for improving legume productivity and soil fertility: A critical assessment. *Plant Soil, 174*(1–2), 143–180.
43. De Lajudie, P., Willems, A., Pot, B., Dewettinck, D., Maestrojuan, G., Neyra, M., Collins, M. D., Dreyfus, B., Kersters, K., Gillis, M. (1994). Polyphasic taxonomy of rhizobia: Emendation of the genus *Sinorhizobium* and description of *Sinorhizobium meliloti* comb. nov., *Sinorhizobium saheli* Sp. Nov., and *Sinorhizobium teranga* Sp. Nov. *International Journal of Systematic Bacteriology, 44*(4), 715–733. doi: 10.1099/00207713-44-4-715
44. Jarvis, B. D. W., Van Berkum, P., Chen, W. X., Nour, S. M., Fernandez, M. P., Cleyet-Marel, J. C., Gillis, M. (1997). Transfer of *Rhizobium loti, Rhizobium huakuii, Rhizobium ciceri, Rhizobium mediterraneum*, and *Rhizobium tianshanense* to *Mesorhizobium* gen. nov. *International Journal of Systematic and Evolutionary Microbiology, 47*(3), 895–898. doi: 10.1099/00207713-47-3-895
45. Stackebrandt, E., Frederiksen, W., Garrity, G. M., Grimont, P. A. D., Kämpfer, P., Maiden, M. C. J., Nesme, X., Rosselló-Mora, R., Swings, J., Trüper, H. G., Vauterin, L. Ward, A. C., Whitman, W. B. (2002). Report of the Ad hoc committee for the re-evaluation of the species definition in bacteriology. *International Journal of Systematic and Evolutionary Microbiology, 52*(3), 1043–1047. doi: 10.1099/00207713-52-3-1043
46. Konstantinidis, K. T., Tiedje, J. M. (2005). Towards a genome-based taxonomy for prokaryotes. *Journal of Bacteriology, 187*(18), 6258–6264. doi: 10.1128/JB.187.18.6258-6264.2005
47. Konstantinidis, K. T., Ramette, A., Tiedje, J. M. (2006). The bacterial species definition in the genomic era. *Philosophical Transactions of the Royal Society B: Biological Sciences, 361*(1475), 1929–1940. doi: 10.1098/rstb.2006.1920
48. Rahi, P., Prakash, O., Shouche, Y. S. (2016). Matrix-assisted laser desorption/ionization time-of-flight mass-spectrometry (MALDI-TOF MS) based microbial identifications: Challenges and scopes for microbial ecologists. *Frontiers in Microbiology, 7*, 1359. doi: 10.3389/fmicb.2016.01359
49. Jaftha, J. B., Strijdom, B. W., Steyn, P. L. (2002). Characterization of pigmented methylotrophic bacteria which nodulate *Lotononis bainesii*. *Systematic and Applied Microbiology, 25*(3), 440–449. doi: 10.1078/0723-2020-00124
50. Bergersen, F. J., Turner, G. L. (1980). Properties of terminal oxidase systems of bacteroids from root nodules of soybean and cowpea and of N2-fixing bacteria grown in continuous culture. *Microbiology, 118*(1), 235–252. doi: 10.1099/00221287-118-1-235
51. Dreyfus, B., Garcia, J. L., Gillis, M. (1988). Characterization of *Azorhizobium caulinodans* gen. nov., sp. nov., a stem-nodulating nitrogen-fixing bacterium isolated from *Sesbania rostrata*. *International Journal of Systematic Bacteriology, 38*(1), 89–98. doi: 10.1099/00207713-38-1-89

52. Zhang, X., Harper, R., Karsisto, M., Lindström, K. (1991). Diversity of *Rhizobium* bacteria isolated from the root nodules of leguminous trees. *International Journal of Systematic Bacteriology*, *41*(1), 104–113. doi: 10.1099/00207713-41-1-104
53. McCray Batzli, J. M., Graves, W. R., Van Berkum, P. (1992). Diversity among rhizobia effective with *Robinia pseudoacacia* L. *Applied and Environmental Microbiology*, *58*(7), 2137–2143. doi: 10.1128/aem.58.7.2137-2143.1992
54. Khbaya, B., Neyra, M., Normand, P., Zerhari, K., Filali-Maltouf, A. (1998). Genetic diversity and phylogeny of rhizobia that nodulate *Acacia* spp. in Morocco assessed by analysis of rRNA genes. *Applied and Environmental Microbiology*, *64*(12), 4912–4917. doi: 10.1128/AEM.64.12.4912-4917.1998
55. Choudhary, S., Meghwal, R. R., Sankhla, I. S., Tak, N., Gehlot, H. S. (2017). Molecular characterization and phylogeny of novel diverse nitrogen fixing microsymbionts associated with *Vachellia* (*Acacia*) *Leucophloea* in arid and semi-arid regions of Rajasthan. *Indian Forester*, *143*(3), 266–278.
56. Elsheikh, E. A. E., Wood, M. (1989). Response of chickpea and soybean rhizobia to salt: Osmotic and specific ion effects of salts. *Soil Biology and Biochemistry*, *21*(7), 889–895. doi: 10.1016/0038-0717(89)90077-1
57. Abdelmoumen, H., Filali-Maltouf, A., Neyra, M., Belabed, A., El Idrissi, M. M. (1999). Effect of high salts concentrations on the growth of rhizobia and responses to added Osmotica. *Journal of Applied Microbiology*, *86*(6), 889–898. doi: 10.1046/j.1365-2672.1999.00727.x
58. Zerhari, K., Aurag, J., Khbaya, B., Kharchaf, D., Filali-Maltouf, A. (2000). Phenotypic characteristics of rhizobia isolates nodulating *Acacia* species in the arid and Saharan regions of Morocco. *Letters in Applied Microbiology*, *30*(5), 351–357. doi: 10.1046/j.1472-765x.2000.00730.x
59. Zabaloy, M. C., Gómez, M. A. (2005). Diversity of rhizobia isolated from an agricultural soil in Argentina based on carbon utilization and effects of herbicides on growth. *Biology and Fertility of Soils*, *42*(2), 83–88.
60. McInroy, S. G., Campbell, C. D., Haukka, K. E., Odee, D. W., Sprent, J. I., Wang, W. J., Young, J. P. W., Sutherland, J. M. (1999). Characterisation of Rhizobia from African acacias and other tropical woody legumes using biolog™ and partial 16S RRNA sequencing. *FEMS Microbiology Letters*, *170*(1), 111–117. doi: 10.1111/j.1574-6968.1999.tb13362.x
61. Yost, C. K., Rath, A. M., Noel, T. C., Hynes, M. F. (2006). Characterization of genes involved in erythritol catabolism in *Rhizobium leguminosarum* bv. viciae. *Microbiology (Reading)*, *152*(7), 2061–2074. doi: 10.1099/mic.0.28938-0
62. de la Cruz, F., Davies, J. (2000). Horizontal gene transfer and the origin of species: Lessons from bacteria. *Trends Microbiology*, *8*(3), 128–133. doi: 10.1016/S0966-842X(00)01703-0
63. Lindström, K., Lehtomäki, S. (1988). Metabolic properties, maximum growth temperature and phage sensitivity of *Rhizobium* sp. (Galega) compared with other fast-growing rhizobia. *FEMS Microbiology Letters*, *50*(2–3), 277–287. doi: 10.1111/j.1574-6968.1988.tb02951.x
64. Graham, P. H., Parker, C. A. (1964). Diagnostic features in the characterisation of the root-nodule bacteria of legumes. *Plant Soil*, *20*(3), 383–396. doi: 10.1007/BF01373828

65. El Idrissi, M. M., Aujjar, N., Belabed, A., Dessaux, Y., Filali-Maltouf, A. (1996). Characterization of rhizobia isolated from carob tree (*Ceratonia siliqua*). *Journal of Bacteriology*, *80*(2), 165–173. doi: 10.1111/j.1365-2672.1996.tb03205.x
66. Sadowsky, M. J., Keyser, H. H., Bohlool, B. B. (1983). Biochemical characterization of fast- and slow-growing rhizobia that nodulate soybeans. *International Journal of Systematic and Evolutionary Microbiology*, *33*(4), 716–722.
67. Kloepper, J. W., Schroth, M. N. (1981). Plant growth-promoting rhizobacteria and plant growth under gnotobiotic conditions. *Phytopathology*, *71*(6), 642–644. doi: 10.1094/Phyto-71-642
68. Wakelin, S. A., Warren, R. A., Harvey, P. R., Ryder, M. H. (2004). Phosphate solubilization by *Penicillium* spp. closely associated with wheat roots. *Biology and Fertility of Soils*, *40*(1), 36–43. doi: 10.1007/s00374-004-0750-6
69. Madhaiyan, M., Poonguzhali, S., Ryu, J., Sa, T. (2006). Regulation of ethylene levels in canola (*Brassica campestris*) by 1-aminocyclopropane-1-carboxylate deaminase-containing *Methylobacterium fujisawaense*. *Planta*, *224*(2), 268–278. doi: 10.1007/s00425-005-0211-y
70. Van Loon, L. C., Bakker, P. A. H. M., Pieterse, C. M. J. (1998). Systemic resistance induced by rhizosphere bacteria. *Annual Review of Phytopathology*, *36*(1), 453–483. doi: 10.1146/annurev.phyto.36.1.453
71. Glick, B. R. (2012). Plant growth-promoting bacteria: Mechanisms and applications. *Scientifica*, *2012*, 1–15. doi: 10.6064/2012/963401
72. Nautiyal, C. S. (1999). An efficient microbiological growth medium for screening phosphate solubilizing microorganisms. *FEMS Microbiology Letters*, *170*(1), 265–270. doi: 10.1111/j.1574-6968.1999.tb13383.x
73. Antoun, H., Beauchamp, C. J., Goussard, N., Chabot, R., Lalande, R. (1998). Potential of *Rhizobium* and *Bradyrhizobium* species as plant growth promoting rhizobacteria on nonlegumes: Effect on radishes (*Raphanus sativus* L.). *Plant Soil*, *204*(1), 57–67.
74. Vessey, J. K. (2003). Plant growth promoting rhizobacteria as biofertilizers. *Plant Soil*, *255*(2), 571–586.
75. Bhattacharyya, P. N., Jha, D. K. (2012). Plant growth-promoting rhizobacteria (PGPR): Emergence in agriculture. *World Journal of Microbiology & Biotechnology*, *28*(4), 1327–1350. doi: 10.1007/s11274-011-0979-9
76. Suman, A., Shasany, A. K., Singh, M., Shahi, H. N., Gaur, A., Khanuja, S. P. S. (2001). Molecular assessment of diversity among endophytic diazotrophs isolated from subtropical Indian sugarcane. *World Journal of Microbiology & Biotechnology*, *17*(1), 39–45. doi: 10.1023/A:1016624701517
77. Rodríguez, H., Fraga, R. (1999). Phosphate solubilizing bacteria and their role in plant growth promotion. *Biotechnology Advances*, *17*(4–5), 319–339. doi: 10.1016/s0734-9750(99)00014-2
78. Afzal, A., Bano, A. (2008). *Rhizobium* and phosphate solubilizing bacteria improve the yield and phosphorus uptake in wheat (*Triticum aestivum*). *International Journal of Agriculture and Biology*, *10*(1), 85–88.
79. Patten, C. L., Glick, B. R. (1996). Bacterial biosynthesis of indole-3-acetic acid. *Canadian Journal of Microbiology*, *42*(3), 207–220. doi: 10.1139/m96-032
80. Senthilkumar, M., Swarnalakshmi, K., Govindasamy, V., Lee, Y. K., Annapurna, K. (2009). Biocontrol potential of soybean bacterial endophytes against charcoal rot

fungus, *Rhizoctonia bataticola*. *Current Microbiology*, *58*(4), 288–293. doi: 10.1007/s00284-008-9329-z
81. Chi, F., Yang, P., Han, F., Jing, Y., Shen, S. (2010). Proteomic analysis of rice seedlings infected by *Sinorhizobium meliloti* 1021. *Proteomics*, *10*(9), 1861–1874. doi: 10.1002/pmic.200900694
82. Arora, N. K., Kang, S. C., Maheshwari, D. K. (2001). Isolation of siderophore-producing strains of *Rhizobium meliloti* and their biocontrol potential against *Macrophomina phaseolina* that causes charcoal rot of groundnut. *Current Science*, *81*(6), 673–677.
83. Siddiqui, I. A., Ehteshamul-Haque, S., Zaki, M. J., Ghaffar, A. (2000). Greenhouse evaluation of rhizobia as biocontrol agent of root-infecting fungi in Okra. *Acta Agrobotanica*, *53*(1), 13–22. doi: 10.5586/aa.2000.002
84. Özkoç, İ., Deliveli, M. H. (2001). In vitro inhibition of the mycelial growth of some root rot fungi by rhizobium leguminosarum biovar phaseoli isolates. *Turkish Journal of Biology*, *25*(4), 435–445.
85. Siddiqui, Z. A., Mahmood, I. (2001). Effects of rhizobacteria and root symbionts on the reproduction of *Meloidogyne javanica* and growth of chickpea. *Bioresource Technology*, *79*(1), 41–45. doi: 10.1016/s0960-8524(01)00036-0
86. Handley, B. A., Hedges, A. J., Beringer, J. E. (1998). Importance of host plants for detecting the population diversity of *Rhizobium leguminosarum* biovar *viciae* in soil. *Soil Biology and Biochemistry*, *30*(2), 241–249. doi: 10.1016/S0038-0717(97)00103-X5
87. Wolde-Meskel, E., Terefework, Z., Lindström, K., Frostegård, Å. (2004). Rhizobia nodulating African *Acacia* spp. and *Sesbania sesban* trees in Southern Ethiopian soils are metabolically and genomically diverse. *Soil Biology and Biochemistry*, *36*(12), 2013–2025. doi: 10.1016/j.soilbio.2004.05.020
88. Sikora, S., Redzepović, S., Bradić, M. (2002). Genomic fingerprinting of *Bradyrhizobium japonicum* isolates by RAPD and Rep-PCR. *Microbiology Research*, *157*(3), 213–219. doi: 10.1078/0944-5013-00153
89. Adiguzel, A., Ogutcu, H., Baris, O., Karadayi, M., Gulluce, M., Sahin, F. (2010). Isolation and characterization of *Rhizobium* Strains from wild vetch collected from high altitudes in Erzurum-Turkey. *Romanian Biotechnological Letters*, *15*, 5017–5024.
90. Jarvis, B. D. W., Gillis, M., De Ley, J. (1986). Intra- and intergeneric similarities between the ribosomal ribonucleic acid cistrons of rhizobium and bradyrhizobium species and some related bacteria. *International Journal of Systematic and Evolutionary Microbiology*, *36*(2), 129–138. doi: 10.1099/00207713-36-2-129
91. Giongo, A., Ambrosini, A., Vargas, L. K., Freire, J. R. J., Bodanese-Zanettini, M. H., Passaglia, L. M. P. (2008). Evaluation of genetic diversity of bradyrhizobia strains nodulating soybean [Glycine max (L.) Merrill] isolated from South Brazilian fields. *Applied Soil Ecology*, *38*(3), 261–269. doi: 10.1016/j.apsoil.2007.10.016
92. Richardson, A. E., Viccars, L. A., Watson, J. M., Gibson, A. H. (1995). Differentiation of *Rhizobium* strains using the polymerase chain reaction with random and directed primers. *Soil Biology and Biochemistry*, *27*(4), 515–524.
93. Thies, J. E., Holmes, E. M., Vachot, A. (2001). Application of molecular techniques to studies in *Rhizobium* ecology: A review. *Australian Journal of Experimental Agriculture*, *41*(3), 299–319. doi: 10.1071/EA99171

94. Willems, A., doignon-Bourcier, F., Goris, J., Coopman, R., de Lajudie, P., De Vos, P., Gillis, M.. (2001). DNA-DNA hybridization study of bradyrhizobium strains. *International Journal of Systematic and Evolutionary Microbiology*, *51*(4), 1315–1322. doi: 10.1099/00207713-51-4-1315
95. Parker, M. A., Lafay, B., Burdon, J. J., Van Berkum, P. (2002). Conflicting phylogeographic patterns in rRNA and *nifD* indicate regionally restricted gene transfer in *Bradyrhizobium*. *Microbiology (Reading)*, *148*(8), 2557–2565. doi: 10.1099/00221287-148-8-2557
96. Stępkowski, T., Moulin, L., Krzyżańska, A., McInnes, A., Law, I. J., Howieson, J. (2005). European origin of *Bradyrhizobium* populations infecting *Lupins* and *Serradella* in soils of Western Australia and South Africa. *Applied and Environmental Microbiology*, *71*(11), 7041–7052. doi: 10.1128/AEM.71.11.7041-7052.2005
97. Vinuesa, P., León-Barrios, M., Silva, C., Willems, A., Jarabo-Lorenzo, A., Pérez-Galdona, R., Werner, D., Martínez-Romero, E. (2005). *Bradyrhizobium canariense* sp. nov., an acid-tolerant endosymbiont that nodulates endemic genistoid legumes (Papilionoideae: Genisteae) from the Canary Islands, along with *Bradyrhizobium japonicum* bv. Genistearum, *Bradyrhizobium* genospecies alpha and *Bradyrhizobium* genospecies beta. *International Journal of Systematic and Evolutionary Microbiology*, *55*(2), 569–575. doi: 10.1099/ijs.0.63292-0
98. Williams, J. G., Kubelik, A. R., Livak, K. J., Rafalski, J. A., Tingey, S. V. D. N. A. (1990). DNA polymorphisms amplified by arbitrary primers are useful as genetic markers. *Nucleic Acids Research*, *18*(22), 6531–6535. doi: 10.1093/nar/18.22.6531
99. Micheli, M. R., Bova, R. (2013). *Fingerprinting Methods Based on Arbitrarily Primed PCR*. Springer Science+Business Media.
100. Law, I. J., Botha, W. F., Majaule, U. C., Phalane, F. L. (2007). Symbiotic and genomic diversity of 'Cowpea' Bradyrhizobia from soils in Botswana and South Africa. *Biology and Fertility of Soils*, *43*(6), 653–663. doi: 10.1007/s00374-006-0145-y
101. Niemann, S., Pühler, A., Tichy, H. V., Simon, R., Selbitschka, W. (1997). Evaluation of the resolving power of three different DNA fingerprinting methods to discriminate among isolates of a natural *Rhizobium meliloti* population. *Journal of Applied Microbiology*, *82*(4), 477–484. doi: 10.1046/j.1365-2672.1997.00141.x
102. Cockerill III, F. R., Uhl, J. R., Temesgen, Z., Zhang, Y., Stockman, L., Roberts, G. D., Williams, D. L., Kline, B. C. (1995). Rapid identification of a point mutation of the *Mycobacterium tuberculosis* Catalase-peroxidase (katG) gene associated with isoniazid resistance. *Journal of Infectious Diseases*, *171*(1), 240–245. doi: 10.1093/infdis/171.1.240
103. Versalovic, J., Koeuth, T., Lupski, J. R. (1991). Distribution of repetitive DNA sequences in eubacteria and application to fingerprinting of bacterial genomes. *Nucleic Acids Research*, *19*(24), 6823–6831. doi: 10.1093/nar/19.24.6823
104. de Lajudie, P., Laurent-Fulele, E., Willems, A., Torck, U., Coopman, R., Collins, M. D., Kersters, K., Dreyfus, B., Gillis, M. (1998). Allorhizobium undicola gen. nov., sp. nov., nitrogen-fixing bacteria that efficiently nodulate neptunia natans in senegal. *International Journal of Systematic Bacteriology*, *48*(4), 1277–1290. doi: 10.1099/00207713-48-4-1277
105. Martens, M., Dawyndt, P., Coopman, R., Gillis, M., De Vos, P., Willems, A. (2008). Advantages of multilocus sequence analysis for taxonomic studies: A case study using 10

housekeeping genes in the genus *Ensifer* (including former *Sinorhizobium*). *International Journal of Systematic and Evolutionary Microbiology*, *58*(1), 200–214. doi: 10.1099/ijs.0.65392-0
106. Hansen, M. C., Tolker-Nielsen, T., Givskov, M., Molin, S. (1998). Biased 16S rDNA PCR amplification caused by interference from DNA flanking the template region. *FEMS Microbiology Ecology*, *26*(2), 141–149. doi: 10.1111/j.1574-6941.1998.tb00500.x
107. Botstein, D., White, R. L., Skolnick, M., Davis, R. W. (1980). Construction of a genetic linkage map in man using restriction fragment length polymorphisms. *American Journal of Human Genetics*, *32*(3), 314–331.
108. Vandamme, P., Pot, B., Gillis, M., De Vos, P., Kersters, K., Swings, J. (1996). Polyphasic taxonomy, a consensus approach to bacterial systematics. *Microbiology Reviews*, *60*(2), 407–438. doi: 10.1128/mr.60.2.407-438.1996
109. Benchimol, L. L., de Souza, C. L., Garcia, A. A. F., Kono, P. M. S., Mangolin, C. A., Barbosa, A. M. M., Coelho, A. S. G., de Souza, A. P. (2000). Genetic diversity in tropical maize inbred lines: Heterotic group assignment and hybrid performance determined by RFLP markers. *Plant Breed*, *119*(6), 491–496. doi: 10.1046/j.1439-0523.2000.00539.x
110. Jarabo-Lorenzo, A., Velázquez, E., Pérez-Galdona, R., Vega-Hernández, M. C., Martínez-Molina, E., Mateos, P. F., Vinuesa, P., Martínez-Romero, E., León-Barrios, M. (2000). Restriction fragment length polymorphism analysis of 16S rDNA and low molecular weight RNA profiling of rhizobial isolates from shrubby legumes endemic to the Canary Islands. *Systematic and Applied Microbiology*, *23*(3), 418–425. doi: 10.1016/S0723-2020(00)80073-9
111. Willems, A., Doignon-Bourcier, F., Coopman, R., Hoste, B., de Lajudie, P., Gillis, M. (2000). AFLP fingerprint analysis of bradyrhizobium strains isolated from Faidherbia Albida and Aeschynomene species. *Systematic and Applied Microbiology*, *23*(1), 137–147. doi: 10.1016/S0723-2020(00)80055-7
112. Deng, W., Xi, D., Mao, H., Wanapat, M. (2008). The use of molecular techniques based on ribosomal RNA and DNA for rumen microbial ecosystem studies: A review. *Molecular Biology Reports*, *35*(2), 265–274. doi: 10.1007/s11033-007-9079-1
113. Smit, E., Leeflang, P., Wernars, K. (1997). Detection of shifts in microbial community structure and diversity in soil caused by copper contamination using amplified ribosomal DNA restriction analysis. *FEMS Microbiology Ecology*, *23*(3), 249–261. doi: 10.1111/j.1574-6941.1997.tb00407.x
114. Diouf, A., de Lajudie, P., Neyra, M., Kersters, K., Gillis, M., Martinez-Romero, E., Gueye, M. (2000). Polyphasic characterization of rhizobia that nodulate phaseolus vulgaris in West Africa (Senegal and Gambia). *International Journal of Systematic and Evolutionary Microbiology*, *50*(1), 159–170. doi: 10.1099/00207713-50-1-159
115. Alberton, O., Kaschuk, G., Hungria, M. (2006). Sampling effects on the assessment of genetic diversity of rhizobia associated with soybean and common bean. *Soil Biology and Biochemistry*, *38*(6), 1298–1307. doi: 10.1016/j.soilbio.2005.08.018
116. Terefework, Z., Nick, G., Suomalainen, S., Paulin, L., Lindström, K. (1998). Phylogeny of rhizobium galegae with respect to other rhizobia and agrobacteria. *International Journal of Systematic Bacteriology*, *48*(2), 349–356. doi: 10.1099/00207713-48-2-349

117. van Berkum, P., Eardly, B. D. (1998). Molecular evolutionary systematics of the rhizobiaceae. In: *The Rhizobiaceae* (pp. 1–24). Springer: the Netherlands.
118. Young, J. P., Downer, H. L., Eardly, B. D. (1991). Phylogeny of the phototrophic rhizobium strain BTAi1 by polymerase chain reaction-based sequencing of a 16S RRNA gene segment. *Journal of Bacteriology*, *173*(7), 2271–2277. doi: 10.1128/jb.173.7.2271-2277.1991
119. Jarvis, B. D. W., Downer, H. L., Young, J. P. W. (1992). Phylogeny of fast-growing soybean-nodulating rhizobia support synonymy of sinorhizobium and rhizobium and assignment to rhizobium fredii. *International Journal of Systematic Bacteriology*, *42*(1), 93–96. doi: 10.1099/00207713-42-1-93
120. Laguerre, G., Mavingui, P., Allard, M. R., Charnay, M. P., Louvrier, P., Mazurier, S. I., Rigottier-Gois, L., Amarger, N. (1996). Typing of rhizobia by PCR DNA fingerprinting and PCR-restriction fragment length polymorphism analysis of chromosomal and symbiotic gene regions: Application to rhizobium leguminosarum and its different biovars. *Applied and Environmental Microbiology*, *62*(6), 2029–2036. doi: 10.1128/aem.62.6.2029-2036.1996
121. Woese, C. R. (1987). Bacterial evolution. *Microbiology Reviews*, *51*(2), 221–271. doi: 10.1128/mr.51.2.221-271.1987
122. Stackebrandt, E., Goebel, B. M. (1994). Taxonomic note: A place for DNA-DNA reassociation and 16S RRNA sequence analysis in the present species definition in bacteriology. *International Journal of Systematic and Evolutionary Microbiology*, *44*(4), 846–849. doi: 10.1099/00207713-44-4-846
123. Owen, R. J., Pitcher, D. (1985). Current methods for estimating DNA base composition and levels of DNA-DNA hybridization. *Society for Applied Bacteriology Technical Series*, *20*, 67–93.
124. Rivas, R., Velázquez, E., Willems, A., Vizcaíno, N., Subba-Rao, N. S., Mateos, P. F., Gillis, M., Dazzo, F. B., Martínez-Molina, E. (2002). A new species of devosia that forms a unique nitrogen-fixing root-nodule symbiosis with the aquatic legume neptunia natans (L.f.) Druce. *Applied and Environmental Microbiology*, *68*(11), 5217–5222. doi: 10.1128/AEM.68.11.5217-5222.2002
125. Ngom, A., Nakagawa, Y., Sawada, H., Tsukahara, J., Wakabayashi, S., Uchiumi, T., Nuntagij, A., Kotepong, S., Suzuki, A., Higashi, S., Abe, M. (2004). A novel symbiotic nitrogen-fixing member of the ochrobactrum clade isolated from root nodules of Acacia mangium. *Journal of General and Applied Microbiology*, *50*(1), 17–27. doi: 10.2323/jgam.50.17
126. Chen, Y. P., Rekha, P. D., Arun, A. B., Shen, F. T., Lai, W.-A., Young, C. C. (2006). Phosphate solubilizing bacteria from subtropical soil and their tricalcium phosphate solubilizing abilities. *Applied Soil Ecology*, *34*(1), 33–41. doi: 10.1016/j.apsoil.2005.12.002
127. Eardly, B. D., Wang, F.-S., van Berkum, P. V. (1996). Corresponding 16S RRNA gene segments in rhizobiaceae and aeromonas yield discordant phylogenies. *Plant Soil*, *186*(1), 69–74. doi: 10.1007/BF00035057
128. van Berkum, P., Eardly, B. D. (2002). The aquatic budding bacterium blastobacter denitrificans is a nitrogen-fixing symbiont of aeschynomene indica. *Applied and Environmental Microbiology*, *68*(3), 1132–1136. doi: 10.1128/AEM.68.3.1132-1136.2002

129. Nichols, R. (2001). Gene trees and species trees are not the same. *Trends in Ecology & Evolution, 16*(7), 358–364. doi: 10.1016/s0169-5347(01)02203-0
130. Wertz, J. E., Goldstone, C., Gordon, D. M., Riley, M. A. (2003). A molecular phylogeny of enteric bacteria and implications for a bacterial species concept. *Journal of Evolutionary Biology, 16*(6), 1236–1248. doi: 10.1046/j.1420-9101.2003.00612.x
131. Naser, S. M., Thompson, F. L., Hoste, B., Gevers, D., Dawyndt, P., Vancanneyt, M., Swings, J. (2005). Application of Multilocus Sequence Analysis (MLSA) for rapid identification of enterococcus species based on rpoA and pheS genes. *Microbiology (Reading), 151*(7), 2141–2150. doi: 10.1099/mic.0.27840-0
132. Thompson, F. L., Gevers, D., Thompson, C. C., Dawyndt, P., Naser, S., Hoste, B., Munn, C. B., Swings, J. (2005). Phylogeny and molecular identification of vibrios on the basis of multilocus sequence analysis. *Applied and Environmental Microbiology, 71*(9), 5107–5115. doi: 10.1128/AEM.71.9.5107-5115.2005
133. Zeigler, D. R. (2003). Gene sequences useful for predicting relatedness of whole genomes in bacteria. *International Journal of Systematic and Evolutionary Microbiology, 53*(6), 1893–1900. doi: 10.1099/ijs.0.02713-0
134. Gevers, D., Cohan, F. M., Lawrence, J. G., Spratt, B. G., Coenye, T., Feil, E. J., Stackebrandt, E., Van de Peer, Y., Vandamme, P., Thompson, F. L., Swings, J. (2005). Opinion: Re-evaluating prokaryotic species. *Nature Reviews Microbiology* Opinion, *3*(9), 733–739. doi: 10.1038/nrmicro1236
135. Turner, S. L., Young, J. P. W. (2000). The glutamine synthetases of rhizobia: Phylogenetics and evolutionary implications. *Molecular Biology and Evolution, 17*(2), 309–319. doi: 10.1093/oxfordjournals.molbev.a026311
136. Wernegreen, J. J., Riley, M. A. (1999). Comparison of the evolutionary dynamics of symbiotic and housekeeping loci: A case for the genetic coherence of rhizobial lineages. *Molecular Biology and Evolution, 16*(1), 98–113. doi: 10.1093/oxfordjournals.molbev.a026041
137. Gaunt, M. W., Turner, S. L., Rigottier-Gois, L., Lloyd-Macgilp, S. A., Young, J. P. (2001). Phylogenies of atpD and recA support the small subunit rRNA-based classification of rhizobia. *International Journal of Systematic and Evolutionary Microbiology, 51*(6), 2037–2048. doi: 10.1099/00207713-51-6-2037
138. Hernández-Lucas, I., Rogel-Hernández, M. A., Segovia, L., Rojas-Jiménez, K., Martínez-Romero, E. (2004). Phylogenetic relationships of rhizobia based on citrate synthase gene sequences. *Systematic and Applied Microbiology, 27*(6), 703–706. doi: 10.1078/0723202042369893
139. Konstantinidis, K. T., Tiedje, J. M. (2007). Prokaryotic taxonomy and phylogeny in the genomic era: Advancements and challenges ahead. *Current Opinion in Microbiology, 10*(5), 504–509. doi: 10.1016/j.mib.2007.08.006
140. Rivas, R., Martens, M., De Lajudie, P., Willems, A. (2009). Multilocus sequence analysis of the genus bradyrhizobium. *Systematic and Applied Microbiology, 32*(2), 101–110. doi: 10.1016/j.syapm.2008.12.005
141. Ribeiro, R. A., Barcellos, F. G., Thompson, F. L., Hungria, M. (2009). Multilocus sequence analysis of Brazilian *Rhizobium* microsymbionts of common bean (*Phaseolus vulgaris* L.) reveals unexpected taxonomic diversity. *Research in Microbiology, 160*(4), 297–306. doi: 10.1016/j.resmic.2009.03.009

142. Degefu, T., Wolde-Meskel, E., Frostegård, A. (2011). Multilocus sequence analyses reveal several unnamed *Mesorhizobium* genospecies nodulating acacia species and sesbania sesban trees in southern regions of Ethiopia. *Systematic and Applied Microbiology*, *34*(3), 216–226. doi: 10.1016/j.syapm.2010.09.006
143. Margaret, I., Becker, A., Blom, J., Bonilla, I., Goesmann, A., Göttfert, M., Lloret, J., Mittard-Runte, V., Rückert, C., Ruiz-Sainz, J. E., Vinardell, J. M., Weidner, S. (2011). Symbiotic properties and first analyses of the genomic sequence of the fast growing model strain sinorhizobium fredii HH103 nodulating soybean. *Journal of Biotechnology*, *155*(1), 11–19. doi: 10.1016/j.jbiotec.2011.03.016
144. Hirsch, A. M., Lum, M. R., Downie, J. A. (2001). What makes the rhizobia-legume symbiosis so special? *Plant Physiology*, *127*(4), 1484–1492. doi: 10.1104/pp.010866
145. Haukka, K., Lindström, K., Young, J. P. W. (1998). Three phylogenetic groups of *nodA* and *nifH* genes in *Sinorhizobium* and *Mesorhizobium* isolates from leguminous trees growing in Africa and Latin America. *Applied and Environmental Microbiology*, *64*(2), 419–426. doi: 10.1128/AEM.64.2.419-426.1998
146. Zurdo-Piñeiro, J. L., García-Fraile, P., Rivas, R., Peix, A., León-Barrios, M., Willems, A., Mateos, P. F., Martínez-Molina, E., Velázquez, E., van Berkum, P. (2009). Rhizobia from lanzarote, the Canary Islands, that nodulate *Phaseolus vulgaris* have characteristics in common with *Sinorhizobium meliloti* Isolates from Mainland Spain. *Applied and Environmental Microbiology*, *75*(8), 2354–2359. doi: 10.1128/AEM.02811-08
147. García-Fraile, P., Mulas-García, D., Peix, A., Rivas, R., González-Andrés, F., Velázquez, E. (2010). *Phaseolus vulgaris* is Nodulated in Northern Spain by *Rhizobium leguminosarum* strains harboring two nodC alleles present in American *Rhizobium etli* strains: Biogeographical and evolutionary implications. *Canadian Journal of Microbiology*, *56*(8), 657–666. doi: 10.1139/w10-048
148. Bailly, X., Olivieri, I., Brunel, B., Cleyet-Marel, J. C., Béna, G. (2007). Horizontal gene transfer and homologous recombination drive the evolution of the nitrogen-fixing symbionts of *Medicago* species. *Journal of Bacteriology*, *189*(14), 5223–5236. doi: 10.1128/JB.00105-07
149. Rincón-Rosales, R., Villalobos-Escobedo, J. M., Rogel, M. A., Martinez, J., Ormeño-Orrillo, E., Martínez-Romero, E. (2013). *Rhizobium calliandrae* sp. nov., *Rhizobium mayense* sp. nov. and *Rhizobium jaguaris* sp. nov., rhizobial species nodulating the medicinal legume *Calliandra grandiflora. International Journal of Systematic and Evolutionary Microbiology*, *63*(9), 3423–3429. doi:10.1099/ijs.0.048249-0
150. Stępkowski, T., Czaplińska, M., Miedzinska, K., Moulin, L. (2003). The variable part of the *dnaK* Gene as an alternative marker for phylogenetic studies of rhizobia and related alpha proteobacteria. *Systematic and Applied Microbiology*, *26*(4), 483–494. doi: 10.1078/072320203770865765.
151. Richardson, A. E., Viccars, L. A., Watson, J. M., Gibson, A. H. (1995). Differentiation of rhizobium strains using the polymerase chain reaction with random and directed primers. *Soil Biology and Biochemistry*, *27*(4–5), 515–524. doi: 10.1016/0038-0717(95)98626-Y
152. Normand, P., Cournoyer, B., Simonet, P., Nazaret, S. (1992). Analysis of a ribosomal RNA operon in the actinomycete frankia. *Gene*, *111*(1), 119–124. doi: 10.1016/0378-1119(92)90612-s.

153. Weisburg, W. G., Barns, S. M., Pelletier, D. A., Lane, D. J. (1991). 16S ribosomal DNA amplification for phylogenetic study. *Journal of Bacteriology*, *173*(2), 697–703. doi: 10.1128/jb.173.2.697-703.1991.
154. Yanagi, M., Yamasato, K. (1993). Phylogenetic analysis of the family rhizobiaceae and related bacteria by sequencing of 16S RRNA gene using PCR and DNA sequencer. *FEMS Microbiology Letters*, *107*(1), 115–120. doi: 10.1111/j.1574-6968.1993.tb06014.x
155. Haukka, K., Lindström, K., Young, J. P. W. (1998). Three phylogenetic groups of nodA and nifH genes in sinorhizobium and mesorhizobium isolates from leguminous trees growing in Africa and Latin America. *Applied and Environmental Microbiology*, *64*(2), 419–426. doi: 10.1128/AEM.64.2.419-426.1998.
156. Zhang, X. X., Turner, S. L., Guo, X. W., Yang, H. J., Debellé, F., Yang, G. P., Dénarié, J., Young, J. P. W., Li, F. D. (2000). The common nodulation genes of astragalus sinicus rhizobia are conserved despite chromosomal diversity. *Applied and Environmental Microbiology*, *66*(7), 2988–2995. doi: 10.1128/AEM.66.7.2988-2995.2000.
157. Laguerre, G., Nour, S. M., Macheret, V., Sanjuan, J., Drouin, P., Amarger, N. (2001). Classification of rhizobia based on nodC and nifH gene analysis reveals a close phylogenetic relationship among phaseolus vulgaris symbionts. *Microbiology (Reading)*, *147*(4), 981–993. doi: 10.1099/00221287-147-4-981.
158. Bhoi, T. K., Samal, I., Majhi, P. K., Komal, J., Mahanta, D. K., Pradhan, A. K., Saini, V., Nikhil Raj, M., Ahmad, M. A., Behera, P. P., Ashwini, M. (2022). Insight into aphid mediated Potato Virus Y transmission: A molecular to bioinformatics prospective. *Frontiers in Microbiology*, *13*, 1001454.
159. Mahanta, D. K., Jangra, S., Ghosh, A., Sharma, P. K., Iquebal, M. A., Jaiswal, S., Baranwal, V. K., Kalia, V. K., Chander, S. (2022). Groundnut bud necrosis virus modulates the expression of innate immune, endocytosis, and cuticle development-associated genes to circulate and propagate in its vector, Thrips palmi. *Frontiers in Microbiology*, *13*, 773238.
160. Bhatnagar, S., Mahanta, D. K., Vyas, V., Samal, I., Komal, J., Bhoi, T. K. (2023). Storage pest management with nanopesticides incorporating silicon nanoparticles: a novel approach for sustainable crop preservation and food security. *Silicon*, *16*, 471–483.
161. Mahanta, D. K., Komal, J., Samal, I., Bhoi, T. K., Dubey, V. K., Pradhan, K., Nekkanti, A., Gouda, M. N. R., Saini, V., Negi, N., Bhateja, S., Jat, H. K., Jeengar, D. (2023a). Nutritional aspects and dietary benefits of "Silkworms": Current scenario and future outlook. *Frontiers in Nutrition*, *10*, 1121508.
162. Samal, I., Bhoi, T. K., Vyas, V., Majhi, P. K., Mahanta, D. K., Komal, J., Singh, S., Kumar, P. V. D., Acharya, L. K. (2023a). Resistance to fungicides in entomopathogenic fungi: Underlying mechanisms, consequences, and opportunities for progress. *Tropical Plant Pathology*, *49*, 5–17.
163. Mahanta, D. K., Bhoi, T. K., Komal, J., Samal, I., Nikhil, R. M., Paschapur, A. U., Singh, G., Kumar, P. V. D., Desai, H. R., Ahmad, M. A., Singh, P. P., Majhi, P. K., Mukherjee, U., Singh, P., Saini, V., Shahanaz, Srinivasa, N., Yele, Y. (2023b). Insect-pathogen crosstalk and the cellular-molecular mechanisms of insect immunity: Uncovering the underlying signaling pathways and immune regulatory function of non-coding RNAs. *Frontiers in Immunology*, *14*, 1169152.

164. Samal, I., Bhoi, T. K., Majhi, P. K., Murmu, S., Pradhan, A. K., Kumar, D., Nikhil Raj, M., Ankur, Manik, S., Pratim Behera, P., Balawi, T. A. (2023b). Combatting insects mediated biotic stress through plant associated endophytic entomopathogenic fungi in horticultural crops. *Frontiers in Plant Science*, *13*, 1098673.
165. Komal, J., Desai, H. R., Samal, I., Mastinu, A., Patel, R. D., Kumar, P. D., Majhi, P. K., Mahanta, D. K., Bhoi, T. K. (2023). Unveiling the genetic symphony: Harnessing CRISPR-Cas genome editing for effective insect pest management. *Plants*, *12*(23), 3961.
166. Samal, I., Bhoi, T. K., Mahanta, D. K., Komal, J. (2023c). Establishing the role of Silicon (Si) in plant resistance to insects: A bibliometric approach. *Silicon*, 1–10.
167. Majhi, P. K., Samal, I., Bhoi, T. K., Pattnaik, P., Pradhan, C., Gupta, A., Mahanta, D. K., Senapati, S. K. (2023). Climate-smart agriculture: An integrated approach to address climate change and food security. In: *Climate Change and Insect Biodiversity* (pp. 221–245). CRC Press.
168. Mahanta, D. K., Samal, I., Komal, J., Bhoi, T. K., Majhi, P. K., Ahmad, M. A. (2023c). Understanding anthropogenic climate change, its consequences on insect pests, and techniques in forecasting and monitoring pest dynamics: A contemporary scenario. In: *Climate Change and Insect Biodiversity* (pp. 44–67). CRC Press.
169. Bhoi, T. K., Samal, I., Mahanta, D. K., Komal, J., Jinger, D., Sahoo, M. R., Achary, G. C., Nayak, P., Sunani, S. K., Saini, V., Raghuraman, M., Singh, S. (2023). Understanding how silicon fertilization impacts chemical ecology and multitrophic interactions among plants, insects and beneficial arthropods. *Silicon*, *15*(6), 2529–2549.
170. Bhoi, T. K., Samal, I., Mahanta, D. K., Komal, J., Majhi, P. K., Ankur. (2024). Exploring Melatonin's potential as an alternative strategy for protecting plants from biotic stresses. In: *Melatonin in Plants: A Pleiotropic Molecule for Abiotic Stresses and Pathogen Infection* (pp. 223–242). Springer Nature Singapore: Singapore.
171. Samal, I., Bhoi, T. K., Pradhan, A. K., Mahanta, D. K. (2023d). Plantix app: A success story of artificial intelligence in plant protection. *Van Sangyan, 24*.
172. Mahanta, D. K., Bhoi, T. K., Komal, J., Samal, I., Mastinu, A. (2024). Spatial, spectral and temporal insights: Harnessing high-resolution satellite remote sensing and artificial intelligence for early monitoring of wood boring pests in forests. *Plant Stress*, *11*, 100381.
173. Samal, I., Mahanta, D. K., Bhoi, T. K., Komal, J., Jatav, H. S., Jatav, S. S., Sathyanarayana, E. (2024). *Approaches of Biochar in Ecosystem Management: Current Scenario and Future Perspectives*. IntechOpen
174. Vyas, V., Bhoi, T. K., Samal, I., Singh, S., Mahanta, D. K. (2024). Revitalizing degraded soils with agroforestry interventions: Opportunities, challenges, and future direction. In: *Agroforestry to Combat Global Challenges: Current Prospects and Future Challenges* (pp. 529–549).
175. Samal, I., Bhoi, T. K., Mahanta, D. K., Komal, J., Singh, S. (2024). Biorational pest management: Potentials, unintended consequences, and future concerns. In: *Biorationals and Biopesticides: Pest Management* (pp. 47–75).

CHAPTER 5

In Multitrophic Agroforestry, Integrated Pest and Disease Management

MANOJ K. TRIPATHY,[1] SASHANKA S. DASH,[1*] SUBASH C. MAHAPATRA,[2] and SAJIYA QUADRI[1]

[1]*Department of Entomology, College of Agriculture, OUAT, Bhubaneswar, Odisha, India*

[2]*AICRP on Agroforestry, OUAT, Bhubaneswar, Odisha, India*

*Corresponding author

ABSTRACT

Agroforestry is a sustainable land-use system that integrates trees, shrubs, or other woody perennials with agricultural crops to create a diverse and productive agricultural landscape. A specialized form, multitrophic agroforestry (MAF), incorporates trees, shrubs, vines, and crops, forming a dynamic ecosystem with multiple trophic levels. By fostering biodiversity, nutrient cycling, and ecological balance, MAF enhances productivity while reducing dependence on synthetic inputs. However, due to dense plantings, high humidity, poor drainage, and the presence of alternative hosts, pests and diseases are often more prevalent in tropical and subtropical regions. Key insect pests include defoliators, root feeders, sap-sucking insects such as aphids, whiteflies, leafhoppers, psyllids, mealybugs, scales, and borers. Common diseases include fungal problems like damping-off, powdery mildew, and rust; bacterial diseases such as wilt, fire blight, and crown gall; viral infections; and nematode infestations. Effective management of these challenges requires Integrated Pest and Disease Management (IPDM) strategies. These include cultural and silvicultural practices, mechanical control, biological

Multitrophic Interactions for Sustainable Agroforestry: Harnessing the Power of Beneficial Insects and Ecological Dynamics. Hanuman Singh Jatav, Vishnu D. Rajput, Ipsita Samal, and Tanmaya Kumar Bhoi (Eds).

DOI: 10.1201/9781779641250-5

methods such as natural enemies, and judicious pesticide application. Together, these approaches maintain productivity, improve soil health, reduce pesticide and fertilizer use, and enhance biodiversity and water quality. MAF thus provides ecological, social, and economic benefits, making it a holistic and sustainable farming model.

5.1 INTRODUCTION

Multitrophic agroforestry (MAF) is an innovative approach to agriculture that mimics the ecological complexity of natural ecosystems. By integrating crops, trees, and shrubs in a single system, MAF fosters beneficial interactions between different trophic levels, creating a more resilient and productive agricultural landscape [1].

5.1.1 THE POWER OF DIVERSITY

At the heart of MAF lies the concept of biodiversity. By incorporating a diverse range of plant species, MAF systems create a web of complex ecological relationships. Trees provide shade and shelter for crops, while their roots help to fix nitrogen and improve soil fertility. Additionally, the presence of beneficial insects and predators helps to keep pest populations in check.

5.1.2 BENEFITS OF MAF

The benefits of MAF are numerous and far reaching. These systems can:

Multitrophic agroforestry (MAF) systems emerge as a multifaceted solution to the challenges facing contemporary agriculture, offering an array of benefits that extend across ecological, economic, and social dimensions. At the forefront of these advantages is the potential for significant increases in crop yields and improvements in crop quality. This is achieved through the intricate interactions between various trophic levels within the agroforestry ecosystem, where the synergy between plants and animals enhances nutrient cycling, pest control, and overall productivity.

Beyond the immediate agricultural outcomes, MAF systems contribute to the enhancement of soil health. The diverse array of plant species and the introduction of animals into the system foster improved soil fertility, heightened organic matter content, and enhanced water infiltration.

These improvements not only fortify the foundation of agricultural productivity but also mitigate the risks of soil degradation, ensuring the long-term sustainability of farming practices.

The ecological resilience of MAF systems further manifests in their ability to curtail pest and disease pressures. The presence of natural predators, coupled with the complexity of the ecosystem, creates an environment less susceptible to outbreaks. This not only reduces the dependence on chemical inputs but also promotes the development of a balanced and self-regulating agricultural ecosystem.

Moreover, MAF systems play a pivotal role in climate change mitigation and adaptation. The inclusion of trees in agroforestry not only sequesters carbon but also mitigates soil erosion, acts as a windbreak, and provides shade and shelter. These attributes collectively contribute to the resilience of the agricultural landscape in the face of climate uncertainties.

The positive environmental impacts of MAF systems extend to the promotion of biodiversity. By creating diverse habitats for a wide range of plant and animal species, agroforestry enhances overall landscape biodiversity. This not only safeguards ecosystems but also contributes to the conservation of indigenous flora and fauna.

On the socioeconomic front, the adoption of MAF systems can transform the livelihoods of farmers. The diversified nature of these systems provides farmers with a resilient and varied source of income. Simultaneously, the integration of agroforestry practices improves access to nutritious food, addressing both food security concerns and the nutritional needs of communities.

5.1.3 EXAMPLES OF MAF

There are many different ways to implement MAF, and the specific design of a system will vary depending on the local climate, soil conditions, and market opportunities. Some common examples of MAF systems include:

- Silvopasture: This system combines trees, forage crops, and livestock. The trees provide shade and shelter for the animals, while the forage crops provide food. The animals, in turn, help to fertilize the soil and control weeds.
- Alley cropping: In this system, rows of trees are planted between rows of crops. The trees provide shade and nitrogen fixation for

the crops, while the crops help to suppress weeds and provide income for the farmer.

- Forest gardening: This system mimics the structure of a natural forest, with a diverse mix of trees, shrubs, herbs, and vines. Forest gardens can be very productive and provide a wide range of food, fuel, and medicine for the household.

5.1.4 HISTORICAL ROOTS

The concept of agroforestry itself has ancient roots, with indigenous communities around the world practicing some form of agroforestry for centuries. However, the contemporary understanding of MAF emerged as a response to the negative environmental and ecological consequences associated with intensive, monoculture-based agriculture [2]. Over the past few decades, researchers and practitioners have sought to develop and refine agroforestry systems that harness the benefits of biodiversity and ecological complexity.

5.1.5 KEY CHARACTERISTICS

MAF is characterized by several key features:

Diversity of Species: The intentional inclusion of diverse tree species, crops, and often livestock in the same system creates a rich and dynamic environment.

Spatial Arrangement: Careful planning of the spatial arrangement of components allows for beneficial interactions, such as shade provision, nutrient cycling, and habitat creation for beneficial organisms.

Ecosystem Services: Agroforestry systems are designed not only to produce crops but also to provide a range of ecosystem services. These services may include soil conservation, water regulation, carbon sequestration, and habitat preservation.

Resilience: The diversity inherent in MAF enhances the system's resilience to pests, diseases, and environmental stressors. This resilience is a key component of sustainable agriculture.

Holistic Approach: Unlike conventional agriculture that may rely heavily on external inputs, MAF takes a holistic approach by utilizing internal ecological processes to enhance productivity and mitigate risks [3, 4].

5.2 IMPORTANCE OF INTEGRATED PEST AND DISEASE MANAGEMENT (IPDM)

5.2.1 PRESERVATION OF BIODIVERSITY

IPDM emphasizes the use of diverse ecological processes to control pests, avoiding heavy reliance on chemical inputs [5]. By minimizing the use of broad-spectrum pesticides, IPDM preserves the biodiversity within MAF systems. Healthy and diverse ecosystems contribute to the conservation of beneficial organisms, including natural predators and pollinators.

5.2.2 RESILIENCE AGAINST PEST OUTBREAKS

MAF inherently fosters a diverse environment, reducing the risk of monoculture-associated pest outbreaks [6]. IPDM strategies, such as companion planting, trap crops, and biological control, enhance the system's resilience by promoting natural checks and balances. The diversity of plants and animals within the agroforestry system acts as a buffer, preventing the rapid proliferation of specific pests.

5.2.3 ENVIRONMENTAL SUSTAINABILITY

IPDM aligns with the principles of sustainable and regenerative agriculture by minimizing the environmental impact of pest and disease control measures. Reduced reliance on synthetic pesticides in MAF systems prevents soil and water contamination, preserving the overall environmental health of the ecosystem [7].

5.2.4 ECONOMIC VIABILITY

IPDM contributes to the economic viability of agroforestry by reducing the costs associated with chemical inputs. Sustainable pest management practices, such as biological control and integrated approaches, provide long-term solutions, potentially increasing yields, and income for farmers.

5.2.5 CULTURAL AND ETHICAL CONSIDERATIONS

IPDM aligns with many traditional and indigenous farming practices that emphasize harmony with nature and a holistic understanding of ecosystems. Integrating IPDM into MAF respects ethical considerations by minimizing harm to nontarget organisms and fostering a balanced coexistence between agriculture and the environment.

5.2.6 *HUMAN HEALTH AND FOOD SAFETY*

Minimizing the use of synthetic pesticides reduces the risk of pesticide residues in food products.

IPDM promotes the use of environmentally friendly pest management practices, contributing to the overall health and safety of farmworkers, consumers, and surrounding communities.

5.2.7 *ADAPTIVE MANAGEMENT IN CHANGING ENVIRONMENTS*

As climate change brings about new challenges, IPDM allows farmers to adopt adaptive management strategies. The dynamic nature of IPDM in MAF systems enables farmers to respond effectively to changing pest and disease dynamics influenced by climate variability [7, 8].

5.2.8 *ENHANCED ECOSYSTEM SERVICES*

By minimizing disruptions to the natural balance of the agroforestry system, IPDM contributes to the provision of ecosystem services such as pollination, nutrient cycling, and soil conservation. Healthy ecosystems within agroforestry systems, fostered by IPDM, contribute to overall landscape resilience.

5.3 ROLE OF MULTITROPHIC INTERACTION IN PEST AND DISEASE SUPPRESSION

MAF, with its layered mix of trees, crops, and sometimes livestock, creates a complex and diverse ecosystem that offers several natural advantages for pest and disease suppression [9, 10].

5.3.1 *PHYSICAL BARRIERS*

Trees provide shade and acts as a windbreaks/shelterbelt to disrupt pest movement and create microclimates, which is less favorable for some pests. Intercropping of different crops with varying heights and growth patterns create physical obstacles, hindering pest migration, and access to preferred hosts.

5.3.2 HABITAT MANIPULATION

Diverse plant communities provide food and shelter for beneficial insects like predators and parasitoids, boosting their populations and aiding in pest control. Apart from that, trees and shrubs offer nesting opportunities for birds and other natural predators, further increasing pest pressure.

5.3.3 RESOURCE COMPETITION

By employing a variety of vertical levels, such as the canopy, the understory, and the soil, plants are able to compete less for resources such as light and nutrients, hence reducing the amount of resources that are accessible for pests to flourish in. The presence of multiple plant communities also results in a more diverse soil microbial population, which has the ability to inhibit the growth of hazardous pathogens and improve the health of plants.

5.3.4 NUTRIENT CYCLING

Nutrient cycling plays a pivotal role in the effective management of pests and diseases in agricultural systems. The process, involving the dynamic movement of essential elements through soil, plants, and microorganisms, significantly contributes to the overall health and immunity of crops. Well-nourished plants, supported by balanced nutrient cycling, exhibit enhanced resistance to pests and diseases, producing natural compounds that act as defenses. Additionally, nutrient-rich soils foster diverse microbial communities that engage in biological control, suppressing harmful pathogens. Practices like crop rotation, inherent to nutrient cycling, disrupt pest life cycles and minimize disease build-up in the soil. Furthermore, the organic matter decomposition associated with nutrient cycling contributes to soil structure, creating an environment less conducive to certain pests and pathogens. In essence, a sustainable nutrient cycling system not only ensures optimal plant nutrition but also establishes a resilient ecosystem that naturally mitigates the impact of pests and diseases.

5.3.5 ALLELOPATHY

Allelopathy, a fascinating ecological phenomenon, involves certain plants releasing chemicals that hinder the growth or reproduction of pests and diseases, offering a natural form of control. Through this process, these plants create a protective shield, utilizing chemical compounds to deter or inhibit the development of harmful organisms in their vicinity. By doing so, allelopathic plants contribute to the establishment of a balanced and resilient ecosystem where pest populations struggle to proliferate, and diseases find it challenging to spread. This natural suppression of pests and diseases not only enhances the overall health of the ecosystem but also reduces the dependence on harmful pesticides. Consequently, embracing allelopathy in agriculture fosters a more sustainable and environmentally friendly approach, promoting harmony between crops and their surroundings, while minimizing the ecological impact of conventional pest management strategies (Figure 5.1).

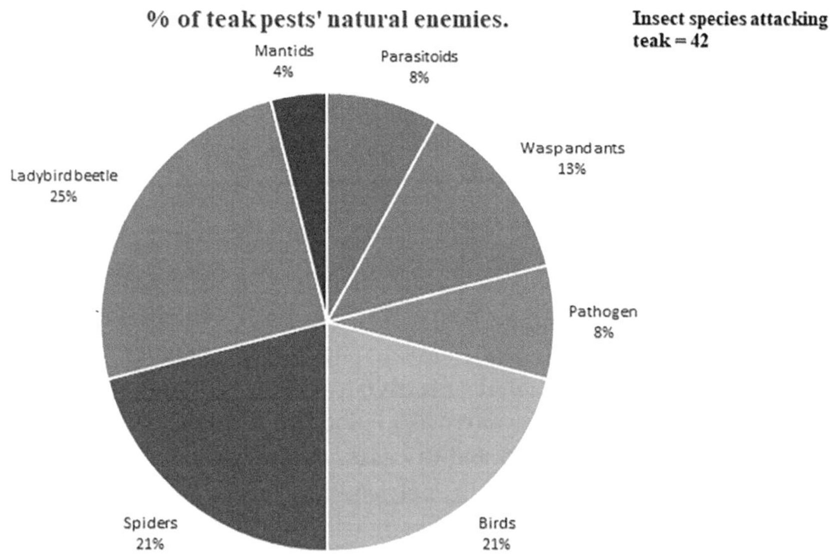

FIGURE 5.1 Insect pests and their natural enemies present in tropical agroforestry system of Odisha.

5.4 IMPORTANT PESTS PREVALENT IN MAF

MAF is divided into three types.

1. Upper story

2. Middle story
3. Under story [11]

Important pests present in the Upper story.

1. Teak [12–14]

Sl. No.	Common Name	Scientific Name	Family	Order	Damage Symptoms	Season of Occurrence
1	Teak defoliator	*Hyblaea puera*	Hyblaedae	Lepidoptera	Feeds upon the leaves. More particularly in green matters.	Early monsoon season
2	Teak skeletoniser	*Eutectona machaeralis*	Pyralidae	Lepidoptera	Skeletonizes the leaves, devours green matters and reduces growth.	Late monsoon
3	Teak sapling borer	*Sahyadrassus malabaricus*	Hapialidae	Lepidoptera	Bore the stem of seedling up to 1 year growth thus causes mortality.	Throughout the year except Dec and Jan
4	White grub	*Lacnosterna consanguinea*	Melolonthidae	Coleoptera	Feeds on roots of seedlings and small plants resulted in mortality.	Early monsoon season
5	Babloo shoot and root borer	*Celosterna scabrator*	Cerambycidae	Coleoptera	Bores the stems and branches which are succumb to wind or frost damage.	Throughout the year

2. Bamboo [15, 16]

Sl. No.	Common Name	Scientific Name	Family	Order	Damage Symptoms	Season of Occurrence
1	Hispine borer	*Estigma chinensis*	Chrysomelidae	Coleoptera	Bore young shoots causing holes made them susceptible to wound.	During rainy season
2	Ghoon borer	*Denoderis minutus*	Bostrichidae	Coleoptera	Bores mature and stored bamboos, cause complete damage	During rainy and winter season
3	Leaf folder		Pyrallidae	Lepidoptera	Folds the leaves. Larvae remains inside and feed on the green matter. Stunted growth	During rainy season
4	Termites	-	Termitidae	Isoptera	Feeds on the roots, stem. Cause mortality and reduces the usefulness	Throughout the year

3. Casuarina [17]

Sl. No.	Common Name	Scientific Name	Family	Order	Damage Symptoms	Season of Occurrence
1	Bark feeder	*Inderbela quadrinotata*	Inderbelidae	Lepidoptera	Feeds on the bark. Makes small holes on branches.	During winter and summer
2	Bag worm	*Pteroma plajiopes*	Psychidae	Lepidoptera	Feed on leaves and cause defoliation.	During winter and summer

4. Gambhar [18, 19]

Sl. No.	Common Name	Scientific Name	Family	Order	Damage Symptoms	Season of Occurrence
1	Gambhar skeletonizer	*Craspedonta leayana*	Chrysomelidae	Coleoptera	Completely skeletonizes the leaves	Rainy season
2	Gambhar shoot weevil	*Pulsidosis ludrivicatus*	Curculionidae	Coleoptera	Bores the tip, causes mortality.	Winter and summer seasons

5. Poplar [20]

Sl. No.	Common Name	Scientific Name	Family	Order	Damage Symptoms	Season of Occurrence
1	Poplar Twig Borer	*Gypsonoma aceriana*	Tortricidae	Lepidoptera	Boring into young shoots and twigs	Throughout the year
2	Poplar Clearwing Moth	*Paranthrene tabaniformis*	Sesiidae	Lepidoptera	Larvae tunnel into trunk and branches	Summer to early autumn
3	Eastern Tent Caterpillar	*Malacosoma americanum*	Lasiocampidae	Lepidoptera	Builds silk tents, defoliation	Spring
4	Aphids	Various species	Aphididae	Hemiptera	Sap feeding, leaf distortion, honeydew	Throughout the year
5	Scale Insects	Various species	Coccoidea	Hemiptera	Sucking sap, weakening the tree	Throughout the year

6. Mango [21]

Sl No.	Common Name	Scientific Name	Family	Order	Damage Symptoms	Season of Occurrence
1	Mango Hopper	*Idioscopus* spp.	Cicadellidae	Hemiptera	Sap-sucking, leaf yellowing, honeydew	Summer
2	Mango Fruit Fly	*Bactrocera dorsalis*	Tephritidae	Diptera	Larvae infest fruit, causing damage	Summer to early autumn

(Continued)

Sl No.	Common Name	Scientific Name	Family	Order	Damage Symptoms	Season of Occurrence
3	Mango Mealybug	*Drosicha mangiferae*	Pseudococcidae	Hemiptera	White, waxy secretions on leaves	Throughout the year
4	Mango Scale	*Aulacaspis tubercularis*	Diaspididae	Hemiptera	Scale insects on leaves and twigs	Throughout the year
5	Mango Shoot Borer	*Chlumetia transversa*	Pyralidae	Lepidoptera	Larvae bore into young shoots and stems	Summer to autumn

7. Eucalyptus [22–24]

Sl No.	Common Name	Scientific Name	Family	Order	Damage Symptoms	Season of Occurrence
1	Eucalyptus Gall Wasp	*Leptocybe invasa*	Eulophidae	Hymenoptera	Formation of galls on leaves and stems	Throughout the year
2	Eucalyptus Longhorned Borer	*Phoracantha* spp.	Cerambycidae	Coleoptera	Larvae bore into the trunk, causing damage	Throughout the year
3	Eucalyptus Psyllid	*Ctenarytaina eucalypti*	Psyllidae	Hemiptera	Sap-sucking, distorted leaves	Throughout the year
4	Eucalyptus Defoliating Beetle	*Paropsis* spp.	Chrysomelidae	Coleoptera	Defoliation, feeding on leaves	Spring to summer
5	Eucalyptus Leaf Beetle	*Chrysophtharta* spp.	Chrysomelidae	Coleoptera	Feeding on eucalyptus leaves, defoliation	Throughout the year

Important pests present in the Middle story.

1. Tea [25]

Sl No.	Common Name	Scientific Name	Family	Order	Damage Symptoms	Season of Occurrence
1	Tea Mosquito Bug	*Helopeltis* spp.	Miridae	Hemiptera	Pierces young tea shoots, causing withering and curling. Feeding results in a characteristic "orange pekoe" appearance in leaves.	Mainly during the growing season

Sl No.	Common Name	Scientific Name	Family	Order	Damage Symptoms	Season of Occurrence
2	Tea Red Spider Mite	*Oligonychus coffeae*	Tetranychidae	Acari	Feeds on the underside of leaves, causing stippling, yellowing, and premature leaf drop.	Hot and dry conditions
3	Tea Tortrix Caterpillar	*Homona* spp.	Tortricidae	Lepidoptera	Larvae feed on tea shoots and young leaves, causing defoliation and reduced yield.	Throughout the year

2. Coffee [26]

Sl. No.	Common Name	Scientific Name	Family	Order	Damage Symptoms	Season of Occurrence
1	Coffee Berry Borer	*Hypothenemus hampei*	Curculionidae	Coleoptera	Larvae bore into coffee berries, reducing yield and quality. Infested beans may develop a characteristic "broca" hole.	Throughout the year, peaks during the rainy season
2	Antestia Bug	*Antestiopsis* spp.	Pentatomidae	Hemiptera	Feeding on coffee cherries, causing discoloration and off-flavors in the beans. May also transmit pathogens.	Mainly during the rainy season

3. Guava [27]

Sl No.	Common Name	Scientific Name	Family	Order	Damage Symptoms	Season of Occurrence
1	Guava Fruit Fly	*Bactrocera correcta*	Tephritidae	Diptera	Larvae infest guava fruits, causing damage by feeding on the pulp. Affected fruits may exhibit premature dropping or rotting.	Warm and humid conditions
2	Guava Whitefly	*Trialeurodes variabilis*	Aleyrodidae	Hemiptera	Feeding on the undersides of leaves, causing yellowing, wilting, and a sticky substance (honeydew) on the foliage.	Throughout the year
3	Red-banded Thrips	*Selenothrips rubrocinctus*	Thripidae	Thysanoptera	Feeding on young leaves, causing silvering, stippling, and deformation. May also transmit plant viruses.	Warm and dry conditions

Important pests present in the Under story.

1. Ginger [28]

Sl. No.	Common Name	Scientific Name	Family	Order	Damage Symptoms	Season of Occurrence
1	Ginger Shoot Borer	*Conogethes punctiferalis*	Crambidae	Lepidoptera	Larvae bore into shoots, wilting of plants	Rainy and winter seasons
2	Aphids	*Pentalonia nigronervosa*	Aphididae	Hemiptera	Sap-sucking, yellowing of leaves	Throughout the year
3	Whiteflies	*Aleurodicus dispersus*	Aleyrodidae	Hemiptera	Whiteflies on the underside of leaves	Throughout the year
4	Red Spider Mite	*Tetranychus ludeni*	Tetranychidae	Acarina	Fine webbing on leaves, stippling	Throughout the year with more intense in summer
5	Ginger Scale	*Aspidiella hartii*	Diaspididae	Hemiptera	Scale insects on stems and leaves	Throughout the year with more intense in winter.

2. Turmeric [29]

Sl No.	Common Name	Scientific Name	Family	Order	Damage Symptoms	Season of Occurrence
1	Turmeric Root Fly	*Neosciara longirostris*	Sciaridae	Diptera	Larvae feed on turmeric rhizomes, causing damage to roots and reducing yield. Adult flies may also cause damage to young shoots.	Rainy season, post-monsoon

3. Pineapple [30]

Sl No.	Common Name	Scientific Name	Family	Order	Damage Symptoms	Season of Occurrence
1	Pineapple Mealybug	*Dysmicoccus brevipes*	Pseudococcidae	Hemiptera	Feeding on plant sap, causing wilting and yellowing of leaves. Honeydew excretion promotes the growth of sooty mold.	Throughout the year
2	Thrips	*Frankliniella schultzei*	Thripidae	Thysanoptera	Feeding on leaves, causing silvering, stippling, and distortion. May transmit plant viruses.	Warm and dry conditions
3	Pineapple Fruit Fly	*Bactrocera dorsalis*	Tephritidae	Diptera	Infestation in fruits, leading to premature ripening, rotting, and decreased market quality.	Warm and humid conditions

5.5 IMPORTANT DISEASE PREVALENT IN MAF [31–33]

Sl. No.	Disease Name	Causal Agent	Host Plants	Symptoms
1	Anthracnose	Various fungi (e.g., *Colletotrichum* spp.)	Various trees and crops	Leaf spots, wilting, cankers, and premature defoliation
2	Rust	Various fungi (e.g., *Puccinia* spp.)	Trees, shrubs, and crops	Yellow or orange powdery pustules on leaves
3	Powdery Mildew	Various fungi (e.g., *Erysiphe* spp.)	Various trees and crops	White powdery growth on leaves and stems
4	Root Rot	Various fungi (e.g., *Phytophthora* spp.)	Trees and shrubs	Wilting, yellowing, and decay of roots
5	Leaf Spot	Various fungi (e.g., *Alternaria* spp.)	Trees, shrubs, and crops	Dark spots on leaves, leading to defoliation
6	Canker	Various fungi and bacteria	Trees and woody plants	Sunken lesions on stems, dieback, and cankers
7	Citrus Canker	Bacterium: *Xanthomonas axonopodis*	Citrus trees	Raised lesions on leaves, fruit, and stems
8	Fusarium Wilt	*Fusarium* spp.	Various trees and crops	Wilting, yellowing, and vascular discoloration

5.6 IPDM STRATEGIES IN MAF

In MAF, IPDM takes advantage of the system's natural pest suppression mechanisms, while employing additional targeted interventions [34]. Here are some key IPDM strategies.

5.6.1 CULTURAL CONTROL

IPDM relies on a nuanced application of cultural control methods to cultivate resilient and sustainable agricultural systems that harmonize with the natural environment. Crop rotation, a linchpin of this strategy, involves the systematic planting of different crops on the same land over time. This disrupts the life cycles of pests, preventing their build-up on specific host plants. Intercropping, the simultaneous cultivation of diverse crops in the same space, creates a dynamic and complex ecosystem. This not only confuses pests but also provides additional benefits, such as nitrogen fixation by legumes and natural weed suppression. Cover cropping introduces fast-growing plants between main crops or during fallow periods, serving as a multifaceted tool by suppressing weeds, offering habitat for beneficial insects, and enhancing soil health through the infusion of organic matter and nitrogen.

Sanitation practices play a pivotal role in IPDM, with meticulous removal of crop residues and infected plant material post-harvest. This serves to curtail overwintering pest populations and reduce disease inoculum, thereby mitigating future issues. Strategic tillage practices, emphasizing minimal soil disturbance, safeguard beneficial soil organisms that act as natural predators of pests and contribute to disease suppression. The adoption of no-till agriculture not only aids in pest control but also promotes improvements in soil organic matter and moisture retention [35].

The effectiveness of these cultural practices stems from their alignment with the natural ecosystem, their relatively low implementation costs, and the enduring benefits they confer on soil health and crop yields. To optimize IPDM strategies, farmers are encouraged to select crop varieties resistant to prevalent pests and diseases, stagger crop maturity dates to break continuous food sources for pests, maintain clean field borders to reduce pest and disease reservoirs, and conduct regular crop monitoring for early identification and intervention. By integrating these cultural practices into an IPDM program, farmers can cultivate a resilient, sustainable, and diverse agroforestry system that not only mitigates pest and disease pressures but also contributes to long-term agricultural viability [36].

5.6.2 SILVICULTURAL PRACTICES

MAF is a land-use system that integrates trees, crops, and livestock in a way that mimics the structure and function of natural ecosystems. This combination creates a complex and beneficial interaction between the different components, leading to increased productivity, improved soil health, and enhanced biodiversity. Silvicultural practices, which are the management of trees in forests, play a crucial role in optimizing the performance of MAF systems.

5.6.2.1 BENEFITS OF SILVICULTURAL PRACTICES IN MAF [33]

5.6.2.1.1 Tree Selection and Planting

The selection of appropriate tree species for a particular MAF system is critical. Trees should be chosen based on their compatibility with the crops and livestock, their ability to fix nitrogen, and their potential to provide shade, shelter, and other benefits. Planting techniques, such as the use of mycorrhizal fungi, can also improve tree establishment and growth.

5.6.2.1.2 Tree Pruning and Thinning

Pruning and thinning trees can help to control their size and shape, allowing more light to reach the crops and livestock below. This can also improve air circulation and reduce the risk of disease.

5.6.2.1.3 Coppicing and Pollarding

Coppicing is the practice of cutting trees back to ground level periodically, while pollarding involves cutting branches to a certain height. These techniques can be used to stimulate the growth of new shoots, which can be used for fodder, fuel, or other purposes.

5.6.2.1.4 Nutrient Cycling

Trees play a vital role in nutrient cycling in MAF systems. Their deep roots can access nutrients from deep in the soil and bring them to the surface, where they can be taken up by crops and livestock. Trees also help to fix nitrogen from the air, making it available to other plants.

5.6.2.1.5 Biodiversity

MAF systems can provide habitat for a wide variety of plants and animals, including beneficial insects, birds, and other predators that can help to control pest populations.

5.6.2.2 CHALLENGES AND CONSIDERATIONS

5.6.2.2.1 Competition for Resources

Trees, crops, and livestock can all compete for resources such as water, nutrients, and sunlight. Careful management is needed to ensure that all components of the system have access to the resources they need.

5.6.2.2.2 *Disease and Pests*

Trees can be susceptible to diseases and pests that can also affect crops and livestock. Integrated pest management (IPM) strategies are needed to control these problems.

5.6.2.2.3 *Labor and Knowledge*

MAF systems can be more labor-intensive than conventional agriculture, and they require a good understanding of ecology and silviculture.

Silvicultural practices play a vital role in the success of MAF systems. By carefully selecting, planting, and managing trees, farmers can create productive, sustainable, and biodiverse systems that provide a wide range of benefits [37].

5.6.3 *BIOLOGICAL CONTROL [1]*

In MAF systems, promoting and enhancing naturally occurring biological control offers a sustainable and effective approach to managing pests and diseases. Here is a breakdown of how multitrophic systems foster biological control:

5.6.3.1 *MECHANISMS OF BIOLOGICAL CONTROL*

Biological control mechanisms within agroforestry systems operate through various strategies aimed at fostering a harmonious ecosystem that naturally regulates pest populations. Diverse tree and plant communities serve as crucial elements by attracting and sustaining populations of natural enemies, including ladybugs, lacewings, and parasitoid wasps, thereby amplifying predation pressure on pests. Habitat manipulation strategies, such as planting flowering strips, providing nesting sites through bird-houses and bat boxes, and maintaining undisturbed areas, contribute to the creation of conducive environments that support and attract beneficial insects, birds, and other predators. Resource compctition comes into play as diverse flora within the agroforestry system competes with pests for essential resources like food and water, thereby restraining their growth and reproductive capabilities. Moreover, the phenomenon of allelopathy, where certain plants release chemicals inhibiting pest growth or reproduction, adds an additional layer of natural control to the agroforestry ecosystem. Collectively, these mechanisms underscore the multifaceted and interconnected strategies employed within biological control to promote a sustainable and self-regulating agroecosystem.

5.6.3.2 STRATEGIES FOR ENHANCING BIOLOGICAL CONTROL

Employing strategies to enhance biological control is paramount for fostering a resilient and sustainable agroecosystem. Intercropping, a fundamental approach, involves strategically combining pest-susceptible crops with either pest-resistant or attractant companion plants. The latter entices beneficial insects, creating a balanced ecosystem that supports natural pest control. Cover cropping, through the cultivation of fast-growing nitrogen-fixing plants between main crops, not only suppresses weeds but also attracts pollinators and offers habitat for beneficial organisms. Establishing flower strips along field edges or within the agroforestry system serves as another effective strategy, providing a haven for pollinators and predatory insects crucial for controlling pest populations. Creating nesting sites, be it through artificial nesting boxes or preserving natural structures, encourages the colonization of birds, bats, and other predators, contributing to effective pest management. Furthermore, a key strategy involves the conservation of natural enemies by minimizing unnecessary pesticide use, as these chemicals can harm beneficial insects alongside pests. Regular monitoring for pest outbreaks allows targeted interventions, minimizing disruptions to natural enemy populations and ensuring the sustained effectiveness of biological control mechanisms within the agroecosystem. These strategies collectively underscore the importance of a holistic and ecologically mindful approach to enhance biological control in agriculture.

5.6.3.3 EXAMPLES OF SUCCESSFUL BIOLOGICAL CONTROL IN MULTITROPHIC SYSTEMS

Several examples showcase the success of biological control within MAF systems, emphasizing the practical application of integrated and sustainable pest management practices. In coffee agroforestry systems across

Central America, the deliberate integration of shade trees, such as Inga species, has proven instrumental in biological control. These shade trees provide a habitat for parasitoids that effectively regulate coffee berry borers, leading to a substantial reduction in the reliance on chemical pesticides and promoting a more environmentally friendly approach to coffee cultivation.

In the cacao agroforestry systems of Indonesia, a diverse array of shade trees plays a pivotal role in biological control. These trees create habitats for ants and spiders, natural predators of cocoa pod borers. The presence of these beneficial organisms helps control pest populations, resulting in higher cacao yields, while concurrently minimizing the need for insecticide applications. This successful example highlights the potential for biological control to enhance productivity and sustainability in cacao cultivation.

In North American fruit tree systems, the integration of flowering herbs and legumes within orchards has demonstrated effective biological control. Interplanting with these diverse companion plants attracts hoverflies and other predators that prey on aphids and other soft-bodied pests affecting fruit trees. By creating a balanced ecosystem, fruit tree growers reduce the incidence of pests and mitigate the use of chemical interventions, promoting a more ecologically sound and economically viable approach to pest management.

These examples underscore the versatility and effectiveness of biological control within MAF systems. By leveraging the interactions between different trophic levels and fostering diverse ecosystems, farmers can successfully manage pests while simultaneously promoting environmental sustainability and resilience within their agricultural landscapes.

5.6.3.4 BENEFITS OF BIOLOGICAL CONTROL

The adoption of biological control strategies in agroforestry systems offers a multitude of benefits, highlighting the ecological and economic advantages of this sustainable pest management approach. First and foremost, the reliance on harmful pesticides is significantly reduced, safeguarding both human health and the environment. By minimizing the use of chemical interventions, biological control mitigates the potential risks associated with pesticide exposure, promoting safer agricultural practices that resonate with broader environmental conservation goals.

Biological control also plays a pivotal role in enhancing the long-term resilience and stability of agroforestry systems. By fortifying natural pest suppression mechanisms through the conservation and augmentation of beneficial organisms, these systems become less vulnerable to pest outbreaks and external disturbances. This intrinsic resilience contributes to the overall sustainability of agricultural practices, aligning with the principles of ecological balance and resource conservation.

Furthermore, the implementation of biological control positively influences crop yields and quality. By effectively managing pest populations, the extent of pest-related damage is minimized, leading to healthier and more robust plants. This, in turn, translates to improved agricultural productivity, ensuring that agroforestry systems can consistently yield high-quality crops, while reducing the economic losses associated with pest damage.

A notable ecological benefit of biological control is its contribution to biodiversity and the creation of healthy ecosystems within agroforestry landscapes. By fostering diverse insect communities, biological control practices support the intricate web of life, promoting a balanced and dynamic environment. This diversity is not only beneficial for the organisms directly involved in pest control but also contributes to overall ecosystem health, including pollination services and the conservation of native species.

5.6.3.5 CHALLENGES AND CONSIDERATIONS

While embracing biological control in agroforestry systems offers numerous benefits, several challenges and considerations must be acknowledged to ensure successful implementation. One notable challenge is the initial establishment and management of diverse systems, which can be more intricate compared to conventional monoculture practices. The integration of various plant species and the careful management of their interactions demand additional expertise and planning, potentially posing challenges for farmers transitioning from more straightforward agricultural systems.

The efficacy of biological control within agroforestry systems may exhibit variability based on several factors. The specific pest species, the composition of natural enemy communities, and local environmental conditions all play crucial roles. Different pests may respond differently to biological control, and the effectiveness of natural enemies in regulating pest populations may vary. Understanding these nuances is essential for optimizing the outcomes of biological control practices within agroforestry contexts.

Monitoring and adaptability are key considerations in the successful implementation of biological control. Agroforestry systems are dynamic environments, and the effectiveness of biological control measures may change over time. Continuous monitoring of pest and natural enemy populations is necessary to assess the system's performance and identify potential challenges. Additionally, interventions may need to be adapted based on shifting ecological dynamics, emphasizing the need for a flexible and responsive management approach.

In conclusion, while the adoption of biological control in agroforestry systems holds great promise, it is essential to recognize and address the challenges associated with this approach. The complexity of establishing diverse systems, the variability in efficacy, and the need for ongoing monitoring and adaptability underscore the importance of informed and adaptive management practices. As agroforestry continues to gain traction as a sustainable agricultural model, addressing these challenges will contribute to the successful integration of biological control, ensuring its effectiveness in promoting resilient and balanced agroecosystems.

5.6.4 CHEMICAL CONTROL

While emphasizing the promotion of naturally occurring biological control and integrating cultural and silvicultural practices, there are circumstances in MAF where the consideration of chemical control becomes a last resort. However, the use of chemical control should be approached with caution and minimized due to its potential drawbacks. Chemical interventions, such as pesticides, can disrupt the delicate balance of the agroecosystem by not only targeting pests but also harming beneficial organisms crucial for biological control. The indiscriminate use of chemicals may lead to the development of resistance in pest populations, rendering future chemical applications less effective. Moreover, chemical residues can have adverse effects on human health, contaminate water sources, and contribute to environmental pollution. Therefore, while chemical control may offer immediate solutions to specific pest issues, its utilization in MAF demands careful evaluation, strict adherence to recommended practices, and a continuous commitment to minimizing its impact on both the ecosystem and human well-being. The judicious integration of chemical control, when absolutely necessary, remains contingent upon sustainable and responsible agricultural practices within the broader framework of MAF [38].

5.6.4.1 REASONS FOR MINIMIZING CHEMICAL CONTROL

The imperative to minimize chemical control in MAF arises from several critical considerations, all of which underscore the potential negative impacts on the ecosystem and long-term agricultural sustainability. One primary concern lies in the disruption of ecological balance. Pesticides, while targeting specific pests, can inadvertently harm beneficial insects, birds, and other predators that contribute to natural pest control within the multitrophic system. This disruption can result in rebound pest outbreaks and a gradual erosion of the system's ability to self-regulate, leading to increased reliance on chemical control in the long run.

Environmental contamination stands out as another compelling reason to minimize chemical control. Pesticide residues have the potential to pollute soil, water, and air, posing significant risks to both human health and wildlife. In multitrophic systems characterized by diverse plant and animal communities, the amplification of these risks becomes a serious concern. The intricate relationships between different trophic levels make the ecosystem more vulnerable to the unintended consequences of chemical applications.

Furthermore, the loss of sustainability is a fundamental consideration. Frequent and indiscriminate reliance on chemical control methods can give rise to pest resistance, rendering the pesticides less effective over time. This phenomenon creates an unsustainable cycle of dependence, where escalating chemical use becomes necessary to achieve the same level of pest control. This not only poses economic challenges for farmers but also contributes to the broader issue of diminishing environmental sustainability [38].

5.6.4.2 WHEN IS CHEMICAL CONTROL CONSIDERED IN MAF?

While the emphasis in MAF remains on sustainable and biological control methods, there are specific situations where the judicious use of chemical control may be considered as a measure of last resort. One such scenario is the consideration of economic thresholds. When pest populations reach or exceed levels where there is a high likelihood of causing economic damage to crops, and other control methods have proven inadequate, chemical interventions might be considered to prevent significant financial losses for farmers.

Protecting public health is another critical consideration warranting the use of chemical control in MAF. In certain cases, when diseases transmitted by pests pose a direct risk to human health, targeted chemical interventions may be necessary. This measure becomes crucial to mitigate the spread of diseases and safeguard the well-being of communities relying on the agroforestry system.

In instances of rapid and severe pest outbreaks, where the immediate threat of catastrophic damage to crops exists, chemical control may be employed as a temporary and emergency measure. This allows for swift intervention to curtail immediate losses while longer-term biological control methods are established and given time to take effect. This strategic use of chemical control can serve as a pragmatic short-term solution to protect crops and prevent extensive damage [39].

5.6.4.3 STRATEGIES FOR MINIMIZING CHEMICAL USE

In MAF systems, prioritizing sustainable and environmentally friendly practices is paramount. IPM emerges as a foundational strategy, where preventative and nonchemical methods take precedence. Cultural practices, such as crop rotation and cover cropping, alongside biological control and habitat manipulation, form the bedrock of pest management, fostering ecological resilience and minimizing the reliance on chemical interventions.

The selection of pesticides is another key strategy, emphasizing the use of selective pesticides with minimal impact on beneficial insects and the overall environment. Opting for targeted approaches that specifically address the identified pest species, rather than employing broad-spectrum pesticides, aligns with the principles of precision and ecological stewardship.

Precise application methods further contribute to minimizing chemical use. Techniques like spot spraying or trunk injection ensure targeted pesticide delivery, reducing the risk of drift and limiting exposure to nontarget organisms. This focused approach enhances the efficacy of chemical interventions while mitigating potential environmental repercussions.

Monitoring and evaluation serve as ongoing components of a comprehensive strategy to minimize chemical use. Regularly assessing pest populations and tracking the effectiveness of implemented control measures enable farmers to make informed decisions. This adaptive approach allows for real-time adjustments, ensuring that chemical interventions are only employed when absolutely necessary, thus avoiding unnecessary pesticide use.

It is crucial to underscore that chemical control should be viewed as a last resort in MAF systems. By steadfastly prioritizing ecological resilience and leveraging natural pest control mechanisms through IPDM, farmers can cultivate sustainable and productive agricultural practices with minimal environmental impact. This holistic approach not only safeguards the health of the agroecosystem but also aligns with the broader goals of promoting biodiversity, environmental conservation, and long-term agricultural viability [2, 40–57].

5.7 CONCLUSION

In conclusion, the multifaceted approach of IPDM in MAF exemplifies a holistic and sustainable strategy for pest control. Cultural practices, such as crop rotation, intercropping, and cover cropping, play a pivotal role in fostering resilient agricultural systems that work in harmony with the natural environment. Silvicultural practices further optimize the performance of agroforestry systems, enhancing productivity, soil health, and biodiversity. The promotion of biological control mechanisms capitalizes on the intricate relationships within the ecosystem, allowing for effective and environmentally friendly pest management. However, the consideration of chemical control is acknowledged as a last resort, with careful evaluation and minimization due to potential drawbacks such as ecological disruption, environmental contamination, and the risk of developing pest resistance.

Strategies for minimizing chemical use, including selective pesticide choices, precise application methods, and continuous monitoring, underscore the commitment to ecological resilience and sustainability. Chemical control is positioned as a temporary and emergency measure, necessitated only in specific situations, such as economic thresholds, public health concerns, or severe pest outbreaks. The overarching message is clear: chemical control should be approached with caution, and every effort should be made to prioritize and integrate nonchemical, environmentally friendly methods.

In embracing these strategies, farmers can create a balanced, biodiverse, and sustainable agroforestry system that not only mitigates pest and disease pressures but also contributes to long-term agricultural viability. By harmonizing with the natural ecosystem and leveraging the benefits of diverse practices, MAF emerges as a promising model for resilient, productive, and environmentally conscious agriculture.

KEYWORDS

- **agroforestry**
- **multitrophic interaction**
- **insect pest diversity**
- **silvicultural practices**
- **integrated pest and disease management**
- **cultural control**
- **biological control**
- **chemical control**

REFERENCES

1. Teodoro, A. V., Tscharntke, T., Klein, A. M. (2009). From the laboratory to the field: Contrasting effects of multi-trophic interactions and agroforestry management on coffee pest densities. *Entomologia Experimentalis et Applicata, 131*(2), 121–129.
2. Beckerman, A. P. (2000). *Some Applications of Multitrophic Level Ecological Theory to Agroforestry Systems*. PhD Thesis. Mark S. Ashton.
3. Marsden, C., Martin-Chave, A., Cortet, J., Hedde, M. Capowiez, Y. (2020). How agroforestry systems influence soil fauna and their functions—A review. *Plant and Soil, 453*, 29–44.
4. Wang, W., Mishra, S., Yang, X. (2023). Seasonal difference in soil health indicators mediating multidiversity-multifunctionality relationship depends on body size of soil organisms: Evidence from rubber plantation agroforestry system. *Soil Biology and Biochemistry, 178*, 108968.
5. Balasubramaniam, M. (2022). Evaluation of frontline demonstrations on integrated pest and disease management (IPDM) practices in paddy. *International Journal of Farm Sciences, 12*(1), 119–121. https://doi.org/10.5958/2250-0499.2022.00027.1

6. Daniel, R., Konam, J. K., Saul-Maora, J. Y., Kamuso, A., Namaliu, Y., Vano, J. T., Wenani, R., N'nelau, P., Palinrungi, R., Guest, D. I. (2011). Knowledge through participation: The triumphs and challenges of transferring Integrated Pest and Disease Management (IPDM) technology to cocoa farmers in Papua New Guinea. *Food Security*, *3*, 65–79.
7. Scudder, M., Wampe, N., Waviki, Z., Applegate, G., Herbohn, J. (2022). Smallholder cocoa agroforestry systems; is increased yield worth the labour and capital inputs. *Agricultural Systems*, *196*, 103350.
8. Konam, J., Namaliu, Y., Daniel, R., Guest, D. (2008). Integrated pest and disease management for sustainable cocoa production: A training manual for farmers and extension workers. Australian Centre for International Agricultural Research (ACIAR).
9. Van der Putten, W. H., Vet, L. E., Harvey, J. A., Wäckers, F. L. (2001). Linking above-and belowground multitrophic interactions of plants, herbivores, pathogens, and their antagonists. *Trends in Ecology & Evolution*, *16*(10), 547–554.
10. Chakraborty, S., Pangga, I. B., Roper, M. M. (2012). Climate change and multitrophic interactions in soil: The primacy of plants and functional domains. *Global Change Biology*, *18*(7), 2111–2125.
11. Devagiri, G. M., Khaple, A. K., Anithraj, H. B., Kushalappa, C. G., Krishnappa, A. K., Mishra, S. B. (2020). Assessment of tree diversity and above-ground biomass in coffee agroforest dominated tropical landscape of India's Central Western Ghats. *Journal of Forestry Research*, *31*(3), 1005–1015.
12. Mathur, R. N. (1961). Pests of teak and their control. *Indian Forest Records (Entomology)*, *10*(3), 43–66.
13. Nascimento, D. A., Anunciação Jr, R. M., Arnhold, A., Ferraz Filho, A. C., dos Santos, A., Zanuncio, J. C. (2016). Expert system for identification of economically important insect pests in commercial teak plantations. *Computers and Electronics in Agriculture*, *121*, 368–373.
14. Sharma, S., Tara, J. S., Bhatia, S. (2013). Bionomics of *Hyblaea puera* (Lepidoptera: Hyblaeidae), a serious pest of Teak (Tectona grandis) from Jammu (India). *Munis Entomology & Zoology*, *8*(1), 139–147.
15. Rishi, R. R., Barthakur, N. D., Borah, R. K., Kumar, R., Pandey, S. (2014). Pest problems of some commercially important bamboo species in Assam, India. *International Journal for Life Sciences and Educational Research*, *2*(4), 113–120.
16. Mathew, G., Varma, R. V. (1990). Occurrence and pest status of some insects attacking bamboos in newly established plantations in Kerala. In: Rao, I. V. R., Gnanaharan, R., Sastry, C. B. (Eds.), *Bamboos—Current Research* (pp. 195–198).
17. Sasidharan, K. R., Varma, R. V. (2008). Seasonal population variations of the bark eating caterpillar (*Indarbela quadrinotata*) in Casuarina plantations of Tamil Nadu. *Tropical Ecology*, *49*(1), 79.
18. Tripathy, M. K., Dandapat, B. (2020). Seasonal incidence and population dynamics of important insect pests infesting gambhar Gmelina arborea Roxb. at coastal agro ecosystem of Odisha, India. *Journal of Entomology and Zoology Studies*, *8*(2), 1767–1773.
19. Tripathy, M. K., Dandapat, B., Panda, N. K. (2020). Studies on insect diversity associated with important tree species at Bhubaneswar, Odisha, India. *International Journal of Current Microbiology and Applied Sciences*, *9*(5), 46–56.

20. Ahmad, M., Faisal, M. (2012). Status of insect pests of poplar in India with special reference to Clostera spp. *Forestry Bulletin*, *12*(1), 105–122.
21. Venkata Rami Reddy, P., Gundappa, B., Chakravarthy, A. K. (2018). Pests of mango. *Pests and Their Management* (pp. 415–440). Springer Nature Singapore: Singapore.
22. Paine, T. D., Steinbauer, M. J., Lawson, S. A. (2011). Native and exotic pests of Eucalyptus: A worldwide perspective. *Annual Review of Entomology*, *56*, 181–201.
23. Wingfield, M. J., Slippers, B., Hurley, B. P., Coutinho, T. A., Wingfield, B. D., Roux, J. (2008). Eucalypt pests and diseases: Growing threats to plantation productivity. *Southern Forests: A Journal of Forest Science*, *70*(2), 139–144.
24. Batish, D. R., Singh, H. P., Kohli, R. K., Kaur, S. (2008). Eucalyptus essential oil as a natural pesticide. *Forest Ecology and Management*, *256*(12), 2166–2174.
25. Shrestha, G., Thapa, R. B. (2015). Tea pests and pesticide problems and integrated management. *International Journal of Agriculture Environment and Biotechnology*, *16*, 188–200.
26. Johnson, M. A., Ruiz-Diaz, C. P., Manoukis, N. C., Verle Rodrigues, J. C. (2020). Coffee berry borer (*Hypothenemus hampei*), a global pest of coffee: Perspectives from historical and recent invasions, and future priorities. *Insects*, *11*(12), 882.
27. Haseeb, M. (2005). Current status of insect pest problems in guava. *International Guava Symposium*, 453–467.
28. Devasahayam, S., Koya, K. A. (2016). Insect pests of ginger. In: *Ginger* (pp. 387–410). CRC Press.
29. Shanmugam, P. S., Indhumathi, K., Sangeetha, M., Tamilselvan, N. (2015). Evaluation of different pest management modules against major insect pests and diseases of turmeric. *Current Biotica*, *9*(1), 17–24.
30. Petty, G. J., Stirling, G. R., Bartholomew, D. P. (2002). Pests of pineapple. In: *Tropical Fruit Pests and Pollinators: Biology, Economic Importance, Natural Enemies and Control* (pp. 157–195). CABI Publishing: Wallingford, UK.
31. Ambele, C. F., Bisseleua, H. D., Djuideu, C. T., Akutse, K. S. (2023). Managing insect services and disservices in cocoa agroforestry systems. *Agroforestry Systems*, *97*, 965–984.
32. Durand-Bessart, C., Tixier, P., Quinteros, A., Andreotti, F., Rapidel, B., Tauvel, C., Allinne, C. (2020). Analysis of interactions amongst shade trees, coffee foliar diseases and coffee yield in multistrata agroforestry systems. *Crop Protection*, *133*, 105137.
33. Octavia, D., Suharti, S., Murniati, Dharmawan, I. W., Nugroho, H. Y., Supriyanto, B., Rohadi, D., Njurumana, G. N., Yeny, I., Hani, A., Mindawati, N. (2022). Mainstreaming smart agroforestry for social forestry implementation to support sustainable development goals in Indonesia: A review. *Sustainability*, *14*(15), 9313.
34. Maredia, K. M., Dakouo, D., Mota-Sanchez, D. (2003). *Integrated Pest Management in the Global Arena*. CABI.
35. Bajwa, W. I., Kogan, M. (2004). Cultural practices: Springboard to IPDM. In: *Integrated Pest Management: Potential, Constraints and Challenges* (pp. 21–38). CABI Publishing.
36. Devi, D. L., Mandal, K. S., Dadaro, S. D., Kumar, V. (2023). Cultural practices: An urgent need in IPDM for sustainable crop production. *Just Agriculture*, *3*, 78–86.
37. Nöldeke, B., Winter, E., Laumonier, Y., Simamora, T. (2021). Simulating agroforestry adoption in rural Indonesia: The potential of trees on farms for livelihoods and environment. *Land*, *10*(4), 385.

38. Song, B., Han, Z. (2020). Assessment of the biocontrol effects of three aromatic plants on multiple trophic levels of the arthropod community in an agroforestry ecosystem. *Ecological Entomology*, *45*(4), 831–839.
39. Teodoro, A., Tscharntke, T., Klein, A. M. (2008). From the lab to the field: Contrasting effects of multitrophic interactions and agroforestry management on coffee pest densities. In: *Agroforestry Management, Seasonal Changes, Biodiversity and Multitrophic Interactions of Coffee Arthropods* (p. 61).
40. Bhoi, T. K., Samal, I., Majhi, P. K., Komal, J., Mahanta, D. K., Pradhan, A. K., Saini, V., Nikhil Raj, M., Ahmad, M. A., Behera, P. P., Ashwini, M. (2022). Insight into aphid mediated potato virus Y transmission: A molecular to bioinformatics prospective. *Frontiers in Microbiology*, *13*, 1001454.
41. Mahanta, D. K., Jangra, S., Ghosh, A., Sharma, P. K., Iquebal, M. A., Jaiswal, S., Baranwal, V. K., Kalia, V. K., Chander, S. (2022). Groundnut bud necrosis virus modulates the expression of innate immune, endocytosis, and cuticle development-associated genes to circulate and propagate in its vector, Thrips palmi. *Frontiers in Microbiology*, *13*, 773238.
42. Bhatnagar, S., Mahanta, D. K., Vyas, V., Samal, I., Komal, J., Bhoi, T. K. (2023). Storage pest management with nanopesticides incorporating silicon nanoparticles: A novel approach for sustainable crop preservation and food security. *Silicon*, *16*, 471–483.
43. Mahanta, D. K., Komal, J., Samal, I., Bhoi, T. K., Dubey, V. K., Pradhan, K., Nekkanti, A., Gouda, M. N. R., Saini, V., Negi, N., Bhateja, S., Jat, H. K., Jeengar, D. (2023a). Nutritional aspects and dietary benefits of "Silkworms": Current scenario and future outlook. *Frontiers in Nutrition*, *10*, 1121508.
44. Samal, I., Bhoi, T. K., Vyas, V., Majhi, P. K., Mahanta, D. K., Komal, J., Singh, S., Kumar, P. V. D., Acharya, L. K. (2023a). Resistance to fungicides in entomopathogenic fungi: Underlying mechanisms, consequences, and opportunities for progress. *Tropical Plant Pathology*, *49*, 5–17.
45. Mahanta, D. K., Bhoi, T. K., Komal, J., Samal, I., Nikhil, R. M., Paschapur, A. U., Singh, G., Kumar, P. V. D., Desai, H. R., Ahmad, M. A., Singh, P. P., Majhi, P. K., Mukherjee, U., Singh, P., Saini, V., Shahanaz, Srinivasa, N., Yele, Y. (2023b). Insect-pathogen crosstalk and the cellular-molecular mechanisms of insect immunity: Uncovering the underlying signaling pathways and immune regulatory function of non-coding RNAs. *Frontiers in Immunology*, *14*, 1169152.
46. Samal, I., Bhoi, T. K., Majhi, P. K., Murmu, S., Pradhan, A. K., Kumar, D., Saini, V., Paschapur, A. U., Raj, M. N., Ankur, Manik, S., Behera, P. P., Mahanta, D. K., Komal, J., Alam, P., Balawi, T. A. (2023b). Combatting insects mediated biotic stress through plant associated endophytic entomopathogenic fungi in horticultural crops. *Frontiers in Plant Science*, *13*, 1098673.
47. Komal, J., Desai, H. R., Samal, I., Mastinu, A., Patel, R. D., Kumar, P. D., Majhi, P. K., Mahanta, D. K., Bhoi, T. K. (2023). Unveiling the genetic symphony: Harnessing CRISPR-cas genome editing for effective insect pest management. *Plants*, *12*(23), 3961.
48. Samal, I., Bhoi, T. K., Mahanta, D. K., Komal, J. (2023c). Establishing the Role of Silicon (Si) in plant resistance to insects: A bibliometric approach. *Silicon*, *16*, 2119–2128.

49. Majhi, P. K., Samal, I., Bhoi, T. K., Pattnaik, P., Pradhan, C., Gupta, A., Mahanta, D. K., Senapati, S. K. (2023). Climate-smart agriculture: An integrated approach to address climate change and food security. In: *Climate Change and Insect Biodiversity* (pp. 221–245). CRC Press.
50. Mahanta, D. K., Samal, I., Komal, J., Bhoi, T. K., Majhi, P. K., Ahmad, M. A. (2023c). Understanding anthropogenic climate change, its consequences on insect pests, and techniques in forecasting and monitoring pest dynamics: A contemporary scenario. In: *Climate Change and Insect Biodiversity* (pp. 44–67). CRC Press.
51. Bhoi, T. K., Samal, I., Mahanta, D. K., Komal, J., Jinger, D., Sahoo, M. R., Achary, G. C., Nayak, P., Sunani, S. K., Saini, V., Raghuraman, M., Singh, S. (2023). Understanding how silicon fertilization impacts chemical ecology and multitrophic interactions among plants, insects and beneficial arthropods. *Silicon*, *15*(6), 2529–2549.
52. Bhoi, T. K., Samal, I., Mahanta, D. K., Komal, J., Majhi, P. K., Ankur. (2024). Exploring Melatonin's potential as an alternative strategy for protecting plants from biotic stresses. In: *Melatonin in Plants*: *A Pleiotropic Molecule for Abiotic Stresses and Pathogen Infection* (pp. 223–242). Springer Nature Singapore: Singapore.
53. Samal, I., Bhoi, T. K., Pradhan, A. K., Mahanta, D. K. (2023d). Plantix app: A success story of artificial intelligence in plant protection. *Van Sangyan*, *10*, 24–26.
54. Mahanta, D. K., Bhoi, T. K., Komal, J., Samal, I., Mastinu, A. (2024). Spatial, spectral and temporal insights: Harnessing high-resolution satellite remote sensing and artificial intelligence for early monitoring of wood boring pests in forests. *Plant Stress*, *11*, 100381.
55. Samal, I., Mahanta, D. K., Bhoi, T. K., Komal, J., Jatav, H. S., Jatav, S. S., Sathyanarayana, E. (2024). *Approaches of Biochar in Ecosystem Management*: *Current Scenario and Future Perspectives*. IntechOpen
56. Vyas, V., Bhoi, T. K., Samal, I., Singh, S., Mahanta, D. K. (2024). Revitalizing degraded soils with agroforestry interventions: Opportunities, challenges, and future direction. In *Agroforestry to Combat Global Challenges*: *Current Prospects and Future Challenges* (pp. 529–549).
57. Samal, I., Bhoi, T. K., Mahanta, D. K., Komal, J., Singh, S. (2024). Biorational pest management: Potentials, unintended consequences, and future concerns. In: *Biorationals and Biopesticides*: *Pest Management* (pp. 47–75).

CHAPTER 6

Multitrophic Pollination Networks: Enhancing Biodiversity Conservation and Crop Yield Through Integrated Ecological Interactions

JASTI SRI VISHNU MURTHY,[1] ALAGESAN KEERTHANA,[2] LOGESWARAN K,[3] KANCHAN KADAWLA,[4] FEBY ATEE,[2] BINDU GUDI RAMAKRISHNA,[1] DEEPAK KUMAR MAHANTA,[5] ANIL KUMAR,[6] and VINOD KUMAR DUBEY[7*]

[1]Department of Agricultural Entomology, College of Agriculture, Kerala Agricultural University, Thrissur, Kerala, India

[2]Department of Entomology, Tamil Nadu Agricultural University, Coimbatore, Tamil Nadu, India

[3]Division of Entomology and Nematology, Indian Institute of Horticultural Research, Bengaluru, Karnataka, India

[4]Department of Entomology, College of Agriculture, Chaudhary Charan Singh Haryana Agricultural University, Hisar, Haryana, India

[5]Forest Entomology Discipline, Forest Protection Division, Forest Research Institute, (ICFRE-FRI), Indian Council of Forestry Research and Education, Dehradun, Uttarakhand, India

[6]Sugarcane Research Institute, Department of Entomology, Post Graduate College of Agriculture, Dr. Rajendra Prasad Central Agricultural University, Pusa, Bihar, India

[7]Department of Genetics and Plant Breeding, College of Agriculture, Acharya Narendra Deva University of Agriculture and Technology, Kumarganj, Ayodhya, Uttar Pradesh, India

*Corresponding author

Multitrophic Interactions for Sustainable Agroforestry: Harnessing the Power of Beneficial Insects and Ecological Dynamics. Hanuman Singh Jatav, Vishnu D. Rajput, Ipsita Samal, and Tanmaya Kumar Bhoi (Eds).

DOI: 10.1201/9781779641250-6

ABSTRACT

One major global challenge facing humanity is the need to address ongoing crises related to climate change and declining biodiversity. Ecological intensification (EI) refers to a strategy that aims to maximize crop yields while minimizing environmental harm. It achieves this by harnessing the benefits of biodiversity to enhance ecosystem functions and services, particularly through interactions such as multitrophic pollination networks. Understanding the interactions among multiple trophic levels of biodiversity can advance agriculture and forestry. Various factors, including land restrictions, limits, and complex habitats, influence multitrophic ecosystem diversity and services, and also the relationship between multitrophic diversity and ecosystem services. Biodiversity has positive effects on ecosystem functions within production systems. This includes various dimensions of ecosystem function, hierarchy levels, research approaches, and specific functions within ecosystems. However, contextual factors can influence these effects. Additionally, land use and management practices have significant impacts on ecosystem services and multitrophic biodiversity. Several acquaintances remain, among them a lack of data from certain production systems, geographical areas, measurements of functional diversity, and organism groups. Furthermore, these areas require more attention in future research. These areas include trade-offs between numerous ecosystem functions, sequential dynamics, the influence of climate change, the spatial scale of intervention measures, and their practical implementation. These statistics are crucial for ensuring that forest and agricultural landscapes can sustainably provide resources for the growing human population. This is done while staying within the planet's environmental limits.

6.1 INTRODUCTION

Major challenges facing us in the 21st century involve the massive loss of biodiversity and providing food to the huge, growing population amid global changes [1–4]. Current agricultural practices often prioritize maximizing crop yields or ensuring profitability over minimizing the quality of the agricultural product. Inopportunely, in many countries, agricultural productivity has already peaked, relying heavily on extensive inputs such as chemical fertilizers and insecticides, leading to the degradation of ecosystem functions [5, 6]. Furthermore, studies indicate that the productivity of agroecosystems declines with the intensification of agriculture, and these low yields persisted for centuries, even with the subsequent abandonment of agricultural activities [7]. The decline in ecosystem services is largely attributed to biodiversity loss [8]. Until now, the emphasis on sustainability and stability of agricultural production systems and the delivery of ecosystem functions beyond production, and biodiversity conservation has been infrequently regarded as essential goals in agricultural landscapes [4]. Therefore, comprehensive objectives such as sustainable agricultural production, resilient production systems, the provision of diverse ecosystem services beyond mere production, and the conservation of biodiversity often find themselves on the periphery of agricultural priorities. It is within this context that multitrophic pollination networks come to the forefront, offering a nuanced and interconnected perspective on how biodiversity and crop yield enhancement can coalesce for mutual benefit [9]. Multitrophic pollination networks refer to complex interactions between multiple trophic levels (organisms occupying different positions in a food chain) involved in the pollination process [9, 10]. These networks play a major role in supporting biodiversity and enhancing crop yields. The key components of multitrophic pollination networks include plants, pollinators, and natural enemies of herbivores. Multitrophic networks involve a variety of interactions, not just between plants and pollinators, but also with predators and parasitoids of herbivores [4]. Multitrophic pollination networks are crucial for maintaining ecological balance and stability, as they provide diverse ecological niches and reduce competition for resources. They also enhance pollination services by controlling herbivore populations, reducing the need for chemical pesticides, and promoting a healthier environment for pollinators. Maintaining diverse habitats supports a wide range of species in these networks, and restoring ecosystems can help re-establish diverse multitrophic interactions [11]. Regular monitoring of pollinator populations, natural enemies, and plant–pollinator interactions is essential for assessing the health and functionality of these networks, guiding management strategies for sustainable agriculture and ecosystem conservation. This exploration seeks to unravel the intricate layers of multitrophic pollination networks, elucidating their critical role in bolstering ecological resilience, supporting agricultural productivity, and contributing substantively to the broader goals of biodiversity conservation. In this chapter, we thoroughly discussed how the multitrophic pollination network enhances biodiversity and crop yield.

6.2 INSECT POLLINATORS: MULTITROPHIC INTERACTION NETWORKS

In agroforestry ecosystems, the interaction between plants, insects, and other organisms forms complex multitrophic networks. Insect pollinators play a crucial role in these networks, contributing to the reproductive success of many plant and tree species [12]. Here is a description of insect pollinators in multitrophic networks within agroforestry ecosystems, and it is also depicted in Figure 6.1.

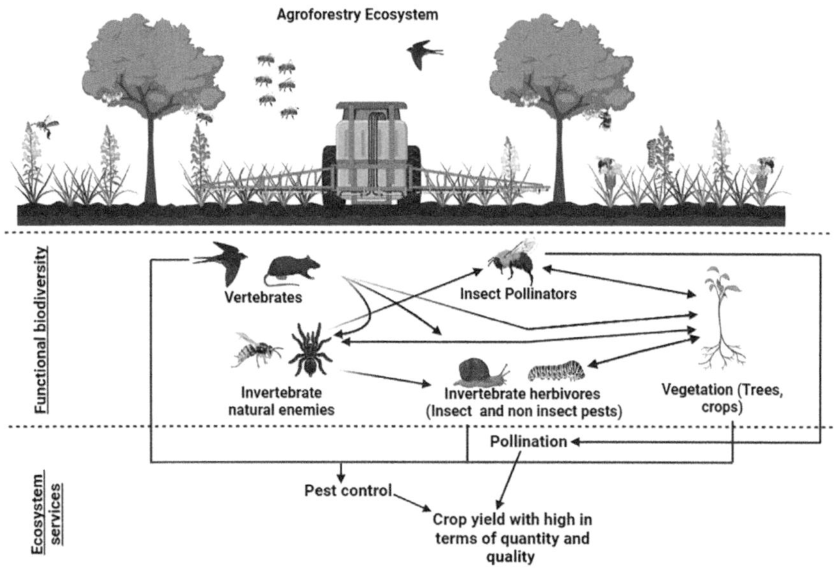

FIGURE 6.1 Multitrophic interaction networks of vertebrates and invertebrates in an agroforestry ecosystem.

6.2.1 *DIVERSITY OF INSECT POLLINATORS*

Agroforestry systems often harbor a diverse community of insect pollinators. This diversity can include bees, butterflies, beetles, flies, moths, wasps, and other insect pollinators. In particular, research indicates that wild bee species frequently exhibit superior pollination capabilities for crop plants when compared to honeybees [13]. Different insect species of pollinators have varying preferences for certain plant and tree species, creating intricate interactions within the ecosystem.

6.2.2 POLLINATION SERVICES

Insect pollinators provide essential pollination services by transferring pollen from male (anthers) to female (stigmas) flowers [14]. This process is fundamental for the next generation of fruits and seeds in many flowering plants, including trees within agroforestry systems. In 2014, Bailey et al. [15] observed that the closer proximity to forestry boundaries, which offer mating or nesting space, results in increased richness of pollinators in oilseed rape. However, the impact is contingent on the size of bees, as larger bees tend to have longer dispersal distances. In general, modifying habitat characteristics and the arrangement of habitats on a broader landscape scale appears to be the most effective strategy for encouraging pollinators to pollinate and enhancing biodiversity and crop yields. But there is a problem, especially with smaller scale projects: some measures and initiatives might not just benefit pollinators but also their natural enemies, which could offset the yield-boosting effects [16].

6.2.3 PLANT–INSECT RELATIONSHIPS

The insect-pollinators and plant relationships are often specific and coevolved. Plants use various strategies to attract pollinators. This includes producing colorful and fragrant flowers, offering nectar as a reward, and providing landing platforms for insects. These features make it easier for pollinators to locate and access the reproductive parts of the flower. For example, Darwin initially formulated the concept of coevolution, employing it to elucidate how mutual interactions between pollinators and nectar-rewarding flowers could lead to the development of elongated mouthparts over time. Significantly, he accurately forecasted that a Madagascan orchid, *Angraecum sesquipedale,* must rely on pollination by a hawkmoth possessing an abnormally elongated tongue. The concept of coevolutionary "race" proposed by Darwin received endorsement from other naturalists of the same period, such as Alfred Wallace. A hawkmoth that met the expected criteria for tongue length was eventually found in Madagascar during the early twentieth century. However, substantial empirical evidence supporting the fundamental process of Darwin's coevolutionary theory has only been apparent in the last twenty years. Existing information confirms that plants with longer flower tubes are often favored by natural selection when their pollinators have shorter tongues [17].

6.2.4 IMPACT ON CROP PRODUCTIVITY

In agroforestry, the presence of diverse insect pollinators can positively impact crop productivity. Improved pollination can lead to increased fruit and seed set, improving the overall productivity and quality of crops [18–20].

6.2.5 HABITAT PROVISION

Agroforestry systems, with their diverse vegetation structure, can deliver a variety of habitats for different insect species. The presence of trees, shrubs, and herbaceous plants creates niches for both generalist and specialist pollinators, contributing to overall biodiversity in the ecosystem. This diversity in vegetation structure creates a range of niches and micro-habitats that can support various species of insect pollinators, including bees, butterflies, beetles, and flies [21, 22]. Trees in agroforestry systems often provide suitable nesting sites for many pollinators. Hollow branches, deadwood, and other tree structures can serve as nesting locations for solitary bees and other insects. This contributes to the reproductive success and population maintenance of these important pollinators. The presence of diverse plant species in agroforestry systems means there is a continuous and varied supply of floral resources such as nectar and pollen. Different flowering plants bloom at different times, ensuring a sustained food source for pollinators throughout the growing season [23].

The canopy and understory of agroforestry systems offer shelter to pollinators, protecting them from extreme weather conditions and providing overwintering habitat. This is crucial for the survival of certain pollinator species during adverse periods [24]. Agroforestry landscapes can act as ecological corridors, connecting different habitat patches.

These corridors facilitate the movement of pollinators between fragmented habitats, promoting genetic diversity and maintaining healthy populations [25]. Agroforestry practices often involve a more integrated and sustainable approach to pest management, reducing the need for intense pesticide use. This benefits pollinators by minimizing exposure to harmful chemicals, thereby supporting their health and reproductive success [21]. Agroforestry can attract not only pollinators but also beneficial insects that act as natural enemies of crop pests. The presence of diverse vegetation provides resources for predatory insects, fostering a balanced ecosystem that supports both pollinators and natural enemies of pests [11].

6.2.6 TROPHIC INTERACTIONS

Insect pollinators participate in multitrophic interactions within agroforestry ecosystems. They serve as a link between plants and higher trophic levels, such as predators and parasitoids. For instance, certain predatory insects may be attracted to areas with abundant floral resources, indirectly benefiting crop protection [26].

6.2.7 SUSTAINABILITY AND RESILIENCE

The presence of diverse insect pollinators enhances the sustainability and resilience of agroforestry systems. A resilient ecosystem with a variety of pollinators is better equipped to adapt to environmental changes, ensuring the continued productivity of both crops and natural vegetation [27].

6.3 TROPHIC INTERACTIONS IN ECOLOGICAL INTENSIFICATION

EI is the process of gradually increasing the primary production per unit area/productivity of the system by favoring more environmentally friendly practices through effective use of the ecosystem goods and services and cutting down the practices and nonrenewable inputs that create a negative impact on the ecosystem [28]. By effectively harnessing all ecosystem services and reducing the negative impacts on the environment, EI aims to create a sustainable agricultural production system [29]. Comparing the characteristics of traditional agricultural systems and natural ecosystems, it is quite easily understandable that traditional agricultural systems closely resemble natural ecosystems, thereby in a way indirectly establishing EI [30]. Pollinators, NEs, beneficial entomopathogens, plant microbiome, detritivores occupying different trophic levels are very well known to have a crucial role in EI by increasing the pollination and setting rate, improving water use efficiency, and tolerance to biotic and abiotic stresses [31].

6.3.1 POLLINATORS AS AN IMPORTANT AGENT IN ECOLOGICAL INTENSIFICATION

The plant–pollinator interaction is a mighty mutualism fundamental to a healthy, sustainable ecosystem. Among the diverse group of biotic and abiotic pollinators, it is estimated that animal pollinators contribute to 87.5% of reproduction in flowering plants [32]. The pollinator community includes species of moths, butterflies, beetles, bees, wasps, ants, flies, thrips, bugs, springtails, termites, cockroaches, lacewings, caddisflies, cricket, scorpion flies, bark flies, stone flies, earwigs, birds, bats, lizards, marine worms, rodents, and reptiles [33, 34]. Insects dominate the pollination system. Around 75% of food crops depend on insect pollination to a certain degree, and it was reported that 39 of the 57 major crops showed increased fruit or seed quality [18, 34]. Agricultural production and diversity rely heavily on *Apis mellifera* and a number of wild bee species [35]. The absence of pollinators can cause a dramatic decrease in the yield of crops. Without honeybees, a decrease in yield by more than 90% was documented in some fruits, seeds, and nut crops [36]. The pollinators, while directly facilitating the reproduction of the crops, indirectly support animal habitats, the stabilization of soil, and carbon sequestration. Thus, maintaining a healthy and diverse population of pollinators is essential for a healthy ecosystem. The native pollinators, apart from increasing the food production of the plant, also aid the plants to prevent erosion, as a food source for birds, egg laying, feeding, and nesting site for insects and other animals [34].

Despite the well-established and widely known importance of pollinators, evidence is mounting on the decline in pollinator species in all countries except Antarctica. Decline in bee abundance over the years has been documented in several works [37]. In the United States, the International Union for Conservation of Nature designated around 185 insect pollinator species as threatened or extinct. Habitat alteration through agricultural intensification and fragmentation, grazing, introduction of alien pollinators, pesticide poisoning, and climate change are the most common contributors to the decline [38, 39]. By planting native vegetation favorable to the indigenous pollinators, careful use of pollinator-safe pesticides, and providing a clean water source can greatly help the pollinator population to survive and flourish in the natural and agroecosystems, thereby helping in EI [34].

6.3.2 *ROLE OF POLLINATORS IN AGROFORESTRY SYSTEMS*

An agroforestry system is a diverse farming approach compared to conventional agriculture, where the crops are raised in between rows of trees in the alley [27]. By integrating crops into the alleys, the agroforestry system exhibits a greater floral and faunal diversity [40]. Agroforestry-induced improvement in biodiversity has been reported from several studies [41, 42]. This underpins that increased pollinators can increase the crop yield of the agroforestry system. Around 73% of the crops cultivated in the Amazon forest were found to be dependent on pollinators for pollination. Eighteen of 24 pollinator-dependent crops studied were found to be important contributor species for socio-biodiversity, along with exhibiting a high food value [43]. The very basic nature of the agroforestry system provides shelter, nesting, and foraging resources for the pollinators, thereby increasing its diversity and abundance within the agroforestry system [21]. By providing the above key benefits to the pollinators, the agroforestry receives crop pollination services in return from the pollinators. Wild bee species showed richness and flower visitation frequency on the almond flowers where the orchards had an adjacent vegetation strip or seminatural habitat. Also, increased fruit set was related to increased species richness of wild bees and their visitation frequency [44]. Despite knowing the benefits of plant–pollinator mutualisms, very little research has been conducted on whether crop pollination services by the pollinators increase crop yield and other benefits in an agroforestry system.

6.4 DIVERSITY OF POLLINATORS ENHANCES THE SERVICES OF POLLINATION

Pollination diversity plays a significant role in enhancing pollination services across various ecosystems. It involves the participation of diverse species of pollinators benefiting both natural and agricultural systems (Table 6.1).

TABLE 6.1 Pollinator Diversity Significance in Natural and Agricultural Ecosystems

Pollinator Diversity Importance	References
Natural ecosystem	
Increases the pollination and genetic diversity by overcoming pollen limitation	[45, 46]
Increases plant diversity by promoting the seedling recruitment	[47, 48]
Agricultural ecosystem	
Increased fruit setting in fruit crops	[23]
Increases agricultural production quality, including fruit sugar, cottonseed weight, and seed oil	[49–51]
Increases the economic value of crops	[52]
Increases crop stability	[53]

6.4.1 *THE SIGNIFICANCE OF DIVERSE POLLINATORS IN NATURAL-PLANT COMMUNITIES*

Pollinators hold profound importance within natural plant communities, serving as a linchpin for their vitality and sustainability. The diverse array of pollinators, encompassing bees, butterflies, birds, bats, and various insects, plays a vital role in the reproductive success of flowering plants [54, 55]. Their interactions with plant species facilitate pollination, leading to the production of seeds, fruits, and nuts essential for plant propagation and genetic diversity [56]. This diversity not only ensures the survival of individual plant species but also contributes significantly to the overall resilience and adaptability of plant communities in the face of environmental changes [57]. A rich assortment of pollinators within these ecosystems fosters intricate relationships, enabling efficient and effective pollination across a wide range of plants with different flowering patterns, shapes, and sizes. This diversity safeguards the stability of natural plant communities, promotes biodiversity, and provides critical ecosystem services that support other organisms, making it an indispensable component for the health and functionality of these ecosystems. Efforts to conserve and enhance pollinator diversity are imperative to maintain the intricate balance and long-term sustainability of plant communities.

6.4.2 *THE ROLE OF DIVERSE POLLINATORS IN ENHANCING AGRICULTURAL PRODUCTIVITY*

Pollinator diversity stands as a cornerstone for sustaining agricultural productivity worldwide. Pollinators, including bees, butterflies, birds, bats, and various insects, contribute significantly to the cultivation of crops by facilitating the transfer of pollen grains, thereby enabling the production of seeds and fruits through fertilization [18]. The variety of these pollinators holds immense significance within agricultural ecosystems for numerous reasons, as represented in Figure 6.2.

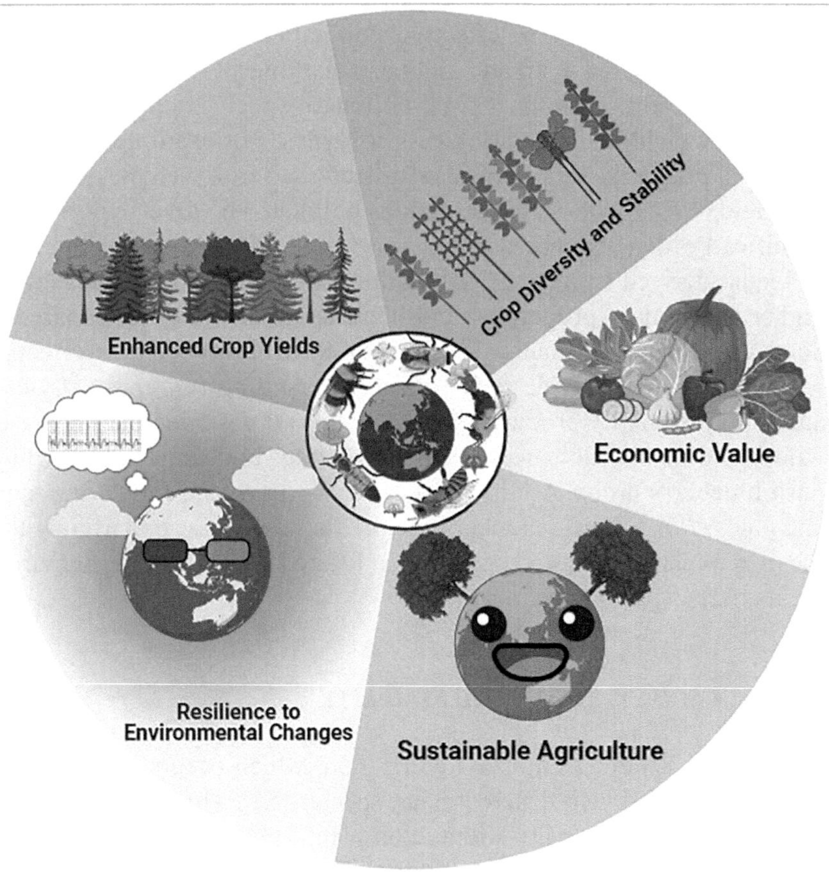

FIGURE 6.2 Importance of pollinator diversity in an agricultural ecosystem.

6.4.2.1 *ENHANCED CROP YIELDS*

Diverse pollinator communities play a pivotal role in enhancing crop yields through the varied foraging behaviors and preferences exhibited by different pollinator species, leading to more effective pollination across various crops. Studies employing pollinator exclusion experiments have demonstrated this phenomenon in diverse agricultural settings. For instance, in a coffee plantation, Philpott et al. [58] observed increased coffee weight due to pollination by dipterans and hymenopterans. Similarly, Bartomeus et al. [50] found higher oil content and lower chlorophyll levels in adequately pollinated oilseed rape, reduced empty seeds in buckwheat, and improved commercial-grade strawberries due to effective pollination. Intercropping mango plantations attracted a greater richness of flower visitors, resulting in increased weight of the fruit [51]. Pollinator species richness may also affect fruit yield and physiological properties. For example, excluding bats from pitaya flowers significantly decreased fruit weight, sweetness, and yield [49]. In cotton, plant sections pollinated by diverse pollinators specifically found to increase the seed weight [59]. Additionally, fruits of loquat derived from open pollination exhibited greater sugar levels, higher antioxidant properties, and increased fruit weight compared to honeybee-pollinated plants. This increased diversity ensures effective pollination across a wide range of crops, resulting in improved yields and superior-quality produce. Rogers et al. [60] observed that increased wild-bee diversity enhances seed set, thereby boosting yields in high bush blueberry crops. Additionally, even conifers like firs (*Abies* spp.) and pines (*Pinus* spp.) attract pollinators for honeydew, underlining that these trees can serve as a sugar source for pollinators in forest environments [61, 62].

6.4.2.2 *CROP DIVERSITY AND STABILITY*

Many crops benefit from cross-pollination, which occurs when diverse pollinator species visit different plant species [52]. This process leads to increased genetic diversity within crop plants, enhancing their resilience to diseases, pests, and environmental stresses [63]. In turn, this diversity contributes to more stable and robust agricultural systems.

6.4.2.3 *ECONOMIC VALUE*

Pollinators provide an invaluable ecosystem service to agriculture, contributing billions of dollars annually to global crop production [64]. Insect-pollinated crops like oilseeds, vegetables, fruits, and nuts derive substantial advantages from diverse groups of pollinators, leading to increased yields and improved economic outcomes for farmers [52]. Beyond food crops, pollinators play a crucial role in supporting fiber crops such as cotton, as well as biofuel sources like canola and palm oils, medicinal plants, and forage for livestock.

6.4.2.4 *SUSTAINABLE AGRICULTURE*

Encouraging diverse pollinator species is essential for fostering sustainable agricultural practices, reducing the dependency on chemical pesticides and fertilizers. These pollinators play a crucial role in natural pest control and maintaining ecological equilibrium within agricultural settings. Certain insect pollinators, such as lacewings (Chrysopidae), hoverflies (Syrphidae), ladybirds (Coccinellidae), and parasitic wasps (Hymenoptera), were reported as natural enemies of pests [65, 66]. Advancements in technology enable the utilization of pollinator species for managing diseases in crops. For instance, *Bombus terrestris* has been successfully employed for strawberry pollination in a greenhouse to control microbial growth [67]. Recent studies have indicated that bats play a significant role in consuming dipterans, which include numerous disease vectors like mosquitoes [68]. Numerous insectivorous bats primarily consume several agroforest pest and disease vectors. In the United States, red bats and big brown bats consume 32.6% and 28% of insect pests, respectively, highlighting their contribution to pest control [69].

6.4.2.5 *RESILIENCE TO ENVIRONMENTAL CHANGES*

A diverse array of pollinators can help mitigate the negative impacts of climate change or other environmental stressors on crop pollination [70]. Conserving and promoting pollinator diversity through habitat preservation, reduced pesticide usage, and sustainable agricultural practices remain imperative for ensuring the long-term viability and productivity of agricultural systems.

6.5 CONTRIBUTION OF POLLINATOR DIVERSITY TO BIODIVERSITY ENHANCEMENT

The following interactions between plants and pollinators are important beneficial relationships in agricultural production as they offer vital ecosystem services that contribute to the improvement of global biodiversity [33].

6.5.1 INCREASED POLLINATOR DIVERSITY IMPROVES POLLINATION SERVICES

6.5.1.1 POLLINATOR DIVERSITY IS IMPORTANT IN NATURAL PLANT COMMUNITIES

Pollination services unquestionably play a major role in maintaining plant communities because they are essential to the reproductive success of plants. For half of all blooming plants, the loss of fertility would be greater than 80%, and for a third, there would be no seed production at all, according to Rodger et al. [54]. Natural pollination is important for ecosystems that exist naturally, as many studies have shown.

6.5.1.2 PLANT PRODUCTION AND RECRUITMENT ARE ENHANCED BY POLLINATOR DIVERSITY

The pollinator diversity is essential for the growth of blooming plant seeds. High-quality and abundant seed yields are produced by plants visited by a diverse pollinator community [45, 56]. Fontaine et al. [47] found that by modifying plant and pollinator functional diversity in trials, boosting pollinator variation increased species richness of the plant community. Lundgren et al. [48] discovered that a four-year restriction on pollinator visits to flowers had an impact on seedling recruitment for the species richness of the plant community. Accordingly, these results demonstrate that diverse pollinators can enrich plant community persistence through promoting the production of seeds and seedling recruitment. At the post-seed development stage, diverse pollinators in plant communities may become apparent. However, to clarify such relationships, longer term research is necessary.

6.5.1.3 *DIVERSITY OF POLLINATORS REDUCES POLLEN LIMITATION IN NATURAL PLANT COMMUNITIES*

Plant species distribution and abundance, as well as ecosystem function, may be impacted by pollen scarcity [71]. A recent meta-analysis found that plants with specialized functions and environments are susceptible to a lack of pollen during the Anthropocene [72]. Pollinator diversity protection is therefore essential. Pollen limitation may also be discouraged by various pollinator assemblages because of variations in pollinator-specific distances of foraging and flower-visiting behaviors. Experiments have shown that pollinator species diversity prevents pollen limitation in natural plant communities, as opposed to the quantity of floral visits [45, 46]. Additionally, alterations in the ambient environmental circumstances leading to alterations in pollinator foraging behavior may enhance pollination in native plant communities [73]. When there is a lot of wind, wild bees visit flowers higher up in the canopy, while honeybees feed at the lowest interior sections of trees [74]. These results demonstrate that pollinator variation can increase pollination under environmental perturbations, potentially circumventing pollen constraint in natural plant communities.

6.5.2 IMPORTANCE OF POLLINATOR DIVERSITY FOR AGRICULTURAL OUTPUT

Pollinator variety may have an impact on the quality of agricultural yields since intercropping mango (*Mangifera indica* Linn.) farms draw a lot of floral visitors, which increases the weight of the mango fruit [51]. Similar findings have been made recently about the possibility that the variety of pollinator species affects the physiological and chemical characteristics of fruit production. Cusser et al. [59] found that heavier cotton seeds were only found in places where a diversity of pollinators was present in a study using cotton, an essential crop for fiber production. When crops are cross-pollinated rather than self-pollinated, the quality of crop yields is higher because of the increased pollinator complementarity that cross-pollinated crops enjoy. This reduces the detrimental effects of various factors on pollination success, including temporal (day or night), environmental, and climatic factors. Therefore, a method for maintaining excellent crop quality in the face of climate change may be provided by taxonomically diverse pollinators.

Agricultural productivity is increased by pollinator variety. Pollinator variety affects fruit set in many crops, including apple [75], pumpkin [76], almond [77], and coffee [78], where honeybee flower visits are frequent. An increase in wild pollinators substantially boosts fruit set [79]. In a similar view, the number of seeds produced by the jalapeño pepper, *Capsicum annuum,* was significantly increased with high pollinator species richness in an urban environment [80]. Senapathi et al. [81] used a global crop data set to determine that temporal pollinator complementarity was essential for producing inter-annual stability in agropollinators worldwide. The proportion of flowers pollinated during crop blooming times increases when different pollinator guilds exhibit spatial and temporal complementarity, hence boosting agricultural productivity.

6.5.3 CONTRIBUTION OF VARIOUS KINDS OF POLLINATORS TO ENVIRONMENTAL AND HUMAN HEALTH SAFETY

6.5.3.1 POLLINATORS ACT AS ENVIRONMENTAL INDICATORS

Monitoring ecological stress is critical to maintaining both environmental and human health. Numerous pollinator species have been assessed as ecological indicators in a range of environments [82]. Because of their extreme sensitivity to anthropogenic contaminants, bees and butterflies have been employed to measure ecological changes. Nymphalid butterflies (*Danaus chrysippus* Linn.) may be biomarkers of cadmium (Cd) and copper (Cu) heavy metal contamination [83–101]. The honeybees (*Apis* spp.) can be used by environmentalists to evaluate seasonal Fe, Zn, Cu, and Mn in urban environments. Research on flowering plants and pollinator groups in various environments could yield more examples of ecological and environmental monitoring. The framework has to be completed with more study on a larger variety of wild pollinators.

6.5.3.2 FUNCTION OF POLLINATORS IN PEST AND DISEASE CONTROL

Some insect pollinators naturally fight these pests [66]. Disease vector eradication is effective with biological control. Bats eat more dipterans, such as mosquitoes, than expected [68]. Whitby et al. [69] discovered that the great brown and red bats, two significant agricultural pollinators, consume 32.6 and 28% of crop pest species, respectively, in a US region dominated by maize and soybean crops. Pollinators can now control agricultural illnesses. Nonhoney bee, *Bombus terrestris* (Linn.), a common crop pollinator, suppressed strawberry microbial growth in a greenhouse [67]. Preserving pollinators that manage pests and disease vectors might significantly decrease the usage of pesticides, which would be advantageous for both humans and the environment.

6.5.4 THE NECESSITY TO ENHANCE BIODIVERSITY

The Council is required by Section 6 of the Environment (Wales) Act 2016 to aim to conserve and increase biodiversity and promote ecosystem resilience. The well-being of Future Generations Act 2015 defines Resilient Wales as "a nation that maintains and improves a biodiverse natural environment with healthy functioning ecosystems that support social, economic, and ecological resilience and the capacity to adapt to change," for instance, "The National Plan for the Future of Wales includes Policy 9: in all cases, action toward securing the maintenance and enhancement of biodiversity (to provide a net benefit), the resilience of ecosystems and green infrastructure assets must be demonstrated as part of development proposals through innovative, nature-based approaches to site planning and the design of the built environment."

6.6 CONCLUSION

Multitrophic pollination networks highlight their crucial role in maintaining a delicate equilibrium between conserving biodiversity and improving crop yields. The complex interactions among plants, pollinators, and natural enemies emphasize the interconnected nature of ecological systems, influencing the stability of production systems and biodiversity conservation. Through the exploration of multifaceted aspects of these networks, opportunities arise to promote ecological resilience, bolster agricultural productivity, and contribute to broader conservation objectives. As we navigate the intricate network of interactions within multitrophic pollination networks, it becomes clear that their sustainable management not only enhances biodiversity but also optimizes crop yields in a mutually beneficial manner. Embracing and giving precedence to these networks in agricultural practices can pave the way for a more resilient and sustainable future, aligning the goals of biodiversity conservation and improved crop productivity for the benefit of both ecosystems and humans.

6.7 FUTURE PROSPECTS

Promoting biodiversity across trophic levels results in multiple ecosystem services, including seed dispersal, biological control, and pollination. However, managing complex ecological networks poses challenges, such as antagonistic interactions (e.g., hyper-parasitism or seed predation by birds), complicating efforts to enhance biocontrol through trophic interactions. Hence, the detailed knowledge of organisms and their intricate interaction networks is essential. Multitrophic pollination results in EI, requires consideration of evolutionary adaptation, and management interventions like manipulating trophic interactions or increasing diversity for long-term stability. A meta-analysis of long-term studies focusing on plant and soil-based measures demonstrates positive impacts on crop yield at the field level. Pollination networks are dynamic and influenced by various environmental conditions. For instance, pollinator activity may decrease during the rainy season. Measures should be implemented to enhance pollination services during such periods of reduced activity. There is a predominant focus on multitrophic functioning at smaller spatial scales; it is crucial to consider all spatial levels, including field, multiple fields and landscape scales. Larger spatial scales gain significance when management aims at multiple ecosystem services. Despite the importance of multitrophic pollination, there is a surprising lack of studies in agroecosystems, and the lack of yield information, a critical factor for farmers, may contribute to the limited adoption of EI and related management types. Previous studies reveal that recommendations for EI often face reluctance due to misalignment with practical needs, consideration of different spatio-temporal scales, or underestimation of logistical challenges. Scientists tend to focus on functions like pollination rather than outcomes such as yield, crop quality, or profits. To promote multitrophic pollination networks, long-term studies incorporating large-scale EI measures are necessary, and there is a need for better transformation of results into the application of specific measures.

KEYWORDS

- **climate change**
- **biodiversity**
- **agroforestry**
- **pollination**
- **sustainable approach**

REFERENCES

1. Barral, M. P., Benayas, J. M. R., Meli, P., Maceira, N. O. (2015). Quantifying the impacts of ecological restoration on biodiversity and ecosystem services in agroecosystems: A global meta-analysis. *Agriculture, Ecosystems & Environment*, *202*, 223–231.
2. Turley, N. E., Bell-Dereske, L., Evans, S. E., Brudvig, L. A. (2020). Agricultural land-use history and restoration impact soil microbial biodiversity. *Journal of Applied Ecology*, *57*(5), 852–863.
3. Doublet, V., Doyle, T., Refoy, I., Hedges, S., Carvell, C., Brown, M. J., Wilfert, L. (2022). Increasing flower species richness in agricultural landscapes alters insect pollinator networks: Implications for bee health and competition. *Ecology and Evolution*, *2*(10), e9442.
4. Buzhdygan, O. Y., Petermann, J. S. (2023). Multitrophic biodiversity enhances ecosystem functions, services and ecological intensification in agriculture. *Journal of Plant Ecology*, *2*, rtad019.
5. Bommarco, R., Kleijn, D., Potts, S. G. (2013). Ecological intensification: Harnessing ecosystem services for food security. *Trends in Ecology & Evolution*, *28*(4), 230–238.
6. Joseph, G. S., Mauda, E. V., Seymour, C. L., Munyai, T. C., Dippenaar-Schoeman, A., Foord, S. H. (2018). Landuse change in savannas disproportionately reduces functional diversity of invertebrate predators at the highest trophic levels: Spiders as an example. *Ecosystems*, *21*, 930–942.
7. Isbell, F., Tilman, D., Reich, P. B., Clark, A. T. (2019). Deficits of biodiversity and productivity linger a century after agricultural abandonment. *Nature Ecology & Evolution*, *3*(11), 1533–1538.
8. Hautier, Y., Tilman, D., Isbell, F., Seabloom, E. W., Borer, E. T., Reich, P. B. (2015). Anthropogenic environmental changes affect ecosystem stability via biodiversity. *Science*, *348*(6232), 336–340.
9. Bascompte, J., Jordano, P. (2007). Plant-animal mutualistic networks: The architecture of biodiversity. *Annual Review of Ecology, Evolution & Systematics*, *38*, 567–593.
10. Bascompte, J., Scheffer, M. (2023). The resilience of plant–pollinator networks. *Annual Review of Entomology*, *68*, 363–380.

11. Arnold, S. E., Elisante, F., Mkenda, P. A., Tembo, Y. L., Ndakidemi, P. A., Gurr, G. M., Darbyshire, I. A., Belmain, S. R., Stevenson, P. C. (2021). Beneficial insects are associated with botanically rich margins with trees on small farms. *Scientific Reports, 11*(1), 15190.
12. Kremen, C. (2020). Ecological intensification and diversification approaches to maintain biodiversity, ecosystem services and food production in a changing world. *Emerging Topics in Life Sciences*, *4*(2), 229–240.
13. Garibaldi, L. A., Steffan-Dewenter, I., Winfree, R., Aizen, M. A., Bommarco, R., Cunningham, S. A., Kremen, C., Carvalheiro, L. G., Harder, L. D., Afik, O., Bartomeus, I. (2013). Wild pollinators enhance fruit set of crops regardless of honey bee abundance. *Science*, *339*(6127), 1608–1611.
14. Allsopp, M. H., De Lange, W. J., Veldtman, R. (2008). Valuing insect pollination services with cost of replacement. *PLoS One*, *3*(9), e3128.
15. Bailey, S., Requier, F., Nusillard, B., Roberts, S. P., Potts, S. G., Bouget, C. (2014). Distance from forest edge affects bee pollinators in oilseed rape fields. *Ecology and Evolution*, *4*(4), 370–380.
16. Shackelford, G., Steward, P. R., Benton, T. G., Kunin, W. E., Potts, S. G., Biesmeijer, J. C. Sait, S. M. (2013). Comparison of pollinators and natural enemies: A meta-analysis of landscape and local effects on abundance and richness in crops. *Biological Reviews*, *88*(4), 1002–1021.
17. Zhang, F., Hui, C., Pauw, A. (2013). Adaptive divergence in Darwin's race: How coevolution can generate trait diversity in a pollination system. *Evolution*, *67*(2), 548–560.
18. Klein, A. M., Vaissière, B. E., Cane, J. H., Steffan-Dewenter, I., Cunningham, S. A., Kremen, C., Tscharntke, T. (2007). Importance of pollinators in changing landscapes for world crops. *Proceedings of the Royal Society B: Biological Sciences*, *274*(1608), 303–313.
19. Ricketts, T. H., Regetz, J., Steffan-Dewenter, I., Cunningham, S. A., Kremen, C., Bogdanski, A., Gemmill-Herren, B., Greenleaf, S. S., Klein, A. M., Mayfield, M. M., Morandin, L. A. (2008). Landscape effects on crop pollination services: Are there general patterns. *Ecology Letters*, *11*(5), 499–515.
20. Garratt, M. P., Coston, D. J., Truslove, C. L., Lappage, M. G., Polce, C., Dean, R., Biesmeijer, J. C., Potts, S. G. (2014). The identity of crop pollinators helps target conservation for improved ecosystem services. *Biological Conservation*, *169*, 128–135.
21. Bentrup, G., Hopwood, J., Adamson, N. L., Powers, R., Vaughan, M. (2021). The role of temperate agroforestry practices in supporting pollinators. In: *Agroforestry and Ecosystem Services* (pp. 275–304).
22. Kay, S., Kühn, E., Albrecht, M., Sutter, L., Szerencsits, E., Herzog, F. (2021). Agroforestry can enhance foraging and nesting resources for pollinators with focus on solitary bees at the landscape scale. *Agroforestry Systems*, *94*, 379–387.
23. Klein, A. M., Steffan-Dewenter, I., Tscharntke, T. (2003). Pollination of Coffea canephora in relation to local and regional agroforestry management. *Journal of Applied Ecology*, *40*(5), 837–845.
24. Bentrup, G., Hopwood, J., Adamson, N. L., Vaughan, M. (2019). Temperate agroforestry systems and insect pollinators: A review. *Forests*, *10*(11), 981.

25. Tscharntke, T., Sekercioglu, C. H., Dietsch, T. V., Sodhi, N. S., Hoehn, P., Tylianakis, J. M. (2008). Landscape constraints on functional diversity of birds and insects in tropical agroecosystems. *Ecology*, *89*(4), 944–951.
26. Klein, A. M., Steffan-Dewenter, I.,Tscharntke, T. (2006). Rain forest promotes trophic interactions and diversity of trap-nesting Hymenoptera in adjacent agroforestry. *Journal of Animal Ecology*, *75*(2), 315–323.
27. Alvarado Sandino, C. O., Barnes, A. P., Sepúlveda, I., Garratt, M. P., Thompson, J., Escobar-Tello, M. P. (2023). Examining factors for the adoption of silvopastoral agroforestry in the Colombian Amazon. *Scientific Reports*, *13*(1), 12252.
28. Gaba, S., Bretagnolle, F., Rigaud, T., Philippot, L. (2014). Managing biotic interactions for ecological intensification of agroecosystems. *Frontiers in Ecology and Evolution*, *2*(29), 1–9.
29. Kleijn, D., Bommarco, R., Fijen, T. P., Garibaldi, L. A., Potts, S. G. Van Der Putten, W. H. (2019). Ecological intensification: Bridging the gap between science and practice. *Trends in Ecology & Evolution*, *34*(2), 154–166.
30. Malézieux, E. (2012). Designing cropping systems from nature. *Agronomy for Sustainable Development*, *32*, 15–29.
31. Ratnadass, A., Barzman, M. S. (2014). Ecological intensification for crop protection. In: Ozier Lafontaine, H., Lesueur Jannoyer, M., (Eds.), *Sustainable Agriculture Reviews 14: Agroecology and Global Change* (pp. 53–81.). Springer: Heidelberg.
32. Ollerton, J., Winfree, R., Tarrant, S. (2011). How many flowering plants are pollinated by animals. *Oikos*, *120*(3), 321–326.
33. Ollerton, J. (2017). Pollinator diversity: Distribution, ecological function, and conservation. *Annual Review of Ecology, Evolution, and Systematics*, *48*, 353–376.
34. Kaur, N., Kaleka, A. S. (2022). *Diversity, Importance and Decline of Pollinating Insects in Present Era* (pp. 77–92). IntechOpen.
35. Aizen, M. A., Garibaldi, L. A., Cunningham, S. A., Klein, A. M. (2009). How much does agriculture depend on pollinators? Lessons from long-term trends in crop production. *Annals of Botany*, *103*(9), 1579–1588.
36. Southwick, E. E., Southwick, L. J. (1992). Estimating the economic value of honey bees (Hymenoptera: Apidae) as agricultural pollinators in the United States. *Journal of Economic Entomology*, *85*, 621–633.
37. Kearns, C. A., Inouye, D. W., Waser, N. M. (1998). Endangered mutualisms: The conservation of plant-pollinator interactions. *Annual Review of Ecology and Systematics*, *29*(1), 83–112.
38. Johansen, C. A. (1997). Pesticides and pollinators. *Annual Review of Entomology*, *22*, 177–192.
39. Memmott, J., Craze, P. G., Waser, N. M., Price, M. V. (2007). Global warming and the disruption of plant–pollinator interactions. *Ecology Letters*, *10*(8), 710–717.
40. Varah, A., Jones, H., Smith, J. Potts, S. G. (2020). Temperate agroforestry systems provide greater pollination service than monoculture. *Agriculture, Ecosystems & Environment*, *301*(10703), 1–13.
41. Torralba, M., Fagerholm, N., Burgess, P. J., Moreno, G., Plieninger, T. (2016). Do European agroforestry systems enhance biodiversity and ecosystem services? A meta-analysis. *Agriculture, Ecosystems & Environment*, *230*, 150–161.

42. Jose, S. (2012). Agroforestry for conserving and enhancing biodiversity. *Agroforestry Systems*, *85*, 1–8.
43. Sabino, W., Costa, L., Andrade, T., Teixeira, J., Araújo, G., Acosta, A. L., Luísa, C., Giannini, T. C. (2022). Status and trends of pollination services in Amazon agroforestry systems. *Agriculture, Ecosystems & Environment*, *335*, 108012, 1–10.
44. Klein, A. M., Brittain, C., Hendrix, S. D., Thorp, R., Williams, N., Kremen, C. (2012). Wild pollination services to California almond rely on semi-natural habitat. *Journal of Applied Ecology*, *49*(3), 723–732.
45. Gómez, J. M., Bosch, J., Perfectti, F., Fernández, J., (2007). Abdelaziz, M. Pollinator diversity affects plant reproduction and recruitment: The tradeoffs of generalization. *Oecologia*, *153*, 597–605.
46. Gomez, J. M., Abdelaziz, M., Lorite, J., Jesús Muñoz-Pajares, A., Perfectti, F. (2010). Changes in pollinator fauna cause spatial variation in pollen limitation. *Journal of Ecology*, *98*(5), 1243–1252.
47. Fontaine, C., Dajoz, I., Meriguet, J., Loreau, M. (2006). Functional diversity of plant–pollinator interaction webs enhances the persistence of plant communities. *PLoS Biology*, *4*(1), e1.
48. Lundgren, R., Lazaro, A., Totland, Ø. (2015). Effects of experimentally simulated pollinator decline on recruitment in two European herbs. *Journal of Ecology*, *103*, 328e337.
49. Tremlett, C. J., Moore, M., Chapman, M. A., Zamora-Gutierrez, V., Peh, K. S. H. (2020). Pollination by bats enhances both quality and yield of a major cash crop in Mexico. *Journal of Applied Ecology*, *57*, 450e459.
50. Bartomeus, I., Potts, S. G., Steffan-Dewenter, I., Vaissiere, B. E., Woyciechowski, M., Krewenka, K. M., Bommarco, R. (2014). Contribution of insect pollinators to crop yield and quality varies with agricultural intensification. *PeerJ*, *2*, e328.
51. Rafique, M. K., Mahmood, R., Stephen, E., Irshad, M., Sarwar, G. (2016). Pollination deficit in mango orchards at Multan, Pakistan. *Pakistan Journal of Zoology*, *48*(1), 35–38.
52. Khalifa, S. A., Elshafiey, E. H., Shetaia, A. A., El-Wahed, A. A. A., Algethami, A. F., Musharraf, S. G., El-Seedi, H. R. (2021). Overview of bee pollination and its economic value for crop production. *Insects*, *12*(8), 688.
53. Westerkamp, C., Gottsberger, G. (2000). Diversity pays in crop pollination. *Crop Science*, *40*(5), 1209–1222.
54. Rodger, J. G., Bennett, J. M., Razanajatovo, M., Knight, T. M., van Kleunen, M., Ashman, T. L., Ellis, A. G. (2021). Widespread vulnerability of flowering plant seed production to pollinator declines. *Science Advances*, *7*(42), eabd3524.
55. Albrecht, M., Schmid, B., Hautier, Y., Müller, C. B. (2012). Diverse pollinator communities enhance plant reproductive success. *Proceedings of the Royal Society B: Biological Sciences*, *279*(1748), 4845–4852.
56. Celep, F., Atalay, Z., Dikmen, F., Doğan, M., Sytsma, K. J., Classen-Bockhoff, R. (2020). Pollination ecology, specialization, and genetic isolation in sympatric bee-pollinated Salvia (Lamiaceae). *International Journal of Plant Sciences*, *181*(8), 800–811.

57. Katumo, D. M., Liang, H., Ochola, A. C., Lv, M., Wang, Q. F., Yang, C. F. (2022). Pollinator diversity benefits natural and agricultural ecosystems, environmental health, and human welfare. *Plant Diversity*, *44*(5), 429–435.
58. Philpott, S. M., Uno, S., Maldonado, J. (2006). The importance of ants and high-shade management to coffee pollination and fruit weight in Chiapas. Mexico. *Biodiversity Conservation*, *15*, 487e501.
59. Cusser, S., Neff, J. L., Jha, S. (2016). Natural land cover drives pollinator abundance and richness, leading to reductions in pollen limitation in cotton agroecosystems. *Agriculture Ecosystem & Environment*, *226*, 33e42.
60. Rogers, S. R., Tarpy, D. R., Burrack, H. J. (2014). Bee species diversity enhances productivity and stability in a perennial crop. *PLoS One*, *9*(5), e97307.
61. Crane, E., Walker, P. (1985). Important honeydew sources and their honeys. *Bee World*, *66*(3), 105–112.
62. Gounari, S., Zotos, Ch., Dafnis, S., Moschidis, G., Papadopoulos, G. (2021). On the impact of critical factors to honeydew honey production: The case of *Marchalina hellenica* and pine honey. *Journal of Apicultural Research*, *45*(2), 23–34.
63. Hajjar, R., Jarvis, D. I., Gemmill-Herren, B. (2008). The utility of crop genetic diversity in maintaining ecosystem services. *Agriculture, Ecosystems & Environment*, *123*(4), 261–270.
64. Porto, R. G., de Almeida, R. F., Cruz-Neto, O., Tabarelli, M., Viana, B. F., Peres, C. A., Lopes, A. V. (2020). Pollination ecosystem services: A comprehensive review of economic values, research funding and policy actions. *Food Security*, *12*(6), 1425–1442.
65. Heimpel, G. E., Jervis, M. A. (2005). Does floral nectar improve biological control by parasitoids? In: Wackers, F. L., van Rijn, P. C. J., Bruin, J. (Eds.), *Plant-derived Food and Plante Carnivore Mutualism* (p. 267e304). Cambridge University Press: Cambridge.
66. Brock, R. E., Cini, A., Sumner, S. (2021). Ecosystem services provided by aculeate wasps. *Biological Review*, *96*, 1645e1675.
67. Mommaerts, V., Put, K., Smagghe, G. (2011). Bombus terrestris as pollinator-and vector to suppress Botrytis cinerea in greenhouse strawberry. *Pest Management Science*, *67*, 1069e1075.
68. Puig-Montserrat, X., Flaquer, C., Gómez-Aguilera, N., Burgas, A., Mas, M., Tuneu, C., López-Baucells, A. (2020). Bats actively prey on mosquitoes and other deleterious insects in rice paddies: Potential impact on human health and agriculture. *Pest Management Science*, *76*(11), 3759–3769.
69. Whitby, M. D., Kieran, T. J., Glenn, T. C., Allen, C. (2020). Agricultural pests consumed by common bat species in the United States corn belt: The importance of DNA primer choice. *Agriculture, Ecosystems & Environment*, *303*, 107105.
70. Nicholson, C. C., Egan, P. A. (2020). Natural hazard threats to pollinators and pollination. *Global Change Biology*, *26*(2), 380–391.
71. Bennett, J. M., Steets, J. A., Burns, J. H., Durka, W., Vamosi, J. C., Arceo-Gómez, G., Knight, T. M. (2018). GloPL, a global data base on pollen limitation of plant reproduction. *Scientific Data*, *5*(1), 1–9.

72. Bennett, J. M., Steets, J. A., Burns, J. H., Burkle, L. A., Vamosi, J. C., Wolowski, M., Ashman, T. L. (2020). Land use and pollinator dependency drives global patterns of pollen limitation in the Anthropocene. *Nature Communications*, *11*(1), 3999.
73. Fründ, J., Dormann, C. F., Holzschuh, A., Tscharntke, T. (2013). Bee diversity effects on pollination depend on functional complementarity and niche shifts. *Ecology*, *94*(9), 2042–2054.
74. Brittain, C., Kremen, C., Klein, A. M. (2013). Biodiversity buffers pollination from changes in environmental conditions. *Global Change Biology*, *19*(2), 540–547.
75. Blitzer, E. J., Gibbs, J., Park, M. G., Danforth, B. N. (2016). Pollination services for apple are dependent on diverse wild bee communities. *Agriculture, Ecosystems & Environment*, *221*, 1–7.
76. Hoehn, P., Tscharntke, T., Tylianakis, J. M., Steffan-Dewenter, I. (2008). Functional group diversity of bee pollinators increases crop yield. *Proceedings of the Royal Society B: Biological Sciences*, *275*(1648), 2283–2291.
77. Norfolk, O., Eichhorn, M. P., Gilbert, F. (2016). Flowering ground vegetation benefits wild pollinators and fruit set of almond within arid smallholder orchards. *Insect Conservation and Diversity*, *9*(3), 236–243.
78. Geeraert, L., Aerts, R., Berecha, G., Daba, G., De Fruyt, N., D'hollander, J., Honnay, O. (2020). Effects of landscape composition on bee communities and coffee pollination in Coffea arabica production forests in southwestern Ethiopia. *Agriculture, Ecosystems & Environment*, *288*, 106706.
79. Garibaldi, L. A., Carvalheiro, L. G., Leonhardt, S. D., Aizen, M. A., Blaauw, B. R., Isaacs, R., Winfree, R. (2014). From research to action: Enhancing crop yield through wild pollinators. *Frontiers in Ecology and the Environment*, *12*(8), 439–447.
80. Cohen, H., Philpott, S. M., Liere, H., Lin, B. B., Jha, S. (2021). The relationship between pollinator community and pollination services is mediated by floral abundance in urban landscapes. *Urban Ecosystems*, *24*, 275–290.
81. Senapathi, D., Fründ, J., Albrecht, M., Garratt, M. P., Kleijn, D., Pickles, B. J., Klein, A. M. (2021). Wild insect diversity increases inter-annual stability in global crop pollinator communities. *Proceedings of the Royal Society B*, *288*(1947), 20210212.
82. Kevan, P. G. (1999). Pollinators as bioindicators of the state of the environment: Species, activity and diversity. In: *Invertebrate Biodiversity as Bioindicators of Sustainable Landscapes* (pp. 373–393). Elsevier.
83. Azam, I., Afsheen, S., Zia, A., Javed, M., Saeed, R., Sarwar, M. K., Munir, B. (2015). Evaluating insects as bioindicators of heavy metal contamination and accumulation near industrial area of Gujrat, Pakistan. *BioMed Research International*, *2015*, 942751.
84. Bhoi, T. K., Samal, I., Majhi, P. K., Komal, J., Mahanta, D. K., Pradhan, A. K., Saini, V., Nikhil Raj, M., Ahmad, M. A., Behera, P. P., Ashwini, M. (2022). Insight into aphid mediated Potato Virus Y transmission: A molecular to bioinformatics prospective. *Frontiers in Microbiology*, *13*, 1001454.
85. Mahanta, D. K., Jangra, S., Ghosh, A., Sharma, P. K., Iquebal, M. A., Jaiswal, S., Baranwal, V. K., Kalia, V. K., Chander, S. (2022). Groundnut bud necrosis virus modulates the expression of innate immune, endocytosis, and cuticle development-associated genes to circulate and propagate in its vector, Thrips palmi. *Frontiers in Microbiology*, *13*, 773238.

86. Bhatnagar, S., Mahanta, D. K., Vyas, V., Samal, I., Komal, J., Bhoi, T. K. (2023). Storage pest management with nanopesticides incorporating silicon nanoparticles: A novel approach for sustainable crop preservation and food security. *Silicon, 16*, 471–483.
87. Mahanta, D. K., Komal, J., Samal, I., Bhoi, T. K., Dubey, V. K., Pradhan, K., Nekkanti, A., Gouda, M. N. R., Saini, V., Negi, N., Bhateja, S., Jat, H. K., Jeengar, D. (2023a). Nutritional aspects and dietary benefits of "Silkworms": Current scenario and future outlook. *Frontiers in Nutrition, 10*, 1121508.
88. Samal, I., Bhoi, T. K., Vyas, V., Majhi, P. K., Mahanta, D. K., Komal, J., Singh, S., Kumar, P. V. D., Acharya, L. K. (2023a). Resistance to fungicides in entomopathogenic fungi: Underlying mechanisms, consequences, and opportunities for progress. *Tropical Plant Pathology*, *49*, 5–17.
89. Mahanta, D. K., Bhoi, T. K., Komal, J., Samal, I., Nikhil, R. M., Paschapur, A. U., Singh, G., Kumar, P. V. D., Desai, H. R., Ahmad, M. A., Singh, P. P., Majhi, P. K., Mukherjee, U., Singh, P., Saini, V., Shahanaz, Srinivasa, N., Yele, Y. (2023b). Insect-pathogen crosstalk and the cellular-molecular mechanisms of insect immunity: Uncovering the underlying signaling pathways and immune regulatory function of non-coding RNAs. *Frontiers in Immunology*, *14*, 1169152.
90. Samal, I., Bhoi, T. K., Majhi, P. K., Murmu, S., Pradhan, A. K., Kumar, D., Saini, V., Paschapur, A. U., Raj, M. N., Ankur, Manik, S., Behera, P. P., Mahanta, D. K., Komal, J., Alam, P., Balawi, T. A. (2023b). Combatting insects mediated biotic stress through plant associated endophytic entomopathogenic fungi in horticultural crops. *Frontiers in Plant Science*, *13*, 1098673.
91. Komal, J., Desai, H. R., Samal, I., Mastinu, A., Patel, R. D., Kumar, P. D., Majhi, P. K., Mahanta, D. K., Bhoi, T. K. (2023). Unveiling the genetic symphony: Harnessing CRISPR-Cas genome editing for effective insect pest management. *Plants*, *12*(23), 3961.
92. Samal, I., Bhoi, T. K., Mahanta, D. K., Komal, J. (2023c). Establishing the role of silicon (Si) in plant resistance to insects: A bibliometric approach. *Silicon*, *16*, 2119–2128.
93. Majhi, P. K., Samal, I., Bhoi, T. K., Pattnaik, P., Pradhan, C., Gupta, A., Mahanta, D. K., Senapati, S. K. (2023). Climate-smart agriculture: An integrated approach to address climate change and food security. In: *Climate Change and Insect Biodiversity* (pp. 221–245). CRC Press.
94. Mahanta, D. K., Samal, I., Komal, J., Bhoi, T. K., Majhi, P. K., Ahmad, M. A. (2023c). Understanding anthropogenic climate change, its consequences on insect pests, and techniques in forecasting and monitoring pest dynamics: A contemporary scenario. In: *Climate Change and Insect Biodiversity* (pp. 44–67). CRC Press.
95. Bhoi, T. K., Samal, I., Mahanta, D. K., Komal, J., Jinger, D., Sahoo, M. R., Achary, G. C., Nayak, P., Sunani, S. K., Saini, V., Raghuraman, M., Singh, S. (2023). Understanding how silicon fertilization impacts chemical ecology and multitrophic interactions among plants, insects and beneficial arthropods. *Silicon*, *15*(6), 2529–2549.
96. Bhoi, T. K., Samal, I., Mahanta, D. K., Komal, J., Majhi, P. K., Ankur. (2024). Exploring Melatonin's potential as an alternative strategy for protecting plants from

biotic stresses. In: *Melatonin in Plants: A Pleiotropic Molecule for Abiotic Stresses and Pathogen Infection* (pp. 223–242). Springer Nature Singapore: Singapore.

97. Samal, I., Bhoi, T. K., Pradhan, A. K., Mahanta, D. K. (2023d). Plantix app: A success story of artificial intelligence in plant protection. *Van Sangyan, 10*, 24–26.
98. Mahanta, D. K., Bhoi, T. K., Komal, J., Samal, I., Mastinu, A. (2024). Spatial, spectral and temporal insights: Harnessing high-resolution satellite remote sensing and artificial intelligence for early monitoring of wood boring pests in forests. *Plant Stress, 11*, 100381.
99. Samal, I., Mahanta, D. K., Bhoi, T. K., Komal, J., Jatav, H. S., Jatav, S. S., Sathyanarayana, E. (2024). *Approaches of Biochar in Ecosystem Management: Current Scenario and Future Perspectives*. IntechOpen
100. Vyas, V., Bhoi, T. K., Samal, I., Singh, S., Mahanta, D. K. (2024). Revitalizing degraded soils with agroforestry interventions: Opportunities, challenges, and future direction. *Agroforestry to Combat Global Challenges: Current Prospects and Future Challenges* (pp. 529–549).
101. Samal, I., Bhoi, T. K., Mahanta, D. K., Komal, J., Singh, S. (2024). Biorational pest management: Potentials, unintended consequences, and future concerns. In: *Biorationals and Biopesticides: Pest Management* (pp. 47–75).

CHAPTER 7

Indigenous Wisdom and Traditional Knowledge in Plant Protection Through Multitrophic Dynamics

SOUVIK ACHARYA

Faculty Centre for Integrated Rural Development and Management, Ramakrishna Mission Vivekananda Educational and Research Institute, Kolkata, West Bengal, India

ABSTRACT

Indigenous wisdom and traditional knowledge have played a vital role for centuries in shaping sustainable practices for agriculture and plant conservation. Rooted in customs, beliefs, and collective experiences passed through generations, these practices reflect a deep understanding of the intricate relationships among plants, animals, insects, and the environment. Traditional methods often include plant-based pest control, where certain species are used to attract beneficial insects or repel harmful ones, thereby fostering ecological balance. For example, Native American tribes developed strategies to protect corn from pests such as borers and weevils, while in Asia, indigenous farmers in countries like China and India devised natural defenses for rice cultivation. Such ethnobotanical practices highlight the effectiveness of multitrophic interactions in maintaining agricultural sustainability. Documenting these plant species and their roles not only preserves cultural heritage but also offers valuable insights for modern sustainable farming. This ethnobotanical study underscores the importance of engaging indigenous communities in knowledge

Multitrophic Interactions for Sustainable Agroforestry: Harnessing the Power of Beneficial Insects and Ecological Dynamics. Hanuman Singh Jatav, Vishnu D. Rajput, Ipsita Samal, and Tanmaya Kumar Bhoi (Eds).

DOI: 10.1201/9781779641250-7

exchange with researchers, policymakers, and environmentalists. By integrating traditional wisdom with contemporary approaches, we can harness multitrophic dynamics to strengthen plant protection, biodiversity conservation, and cultural preservation. Ultimately, these practices promote a more resilient, sustainable, and harmonious relationship between humans and their natural environment.

7.1 INTRODUCTION

Only one particular culture within a society has access to indigenous knowledge. Such statistics acts as a society's foundation of knowledge, supporting way of talking, and judgment. These are the coping procedures that individuals in the area have developed to deal with shifting anthropological and ecological norms. Thinking about the risks that chemicals cause to humans, the environment, and all other living things, it is growing increasingly important to find alternatives to synthetic pesticides. One such moderation is the use of innovative, locally developed insect control approaches, which are cost-effective and draw on the huge experience of the past over time [1]. Indigenous wisdom, rooted in the collective knowledge, customs, and beliefs of ancient cultures, has played a pivotal role in preserving the delicate balance between humans and the environment. Global acceptance of the important role of this age-old wisdom for safeguarding biodiversity and sustainable development is weak to come forward [2]. With less documentation among cultural groups, cultural information is an extensive archive of details and collective wisdom that has been handed down from generation to generation. The various communities' long-standing customs in the areas of farming, customs ceremonies related to religion, and traditional medicine show that resource preserving has always been a key component of local people's lives and has therefore played a role in environmentally friendly growth. An awareness of nature rooted in cultural information illustrates the symbiotic relationship between individuals and the environment as a whole. It presents a regional plan for growth built on respecting ecosystem capacity for support and interacting with the environment. This knowledge assures responsible use and management of the environment and helps indigenous people adjust to changes in the environment. It is based on over time evidence that reflects local conditions.

An understanding of the environment founded on traditional knowledge highlights the symbiotic relationship between individuals and the environment as a whole. It presents a local planning approach built on respecting ecosystem carrying capacity and interacting with the environment. This knowledge assures responsible use and management of the environment and helps indigenous people adjust to changes in the environment. It is based on over time evidence that reflects local conditions [3]. Native populations, in this context, have a great deal of expertise in coordinating plant-based remedies and pest management strategies. They recognize that all living things are interdependent within an ecosystem, which goes beyond conventional understanding.

The significant role in local wisdom and customs will be examined in the chapter and in safeguarding plants through multitrophic dynamics. Topics include the interconnected relationships within ecosystems, exemplification of indigenous practices in crop protection, the importance of ethnobotanical studies, fostering a mutual exchange of knowledge, and the promotion of sustainability. The practices evolved by indigenous communities worldwide form the bedrock of sustainable agriculture, fostering a harmonious relationship between humans and the natural world.

7.2 UNDERSTANDING MULTITROPHIC DYNAMICS

Although interactions between higher and lower levels are an essential component of ecosystems, most scientific studies only examine one trophic level, which makes it more difficult to identify the underlying mechanisms that constitute groups and the consequences of localities [4].

Multitrophic dynamics refer to the interconnected relationships between organisms at different trophic levels within an ecosystem. Trophic levels represent positions in a food chain or web, where energy and nutrients are transferred as organisms feed on each other. Indigenous cultures have, over centuries, developed an intricate understanding of multitrophic dynamics, emphasizing the interconnectedness of animals, plants, insects, and the environment. This holistic perspective recognizes that disruptions in one part of the ecosystem have cascading effects throughout. The act of animals, usually insects, but occasionally vertebrates, birds, or animals, pollinating crop plants is one of the most significant ecological relationships at the core of crop cultivation [5].

7.2.1 INTERCONNECTED RELATIONSHIPS

Multitrophic dynamics recognize that the well-being of one species is actually tied to the health of others. It goes beyond a simplistic view of predator–prey relationships to consider the broader ecological interactions.

7.2.2 ECOSYSTEM HEALTH

Understanding multitrophic dynamics is crucial for maintaining ecosystem health. It involves recognizing the delicate balance between predators and prey, herbivores and plants, and the myriad interactions that contribute to the overall stability of an ecosystem.

7.2.3 INDIGENOUS KNOWLEDGE AND MULTITROPHIC DYNAMICS

Indigenous communities have developed a deep understanding of multitrophic dynamics through generations of observation and practical experience. Their traditional knowledge often includes insights into how different species interact and influence each other within their local ecosystems.

7.2.4 PLANT PROTECTION STRATEGIES

Indigenous practices often involve plant protection strategies that leverage multitrophic dynamics. For example, certain plants may be cultivated strategically to attract beneficial insects or repel pests, contributing to a more balanced and resilient agricultural system.

7.2.5 CULTURAL PRESERVATION AND SUSTAINABILITY

The abstract emphasizes the link between understanding multitrophic dynamics and cultural preservation. Indigenous communities' sustainable practices are often deeply embedded in their cultural identity, emphasizing the need for preserving both ecological and cultural diversity. In the broader context, an understanding of multitrophic dynamics allows for the development of holistic and sustainable approaches to agriculture, pest control, and environmental management. By recognizing the intricate web of relationships in ecosystems, indigenous wisdom becomes a valuable source of knowledge for promoting harmony between humans and their natural surroundings.

7.3 EXEMPLIFYING INDIGENOUS PRACTICES

Native populations have amassed a wealth of experience in plant-based pest control treatments and natural cures. Across diverse regions, indigenous communities have developed ingenious methods to protect crops vulnerable to insect infestations. A case in point is the Native American tribes' cultivation of corn, employing time-tested defenses against pests like corn borers and weevils. Similarly, in Asian nations such as China and India, traditional rice farming practices incorporate knowledge that mitigates threats from pests.

Table 7.1 documents local pest management techniques that were utilized to control white grubs; likewise, the mountain farmers used conventional approaches to manage pests that harm crops besides white grub, like insect pests in rice and the onion. Table 7.1 suggests that the farmers in the region were using six native techniques (Sl. No. 1–6) to manage common pests.

TABLE 7.1 Some Local Traditional Knowledge in Contrast of Pest Control

Sl. No.	Traditional Method of Pest Control	Strategy (Provided by a Professional)
1	Before planting each plant, holes are constructed and cow dung cakes are burned inside to protect the chilli crop from white grub	Cow dung contains lots and lots of nutrients
2	Ash is used in garlic and onion farms to keep eating insects away from the plants	Ash acts as a physical poison, repelling aphids and other insects. Additionally, ash raises the soil's potassium content
3	Spraying of cow urine in the kitchen garden to check wilting symptoms in plants	A solution of cow urine is sprinkled on the plant to prevent insects. Additionally, urine's nitrogen content promotes crop growth
4	In rice, summertime field bund pruning minimizes the pests' and diseases	Eliminating the different insect stages and eliminating the pest alternate hosts
5	(For storage) To avoid pest infestation cereals and pulses are exposed to the sun for two days before to storage	Because moisture content creates a favorable environment for disease growth and insect pest attack, drying the grains before storage lowers their moisture content and lengthens their period of storage
6	Pulses like *Vigna mungo* are combined with chalk powder for storage	In addition to being an insect repellent and antifeedant, chalk powder (calcium carbonate) also has the capacity to stop insects from proliferating

Source: Adapted with permission from Rathore et al. [13].

The general population and scientific community have recently become aware of additional ecosystem benefits and roles, which may be crucial to the emergence of environmentally friendly growth. Apart from the trophic interactions that are primarily mediated by top-down processes, ecological intensification can also be attained by means of promoting relationships from the level upward achieved by plants' defense against diseases and insects. This can be achieved by improving nutrient conditions through the use of manure and fertilizer crops, reducing erosion, and fortifying individual plant defenses. These include ethnic and artistic services like medical care, maintaining support like controlling insects, and supporting services like the process of nitrification [6]. A new investigation built mixed relationships, which together connections of crops and insect pollinators with networks of crops and vegetarians, and proved how intensification of farming affected the durability of the connections based on the likelihood of disappearance [7]. The exception measurements of relationship networks that depend not only on the presence or lack of organisms at a given location, but also on experimentally documented relationship rates, as well as assets. Parasites and aphid ecosystems were measured in a study, and the results suggested that expanding farming might lessen the nutritional chains continuity over time [8]. Not only do intense agroecosystems and production forests have interacting species pairs, but they also have very species-rich food webs.

7.4 GEOGRAPHICAL OVERVIEW: NURTURING MULTITROPHIC DYNAMICS

Indigenous practices, deeply rooted in diverse landscapes, form a mosaic of wisdom that spans continents, ecosystems, and climates. This geographical overview reveals a rich tapestry of traditional ecological knowledge, each thread woven intricately into the fabric of the local environment. These practices, often shaped over generations, exemplify the profound understanding of multitrophic dynamics in different corners of the world.

7.4.1 NORTH AMERICA

Corn farming among Native American tribes: The cultivation of corn by Native American tribes in North America is an example of their traditional ways of living in harmony with multitrophic dynamics. With a deep understanding of plant–insect interactions, these communities have evolved complex defense mechanisms against pests such as corn borers and weevils.

7.4.2 AFRICAN CONTINENT

Ethnobotanical knowledge in agriculture: Native American groups throughout Africa exhibit a keen understanding of the multitrophic dynamics present in their agricultural environments. Sustainable farming systems, which exhibit a harmonious relationship with the environment, are facilitated by practices like intercropping and the strategic use of particular plant species as natural pesticides.

7.4.3 SOUTHEAST ASIA

The customs of rice farming in China and India: The native customs of rice farming in Asia, especially in countries like China and India, demonstrate a sophisticated comprehension of multitrophic dynamics. Conventional techniques demonstrate a comprehensive approach that goes beyond traditional agricultural techniques by carefully balancing planting techniques, water management, and the use of companion plants.

7.4.4 SOUTH AND CENTRAL AMERICA

Amazon Rainforest biodiversity conservation: Indigenous communities in Central and South America, particularly in the Amazon Rainforest, practice methods that support the preservation of biodiversity. The development of sustainable agroforestry systems is guided by the complex understanding of plant–insect relationships, which supports a balanced ecosystem that is advantageous to both people and the various flora and fauna. This geographic investigation makes it clear that native customs are precisely tailored to the peculiarities of regional ecosystems. Indigenous communities provide a wealth of knowledge that cuts across geographic boundaries, whether they are raising rice in Asia, growing corn in the Americas, or protecting biodiversity in the Amazon. This overview highlights the interconnectedness of indigenous practices and their diversity of approaches, which together contribute to a global tapestry of sustainable and peaceful relationships with the natural world.

Again, coming together into indigenous practices as Indigenous communities often practice polyculture, cultivating a variety of crops in proximity. This approach enhances multitrophic dynamics by creating diverse niches, attracting a range of insects, birds, and other organisms. Crop diversity reduces the risk of monoculture vulnerabilities and promotes a more resilient ecosystem. Crop diversity reduces the risk of monoculture vulnerabilities and promotes a more resilient ecosystem, while also enhancing soil fertility, supporting beneficial insects, improving nutrient cycling, and increasing the system's ability to adapt to climate change.

7.5 AGROFORESTRY SYSTEMS

Many indigenous cultures incorporate agroforestry practices, integrating trees and shrubs into agricultural landscapes. This not only provides additional habitat for diverse species but also contributes to nutrient cycling, improved soil structure, and enhanced water retention, fostering a multitrophic balance.

7.5.1 TRADITIONAL SEED SAVING AND EXCHANGE

Indigenous communities often engage in the preservation and exchange of traditional seeds. This practice contributes to maintaining genetic diversity within crops, aligning with multitrophic dynamics by ensuring that plant varieties are adapted to local conditions and resilient against pests.

7.5.2 BENEFICIAL PLANT ASSOCIATIONS

Indigenous agricultural practices often involve the intentional association of certain plants to benefit one another. For instance, nitrogen-fixing plants may be interplanted with food crops, enhancing soil fertility and supporting a multitrophic network that includes beneficial soil microorganisms.

7.5.3 COMPANION PLANTING FOR PEST CONTROL

Indigenous wisdom recognizes specific plant combinations that deter pests or attract natural predators. For example, planting aromatic herbs alongside vegetables can repel certain insects, while flowers may attract pollinators and predatory insects, contributing to a balanced multitrophic system.

7.5.4 FIRE MANAGEMENT PRACTICES

Some indigenous cultures incorporate controlled burning practices as part of land management. These controlled fires not only play a role in promoting certain plant growth but also influence the distribution of species, impacting the multitrophic dynamics of the ecosystem.

7.5.5 COMMUNITY-BASED CONSERVATION PRACTICES

Indigenous communities often engage in communal conservation efforts, collectively managing natural resources. These practices recognize the interdependence of various species within ecosystems, emphasizing the need for collaborative approaches to maintain biodiversity and multitrophic balance.

7.6 ETHNOBOTANICAL STUDY

Engagement ecological investigations aim to encourage people in making choice about the use of available environmental resources and to take various actions that will boost and preserve local culture [9]. Ethnobotanical studies provide a fascinating lens through which we can unfold the comprehensive sectors of traditional ecological knowledge and its application in maintaining multitrophic dynamics. Furthermore, ethnobotany is the study of the close relationships that exist between members of cultural groups and the vegetation that grow in their surroundings, according to de Albuquerque [11]. The analysis of independent communities or the examination of specific activities' conduct are the two defining aspects of ethnobotany, as soon as it becomes evident that the primary emphasis is on individuals who actually interact with or observe the biodiversity around them. People who have to build their shelters and gather or cultivate their food come into direct contact with plants [10]. They need to know when to gather specific plants, how to identify them, and what to anticipate from them. Once it is acknowledged that ethnobotany in so which is clearly defined as belonging either to the study of autonomous societies or to an analysis of the behavior of specific actions, is the domain of those who actually work with or think about the local vegetation. The majority of current ethnobotanical research stays with the conventional method of knowledge records. On the other hand, over the past few decades, methods for choosing the conservation of culturally valuable plant species have been developed by examining the connections between diversity in biology and culture [11].

7.6.1 OVERVIEW OF ETHNOBOTANICAL THEORY

From a field that primarily recorded the variety of uses of plants by the local population, ethnobotany has developed into one that aims to comprehend for what individuals choose specific species considering a broad spectrum of purposes and a variety of applications. This development is a reaction to the ongoing demand considering postulates influenced as well as research with a premise compelled approach facing an increase in a discipline's strictness. Anthropogonic instruction has appropriately concentrated on teaching aspiring ethnobotanists important ideas and techniques like participant observation, livelihood analysis, free-listing, interviews, and emic versus etic methods, according to Martin.

It is essential to first combine the current conceptual study in order to progress the field of cultural botany as a conceptually driven, literature-based science. Recently, a few ideas and concepts that can serve as a basis for formulating inquiry queries that will further improve our knowledge of the relationship between humans and vegetation have been discussed. The following are the primary theories and predictions: the concept of irregular crop choice; the accessibility concept, inclusion, and adaptation of wisdom; the lifespan and gender relationships of knowledge; understanding loss theory; the idea of social connections as a driver of knowledge structure; the most effective searching theory; and the plant's utilization value idea [12].

7.6.2 INDIGENOUS AND ETHANOBOTANY

The majority of indigenous communities and inhabitants hold the belief that they are connected to a specific species of creature, plant, or natural event, and so on in the form of "Totem." The spiritual-cultured aspects were significantly influenced by this idea framework. The folk members also feel that carrying out rites and engaging in medicospiritual activities will help to preserve the regional ecosystem in addition to granting them a life of longevity and prosperity [13]. The clan believes that by worshiping certain plants in front of and around their homes, they will bring wealth and success into their homes. In addition to being utilized for household purposes, customs, holy items, and symbols, parts of green living objects have also been prayed to in sacred sites and grown around them. Certain plants also think they possess unique abilities provided by God, enabling them to ward off wicked spirits, grant health, and bring good fortune [14].

Particular groups use plants like Indian gooseberry, wild tulsi, bidi leaf tree, the *Indica* species of *Azadirachta*, and Indian bael in their health-related spiritualistic rituals [15].

7.6.3 CULTURED-BASED PLANT MATERIALS

The goal of botanical researchers is to investigate the various uses of these plants, including garments, housing, nourishment, animal feed, energy, and household items, as well as the connections between the medicinal applications of these plants and other traits of the type of plant. They employ the cultural method to explain useful plants and gather knowledge about them [16]. A range of cultured-botany indicators are used to assess plant consumption across areas, tribal groups, and organizations. Several custom–agriculture index can be used in the analysis of the scientific information to evaluate the framework of knowledge, its applications, and method of dissemination throughout various regions. Using different group classifications, data analysis is carried out [17] (Table 7.2).

TABLE 7.2 Indigenous Folks Utilize Plants in Their Indigenous Medicine.

Local Name	Family	Parts Used	Treatment
Vasaka	Acanthaceae	Root, leaves, whole plant	Used in curing asthma, chronic bronchitis chest complaints
Wild asparagus	Liliaceae	Roots	Gets rid of diarrhea, spleen disorders
Neem	Meliaceae	Stem bark, leaves, root bark	Prevents from malaria, antiseptic usage, diabetes, skin-related disorders
Indian Sallaki	Burseraceae	Gum	Application to cure dermetic diseases, dysentery, cold and cough
Kurchi Plant	Apocynaceae	Bark and seeds	Better digestion and bowel functioning
Arjuna	Combretaceae	Bark	Purify blood, heart-related disorders
Heart-leaved moonseed plant	Menispermaceae	Leaves and matured stem	Purify the blood circulatory, arthritis cure, heart disorders cure
Fenugreek	Fabaceae	Seeds	Dysentery, swellings cure
Five-leaved chaste	Verbenaceae	Leaves	Cardiac disorders cure, asthma cure
Ginger root	Zingiberaceae	Rhizomes	Treat nausea

Source: Adapted with permission from Manoharachary and Nagaraju [18].

7.7 VISUALIZING SUSTAINABILITY THROUGH INDIGENOUS PRACTICES

Indigenous communities' practices serve as concrete examples of sustainability, which is a concept that goes far beyond theoretical debates. Visible examples of human–environment harmony are provided by the visual representation of sustainability provided by indigenous practices, which is a powerful feature that goes beyond philosophical ideas. The practices that are shown in this section as stepping stones to a future that is more ecologically balanced and sustainable are examined as ways in which indigenous communities act as living examples of sustainability.

7.7.1 ENGAGING WITH INDIGENOUS COMMUNITIES

Interaction with indigenous communities offers an opportunity to observe sustainable practices in action firsthand. Through engaging in daily activities within these communities, observers acquire valuable insights into the realistic implementation of traditional knowledge. This collaboration serves as a visual illustration of sustainability, showing how native knowledge is incorporated into routine tasks like farming, resource management, and biodiversity conservation. For indigenous communities, sustainability is a comprehensive strategy that includes cultural preservation in addition to being an ecological undertaking. The integration of environmentally friendly techniques with customs, ceremonies, and traditions is depicted aesthetically, providing a comprehensive approach in line with the values of sustainable development.

7.7.2 SUSTAINABILITY PILLAR OF CULTURAL PRESERVATION

Alongside environmentally sound practices, indigenous practices frequently involve the preservation of cultural heritage. A testament to the crucial connection between cultural preservation and sustainability is provided by the visual depiction of ceremonies, rituals, and traditions. These images depict communities that are preserving long-standing customs while balancing their relationship with the environment—a comprehensive strategy that aligns with the values of sustainable development.

7.8 CONCLUSION

Thus, the rich fabric of indigenous knowledge bears witness to the continuous interaction that exists between individuals and the natural environment. An invaluable legacy that has been successfully passed down through the ages is the foundation of sustainable coexistence, which is emphasized by oral traditions and rituals as the essence of collective knowledge transmission. Ethnobotanical information, which document specific plant species and their uses, offers light on the advanced farming techniques developed by indigenous people. These methods demonstrate an in-depth comprehension of ecological balance that goes beyond conventional agricultural practices, and they are supported by traditional pest control strategies.

These methods are more than just farming methods; they represent an attitude that shows respect for all the components of the environment. Looking into multitrophic dynamics illustrates how indigenous communities are able to understand and negotiate the complex interactions that exist between plants, animals, insects, and the environment. This chapter, which is based on the fundamental principles of indigenous cultures, ends with a call to action for everyone to work together to encourage harmony with the goal to create a future that is sustainable.

KEYWORDS

- **indigenous wisdom**
- **mutual exchange**
- **traditional knowledge**
- **multitrophic dynamics**
- **plant protection**
- **ethnobotany insights**
- **sustainability**
- **cultural preservation**
- **tribe**

REFERENCES

1. Devanand, Kamala, Merlin. (2017). Indigenous traditional knowledge on crop protection practices. *International Journal of Agricultural Science and Research, 7*(5), 345–352.
2. Gadgil, M., Berkes, F., Folke, C. (1993). Indigenous knowledge for biodiversity conservation. *Ambio, 22*, 151–156.
3. Mazzocchi, F. (2006). Western science and traditional knowledge. Despite their variations, different forms of knowledge can learn from each other. *EMBO Reports, 7*(5), 463–466. doi: 10.1038/sj.embor.7400693.
4. Seibold, S., Cadotte, M. W., MacIvor, J. S., Thorn, S., Müller, J. (2018). The necessity of multitrophic approaches in community ecology. *Trends in Ecology & Evolution, 33*(10), 754–764.
5. Buzhdygan, O. Y., Petermann, J. S. (2023). Multitrophic biodiversity enhances ecosystem functions, services and ecological intensification in agriculture. *Journal of Plant Ecology, 16*(6), rtad019
6. Sileshi, G., Schroth, G., Rao, M. R., Girma, H. (2008). Weeds, diseases, insect pests and tri-trophic interactions in tropical agroforestry. *Ecological Basis of Agroforestry*, pp. 73–94, Boca Raton, FL, USA: CRC Press.
7. Morrison, B. M. L., Brosi, B. J., Dirzo, R. (2020). Agricultural intensification drives changes in hybrid network robustness by modifying network structure. *Ecology Letters, 23*(2), 359–369.
8. Gagic. V., Hänke, S., Thies, C., et al. (2012). Agricultural intensification and cereal aphid-parasitoid-hyperparasitoid food webs: Network complexity, temporal variability and parasitism rates. *Oecologia, 170*, 1099–1109.
9. Sauini, T., Stern da Fonseca-Kruel, V., Baptistela Yazbek, P., Matta, P., Cassas, F., da Cruz, C., Hortal Pereira Barretto, E., Alice Dos Santos, M., Angelica Silva Gomes, M., José Francischetti Garcia, R., Honda, S., Felipe Domingues Passero, L., Esteves Conde, B., Rodrigues, E. (2020). Participatory methods on the recording of traditional knowledge about medicinal plants in Atlantic forest, Ubatuba, São Paulo, Brazil. *PLoS One, 15*(5), e0232288.
10. Singh, H. B., Yaipharembi, N., Huidrom, E., Devi, C. A. (2022). *Traditional Ecological Knowledge of Resource Management in Asia.* ISBN: 978-3-031-16839-0.
11. de Albuquerque and Oliveira, 2007; Cristancho and Vining, 2004; Garibaldi and Turner, 2004.
12. Gaoue, O. G., Coe, M. A., Bond, M., et al. (2017). Theories and major hypotheses in ethnobotany. *Economic Botany, 71*(3), 269–287.
13. Rathore, S., Chandola, M., Raghuvanshi, R., Kaur, M., Singh, K. V. (2021). Indigenous pest management practices of Indian hill farmers: Introspecting their rationale and communication pattern for secure ecosystems. *Sustainability, 13*(21), 11608.
14. Panigrahi, S., Rout, S., Sahoo, G. (2021). Ethnobotany: A strategy for conservation of plant. *Annals of the Romanian Society for Cell Biology, 25*(6), 1370–1377.
15. Cavender, A. P., Alban, M. (2009). The use of magical plants by curanderos in the Ecuador highlands. *Journal of Ethnobiology and Ethnomedicine*.

16. Lozada, M., Ladio, A., Weigandt, M. (2006). Cultural Transmission of ethnobotanical knowledge in a rural community of Northwestern Patagonia, Argentina. *Economic Botany*, *60*(4), 374–385.
17. Heinrich, M., Barnes, J., Gibbons, S., Williamson, E. M. (2004). *Fundamentals of Pharmacognosy and Phytotherapy*. Churchill Livingstone: London.
18. Manoharachary, C., Nagaraju, D. (2016). Medicinal plants for human health and welfare. *Annals of Phytomedicine, 5*, 24–34.

CHAPTER 8

Harnessing Traditional Knowledge for Plant Protection: Exploring Multitrophic Dynamics in Agroecosystems

ANKITA MUKHOPADHYAY BISWAS, PABITRA KUMAR GHOSH*, and SWARNALI DUARY

Department of Agriculture, Brainware University, Barasat, Kolkata, West Bengal, India

*Corresponding author

ABSTRACT

Indigenous wisdom and traditional knowledge represent time-tested approaches that integrate cultural, ecological, and spiritual values into sustainable agriculture and plant protection. Passed down through generations, these practices emphasize biodiversity, soil fertility, erosion control, and resilience to climate change while fostering harmony between humans and nature. In plant protection, Indigenous methods such as plant-based pest control, integrated pest management, and the cultivation of resilient crop varieties reduce reliance on chemical pesticides and safeguard cultural heritage. Incorporating this knowledge into modern strategies enhances sustainability, promotes biodiversity conservation, and strengthens community participation. A multitrophic perspective, which considers interactions among plants, herbivores, predators, and decomposers, further enriches ecological management by highlighting the interconnectedness of ecosystem components. Traditional agricultural practices—including terrace farming, crop rotation, and agroforestry—demonstrate these principles in action, ensuring food security, cultural identity, and environmental preservation.

Multitrophic Interactions for Sustainable Agroforestry: Harnessing the Power of Beneficial Insects and Ecological Dynamics. Hanuman Singh Jatav, Vishnu D. Rajput, Ipsita Samal, and Tanmaya Kumar Bhoi (Eds).

DOI: 10.1201/9781779641250-8

Integrating indigenous knowledge with scientific innovations offers a holistic framework for developing resilient agroecosystems, advancing both ecological sustainability and cultural continuity.

8.1 INTRODUCTION

8.1.1 DEFINITION OF INDIGENOUS WISDOM AND TRADITIONAL KNOWLEDGE

Dependency on synthetic chemical pesticides by growers has an adverse impact on health and the ecological balance [8]. Indigenous wisdom, intricately linked to the land, encompasses the delicate interplay between humanity, nature, and spirituality. Passed down virtually, this holistic understanding permeates facets like agriculture, medicine, navigation, and resource management, showcasing an intimate connection with ecosystems. Spirituality, integral to Indigenous knowledge, manifests through daily ceremonies, emphasizing a profound Earth connection. Elders safeguard community wisdom rooted in conventional ecological understanding, adapting it to modern challenges for resilience, cultural diversity, and sustainability. Efforts to preserve Indigenous customs amid colonization threats underscore the need for intellectual property rights, challenging Western views on climate change and advocating for mainstream policy collaboration.

8.1.2 SIGNIFICANCE OF INCORPORATING THESE INTO MODERN PLANT PROTECTION STRATEGIES

In the ever-evolving field of plant protection, there is a growing recognition of the profound significance of incorporating Indigenous wisdom and traditional knowledge. Indigenous communities, with their deep connections to the land, possess invaluable insights into sustainable and holistic approaches to safeguarding plant ecosystems. Incorporating these tried-and-true methods, which are based on a deep knowledge of nature, into contemporary plant protection strategies not only increases the effectiveness of conservation efforts but also fosters respect for Indigenous heritage and cultural diversity. This symbiotic strategy could completely transform how we think about protecting plants by promoting a peaceful coexistence of conventional knowledge and cutting-edge scientific discoveries. The following points can be highlighted:

- **Preservation of biodiversity:** Traditional knowledge and indigenous wisdom provide a comprehensive understanding of ecosystems, which helps to maintain biodiversity.
- **Sustainable agriculture practices:** By fostering environmental harmony, sustainable agriculture practices are developed by incorporating Indigenous knowledge.
- **Adaptation to climate change:** Indigenous knowledge offers techniques for coping with the effects of climate change on plant life that are based on centuries of observation.
- **Localized solutions:** Native knowledge is frequently place-based and provides context-specific solutions that take into account the special qualities of the land and its resources.
- **Resilient crop varieties:** Traditional knowledge contributes to the identification and cultivation of resilient crop varieties, crucial for food security in changing environmental conditions.
- **Erosion control and soil health:** Native American methods support the preservation of soil, reduce erosion, and improve soil health—all essential elements of long-term plant protection.
- **Integrated pest management:** By developing integrated pest management strategies, traditional knowledge helps lessen the need for hazardous chemical inputs.
- **Cultural significance:** Including Indigenous knowledge acknowledges the importance of plants in culture, which promotes a more considerate and well-rounded approach to their preservation.
- **Community engagement:** By including Indigenous communities in plant protection plans, you can guarantee their active involvement and leverage their generational knowledge to develop practical, culturally appropriate solutions.
- **Ethical framework:** Indigenous knowledge frequently carries moral precepts that emphasize environmental stewardship, reciprocity, and community well-being when it comes to ethical plant protection.

8.1.3 FOCUS ON MULTITROPHIC DYNAMICS AS A HOLISTIC APPROACH

A comprehensive approach that recognizes the complex interactions between different trophic levels in ecosystems is to concentrate on multitrophic dynamics. Seeing the interdependencies between plants, herbivores, predators, and decomposers, this viewpoint goes beyond isolated species interactions. Ecologists and conservationists can develop more sustainable and successful management strategies by adopting a multitrophic lens, which gives them a thorough understanding of ecosystem dynamics. By taking into account the ripple effects of changes at one trophic level on the ecosystem as a whole, this holistic approach sheds light on the complex balance needed for ecological health. By applying a multitrophic lens to our analysis, we can develop strategies that not only target individual species but also take into account the larger ecological web, strengthening resilience and encouraging a more complex and integrated method of ecological management (Figure 8.1).

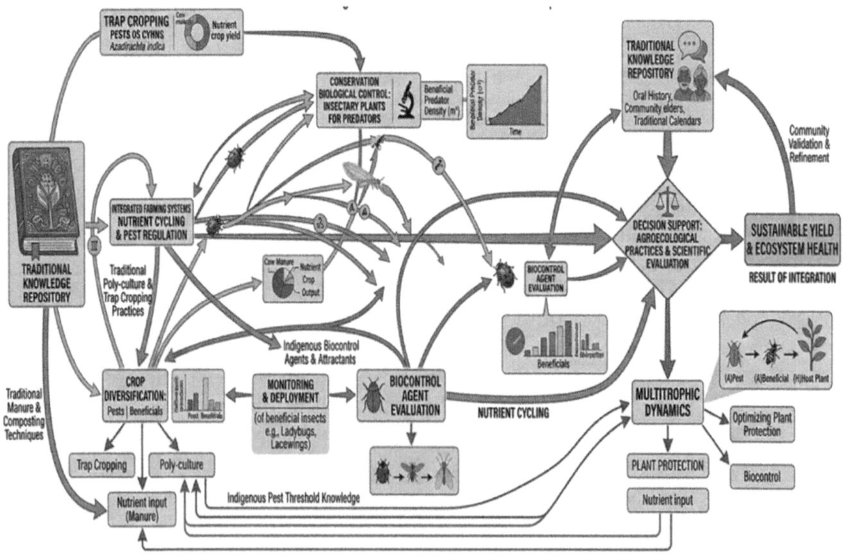

FIGURE 8.1 Conceptual framework of agroecological plant protection integrating Indian indigenous technical knowledge (ITK) and multitrophic dynamics for sustainable agroforestry.

8.2 INDIGENOUS PRACTICES IN PLANT PROTECTION

8.2.1 OVERVIEW OF TRADITIONAL AGRICULTURAL PRACTICES

An extensive array of enduring and sustainable techniques for land cultivation is provided by traditional agricultural practices, which are ingrained in the cultural fabric of many different societies. The harmony between humans and nature is emphasized by these practices, which are frequently characterized by a deep understanding of local ecosystems. To build resilience against environmental challenges, traditional agricultural communities prioritize biodiversity, water conservation, and soil fertility in their techniques. Among the many traditional methods that have helped communities survive for many years are terrace farming, agroforestry, and crop rotation. By combining communal and spiritual values, these practices frequently produce a comprehensive framework that goes beyond simple agricultural production. Looking at a broad overview of traditional agricultural practices makes it clear that these approaches support cultural identity, environmental preservation, and the preservation of priceless agricultural knowledge that has been passed down through the years in addition to ensuring food security.

8.2.2 INDIGENOUS PEST CONTROL METHODS

Indigenous populations all over the world have been using traditional methods of controlling pests for centuries, a sign of their profound knowledge of regional ecosystems and sustainable farming methods. Efficacious substitutes for contemporary chemical-based pest control, these conventional techniques frequently depend on inherited knowledge and natural resources. Native American-style pest management emphasizes the value of utilizing local expertise and balancing with the environment to build sustainable agricultural systems. These methods provide insightful information for contemporary agriculture, promoting a change in pest management strategies toward those that are more ecologically and culturally conscious. Even though chemical pesticides give instantaneous control over pests, conventional wisdom holds a strong position when ecology and sustainability are taken into account. From this angle, these indigenous knowledge systems are mega-preserved by the northeast Himalayan region [3].

Here are some examples of indigenous pest control methods	
Companion planting	Native agricultural knowledge supports companion planting, which involves carefully growing a variety of crops to promote mutual advantages. Marigolds, for example, naturally ward off nematodes when planted next to vegetables, and intercropping particular plants confuses and repels pests in an ingenious way. This tried-and-true method promotes sustainable and pest-resistant agriculture by utilizing the innate relationships between plants, which is indicative of a holistic approach
Biological control	Natural predators are frequently used by indigenous tribes to manage pest populations. One way to manage pests without using chemical pesticides is to introduce beneficial insects like parasitic wasps, predatory beetles, and ladybirds
Crop rotation	Crop rotation is a conventional method that helps keep diseases and pests from growing in the soil. Indigenous farmers can lower the risk of infestations by uprooting their crops and relocating them every year
Traditional traps and barriers	Indigenous groups have created a variety of barriers and traps to keep pests out of their crops. To help control insect populations, physical barriers like fences or sticky traps made of natural materials can be built
Natural repellents	Natural repellents made from nearby plants are frequently used, according to indigenous knowledge. In many conventional pest control techniques, for example, neem oil—which is extracted from the neem tree—is utilized due to its insecticidal qualities
Cultural practices	Certain native communities use traditional methods to combat pests. In order to protect crops or ward off pests, rituals or ceremonies may be carried out at particular times of the year. This reflects a holistic approach that combines agricultural practices with spiritual beliefs
Traditional knowledge sharing	An important part of native pest management is the oral tradition of knowledge transfer. A sustainable agricultural ecosystem is fostered by the knowledge that elders impart about when to plant, which crops go well together, and how to spot and deal with pest problems
Water management	Indigenous communities frequently combine pest control with water management techniques. The prevalence of pests can be influenced by appropriate irrigation methods since certain pests prefer damp environments. Conventional techniques for conserving water, like collecting rainfall, can be incorporated into a comprehensive strategy for managing pests
Selective harvesting	Pest control can benefit from harvesting techniques as well. Using selective crop harvesting, indigenous communities can keep pests from spreading to healthy plants by removing infested ones

8.2.3 ROLE OF CULTURAL BELIEFS IN SHAPING SUSTAINABLE AGRICULTURE

Traditional wisdom and indigenous knowledge frequently highlight a profound reverence for the environment, directing sustainable farming methods based on an all-encompassing comprehension of ecosystems. Cultural values emphasize responsible land stewardship, support crop diversity, and preserve biodiversity, all of which help to create a positive relationship between people and the land. Cultural values can also affect the adoption of community-based agricultural systems that put long-term environmental health ahead of short-term profits, organic farming practices, and traditional seed saving. The importance of cultural beliefs in sustainable agriculture essentially draws attention to the connections that exist between environmental stewardship, cultural heritage, and the development of resilient farming techniques.

8.3 MULTITROPHIC DYNAMICS IN ECOSYSTEMS

8.3.1 EXPLANATION OF MULTITROPHIC INTERACTIONS IN ECOSYSTEMS

Multitrophic interactions in ecosystems are interactions among producers, consumers, and decomposers at various trophic levels. Photosynthesis, the process by which primary producers, like plants, convert sunlight into energy, is the cornerstone of these interactions. Herbivores, which primarily feed on plants, create the second trophic level. When herbivores are eaten by predators or other secondary consumers, a cascade effect occurs. By feeding on secondary consumers, tertiary consumers form a trophic pyramid that maintains ecological balance by regulating population sizes and preventing species dominance. These interactions support biodiversity by promoting species coexistence and specialization. Trophic cascades illustrate the interdependence of these relationships by demonstrating how changes at one trophic level affect those at other levels. Decomposers, which are necessary for the decomposition of organic matter and recycle nutrients, influence the nutrients that plants can access. Human actions that disrupt multitrophic interactions, like deforestation and overfishing, can lead to ecosystem imbalances and degradation. It draws attention to the complex web of life and stresses how important it is to preserve the precarious balance between various ecosystem components. Comprehending these intricate connections is imperative for proficient preservation and ecological administration.

8.3.2 IMPORTANCE OF UNDERSTANDING RELATIONSHIPS BETWEEN PLANTS, PESTS, AND PREDATORS

Understanding the relationships between plants, pests, and predators is crucial for effective agricultural pest management. By assisting in the identification of natural control mechanisms, an understanding of these relationships helps to reduce the need for chemical pesticides. Beneficial predators can play a critical role in controlling pest populations in addition to promoting sustainable and ecologically friendly farming practices. In addition, understanding these dynamics enhances crop resilience, increasing the productivity of agriculture. In summary, this knowledge promotes a balanced ecosystem, thereby reducing the negative effects of pests on crops.

8.3.3 ECOLOGICAL BALANCE AND RESILIENCE IN TRADITIONAL AGRICULTURAL SYSTEMS

By employing a range of plants and animals, conventional agricultural systems promote biodiversity and preserve ecological balance. Maintaining healthy soil and minimizing pest outbreaks are facilitated by crop rotation, polyculture, and the application of natural fertilizers. These methods improve ecosystem resilience, which makes traditional agriculture more adaptable to shifting environmental conditions. Traditional agricultural systems contribute significantly to long-term ecological stability because they maintain ecological knowledge and sustainable practices. The ethnic people primarily embraced two farming systems, namely, terrace or wet cultivation and jhum or shifting cultivation, in addition to general agricultural practices in plain land. Shifting cultivation, also referred to as slash-and-burn agriculture or "Jhum," is a significant farming system in which farmers rotate land instead of crops to make a living [1]. This method is comparable to approaches taken in Southeast Asia, Africa, and certain regions of Europe [2].

8.3.4 BIODIVERSITY AND PLANT PROTECTION

Biodiversity is crucial for plant protection because diverse ecosystems often exhibit greater resilience against pests and diseases. Many types of plants serve as a natural barrier to prevent infections from spreading rapidly. Beneficial organisms like pollinators and predators thrive in biodiverse settings and offer natural pest control. The preservation of a variety of plant species ensures a rich genetic pool, which facilitates the development of resistant cultivars through selective breeding. In the end, protecting biodiversity is essential to preserving the health of plant populations and guaranteeing the availability of food for all.

8.3.5 INDIGENOUS PERSPECTIVES ON THE VALUE OF BIODIVERSITY

Indigenous perspectives emphasize the significance of spiritual health, cultural identity, and biodiversity. Indigenous communities believe that because ecosystems are thought to be interconnected, maintaining harmony and balance within them depends on all species. The symbiotic relationship between humans and the environment is strengthened by biodiversity, which provides traditional medicines, sustenance, and materials for cultural practices. Indigenous knowledge has been passed down through the ages and recognizes the need of preserving a variety of ecosystems for future generations. Indigenous peoples see the preservation of biodiversity as both an ecological need and a reflection of their strong spiritual and cultural ties to the natural world.

8.3.6 HOW DIVERSE PLANT SPECIES CONTRIBUTE TO PEST RESISTANCE

A phenomenon known as "Allelopathy," in which some plant species release chemicals that prevent the growth of possible pests, is one way that a variety of plant species cause pest resistance. Pests find it harder to establish and flourish in polyculture systems because different plant species disrupt the pest population's continuity. There is less chance of infestations because different plant structures and chemical compositions make certain pests' environments less favorable. Added to that, a more robust and balanced ecosystem is promoted by a diversity of beneficial insects and predators that are drawn to diverse plantings and help control pest populations naturally. An important aspect of indigenous knowledge for controlling crop diseases and pests, especially in the jhum system, is the use of plant and animal parts and products [5]. The ecological affordability, social acceptability, economic viability, and environmental sustainability of this knowledge make it unique [7, 11].

8.3.7 TRADITIONAL CROP VARIETIES AND THEIR ROLE IN MULTITROPHIC DYNAMICS

Traditional crop varieties are crucial for multitrophic dynamics because they boost biodiversity in agroecosystems. These crops often have a wide range of properties that support various trophic levels and provide food and habitat for a variety of organisms. Because they contribute to the overall stability of multitrophic interactions, the adaptability and resilience of traditional varieties enhance pest resistance and promote sustainable agricultural practices.

8.4 HOLISTIC APPROACHES TO PEST MANAGEMENT

Integrated strategies that take into account the entire ecosystem are a feature of holistic approaches to pest management. These strategies combine crop diversification, cultural practices, and biological controls to manage pest populations in agriculture in a sustainable manner, increase crop resilience, and lessen reliance on chemical pesticides.

8.4.1 INTEGRATION OF SPIRITUAL AND CULTURAL PRACTICES IN AGRICULTURE

A holistic strategy that recognizes the interdependence of people, the land, and crops is emphasized in agriculture that integrates spiritual and cultural practices. Agricultural practices are often significantly influenced by indigenous knowledge and customs, which show a deep respect for the environment and biodiversity. This integration conserves cultural heritage, encourages sustainable farming methods, and deepens a person's spiritual connection to the land.

8.4.2 RITUALS AND CEREMONIES ASSOCIATED WITH PLANTING AND HARVESTING

Customs and rituals surrounding planting and harvesting can be found in many cultures, and they signify a close relationship between human societies and the natural cycles. Planting ceremonies often involve prayers, blessings, and rituals intended to create the best possible growing conditions for the crops.

Harvest festivals celebrate the accomplishments of agriculture while honoring and recognizing the relationship between human life and the land. During these rituals, traditional foods are shared, music and dance are performed, and community involvement is typical. These ceremonies typically commemorate important agricultural milestones and work to strengthen cultural identity and a sense of unity with the environment. In the district of Nayagarh, women farmers from 18 villages gathered in Gambharikhola to participate in the inaugural festival of indigenous seeds. Every woman farmer wore a tiny, exquisitely decorated earthen pot on her head, carrying native seeds as a symbol of her respect for farming, seeds, and farmland. Every stage of the agricultural cycle involves complex rituals that are essential to the cultivation of the local rice variety, known as "Tinawon." Every activity in the rice fields is accompanied by epics, myths, ballads, and prayer chants. A deeper comprehension of Ifugao history, religion, spirituality, and culture is possible thanks to this intangible heritage.

8.4.3 THE CONCEPT OF BALANCE AND HARMONY IN TRADITIONAL FARMING

The concepts of harmony and balance are central to traditional farming, wherein an all-encompassing strategy is employed to attain equilibrium among farming methods, the environment, and the human race. Traditional farmers often use sustainable farming methods that respect natural cycles and ecological processes in order to maintain harmony. Crop rotation, polyculture, and fallow periods are common practices to maintain soil fertility and prevent imbalances. By embracing this interconnected perspective, traditional farming practices support long-term environmental sustainability as well as community livelihoods.

8.5 ETHNOBOTANICAL KNOWLEDGE

Ethnobotanical knowledge is the customary plant knowledge that various human cultures have amassed over time. This knowledge is often passed down through the generations and includes the identification, classification, and use of plants for culinary, medicinal, and cultural purposes. Ethnobotanical practices highlight the intricate relationship between humans and plants, and they also highlight how important it is to preserve indigenous knowledge in order to preserve biodiversity and ensure sustainable resource management.

8.5.1 MEDICINAL PLANTS USED IN PEST MANAGEMENT

- **Botanical insecticides:** A plethora of therapeutic plants contain insecticidal compounds. Plant extracts, such as garlic (*Allium sativum*), pyrethrum (*Chrysanthemum cinerariifolium*), and neem (*Azadirachta indica*), have been used as natural insecticides. Broad-range insecticidal actions disrupt insect growth, support oviposition, and have sanitizing, sterilizing, and antifeedant properties [4].
- Some medicinal plants, such as basil (*Ocimum* spp.) and citronella (*Cymbopogon nardus*), are known to have insect-repelling properties. When grown near crops or when their extracts are applied, they aid in providing insect protection for plants.
- **Companion planting:** Companion planting schemes often include medicinal plants to ward off or divert pests. Planting marigolds (*Tagetes* spp.) next to vegetables, for example, can repel nematodes. The active ingredients in turmeric leaves, namely, phenolics, terpenoids, tannins, and alkaloids, have a repulsive effect [9].
- **Biopesticides:** Biopesticides, which are less harmful to the environment than chemical pesticides, can be found in medicinal plants. Plant extracts, such as neem and eucalyptus (*Eucalyptus* spp.), are used in biopesticide formulations to control pest populations.
- **Allelopathy:** Allelopathic compounds, secreted by some medicinal plants, prevent pests from growing and developing. For instance, the allelopathic characteristics of sunflower (*Helianthus annuus*) can aid in the management of certain soil-borne pests.

8.5.2 WISDOM PASSED DOWN THROUGH GENERATIONS REGARDING PLANT–PEST RELATIONSHIPS

Plant–pest relationships have been enacted down through the generations, and with it comes knowledge of the intricate relationships between crops and their pests. Plants with certain characteristics that either attract or repel pests are chosen based on customary knowledge to guide planting practices for natural pest control. This generational wisdom centers on sustainable, holistic agricultural practices, which support the harmony between plants and the ecosystems they live in.

8.6 COMMUNITY INVOLVEMENT AND KNOWLEDGE TRANSFER

Community involvement in agriculture facilitates the transfer of invaluable knowledge from seasoned residents to the next generation. Through active participation in hands-on learning, community members ensure the transmission and preservation of traditional practices that support resilient and sustainable farming systems.

8.6.1 IMPORTANCE OF COMMUNITY-BASED APPROACHES

- **Preservation of local knowledge:** Community-based approaches ensure that traditional and indigenous knowledge about farming, biodiversity, and eco-friendly practices is preserved.
- **Cultural relevance:** These strategies promote culturally sensitive solutions that foster a sense of ownership and participation by recognizing the unique needs and values of local communities.
- **Social cohesion:** By incorporating the community in decision-making and execution, community-based approaches enhance social cohesion. This strengthens resilience and promotes sustainable development by fostering a collaborative environment.

8.7 CASE STUDIES

8.7.1 EXAMPLES OF SUCCESSFUL INTEGRATION OF INDIGENOUS WISDOM IN MODERN AGRICULTURE

Indigenous farming communities have a wealth of agricultural knowledge, especially among tribal members. These traditional farming methods are the product of centuries of knowledge. Using local resources, farmers have developed their farming systems all over the world. These Indigenous technical knowledge (ITK) have accompanying descriptions and images.

- **Crow screw or bird scarecrow:** Used to deter birds from eating crops. Its construction is made of straw and similar materials. It appears to be a man standing in a field wearing an outdated dress. Its face is a drawing of clay pot eyes and mouth (Agartala, Noabari, Chandrarapur, Killa).
- **Field house:** A hut used to keep an eye on the monkeys and protect crops from their attacks. It is built at a height above the field's surface out of straw or a similar material. People used to hide inside this house and make noise by firing rubber bullets or anything else when monkeys attacked the field (Belonia).
- **Water-conservation vegetable gardening:** Reusing banana trash to save water. Banana leaves that have been dried out are used as a sunblock. Bamboo or rope is used at the base of the crop on the east and south faces, where sunlight is received, to keep the dry leaves hanging (Belonia, Rajnagar).
- **Honey bee dry hives:** Honey bee dry hives are used to treat human wounds. After being extracted from the beehive, honey is either sun-dried or naturally dried. This hive is used by traditional healers for their purposes (Belonia, Chittamara).
- **Unidentified shrub that repels insects:** This leaf is used to keep insects away from paddy fields. In rice fields, leaves are used by being transplanted and then placed in different parts of the field. The fragrance of leaves deters insects (Belonia, Chittamara).
- **Hand-held sprayer made of bamboo:** In the absence of a sprayer, pesticides are applied manually. It has two nodes and is made of bamboo. On one side of the bamboo, there are two holes—one small and the other larger—where a through valve is intended to be inserted. Pressure at the other end creates a tiny hole where water or spray material emerges when it passes through a manually built bamboo valve-style component (Belonia, Chittamara).
- **Banana stem for weed control:** For increasing aeration, controlling weeds, clearing rows, and so on. In standing pulse crops, the banana pseudo-stem is used to remove weeds, clean the rows in

between the lines, and eventually improve aeration (Killa Gomati) (Figure 8.5c).

- **Seed preservation with banana pseudo-stem:** To enhance bitter gourd seed germination. Before seeding, bitter gourd seeds are stored for a few days by wrapping them in a pseudo-stem made of bananas. The germination percentage of seeds will rise, and their overall health will remain intact (Killa Gomati).
- **Rice with neem leaf:** To shield rice from infestation by stored grain insects (Chandrapur, Agartala).
- **Ash spreading to plants:** To shield rice from insect attacks on grains that are stored. Chandrapur, Agartala).
- **Seed preservation in bamboo (Wasung):** For planting the following year, store the seeds. For an entire year, seeds are stored in the bamboo. It will keep seeds healthy and at a high rate of germination. It serves to keep seeds stored (Killa, Gomati).
- **Rat trap by bamboo (rat killing Kaman—local name):** Rat attacks on crops can be decreased by trapping and killing rats. The building is situated where the rat hole opens. It is divided into three parts: two bamboo structures and an arrow. Two parts of bamboo, one narrower than the other. The rat food item is placed in the narrower section using a jute rope so that it is positioned such that when the rat enters the bamboo to eat, the rope needs to be cut. Cut to length, the narrower section of the rope is forced into the upper part of the bigger one, where it is tight with the arrow, killing the rat (East R. C. Ghat, Khowai).
- **Bird-trap (run-away of birds, local name—Dongki):** To keep avian predators away from the crop, elude the birds and flee from the crop. Two small pieces of bamboo are splinted together, with one end fixed. You grasp it by the fixed side end with your hand. It produces a characteristic sound when shaken after being held, keeping birds away from crops (East R. C. Ghat, Khowai).
- **Cultivation of Bitter gourd in brinjal as relay cropping:** To cultivate bitter gourd using the brinjal stem as "Machha." Two weeks before the brinjal harvest, the bitter gourd crop is planted as a relay crop. Herbicides are sprayed on the brinjal plants after they

have been harvested, leaving only the stems in the field. These stems are employed to provide trailing bitter gourd crop support (East R. C. Ghat, Khowai).

- **Use of cow urine for pest control:** To ward off termite attacks on the bamboo structure. After being diluted 1:10 with water, cow urine that is collected in the morning at the cowshed is used at the base of the bamboo structure. It has lengthened the structure's lifespan and decreased termite attacks. To store the cow pee for a long time, the base is covered with a cemented poly bag (East R. C. Ghat, Khowai).

8.7.2 OTHER TRADITIONAL TECHNIQUES FOR CROP PROTECTION

To keep moths, weevils, and beetles away, storage rooms can be frequently smoked with neem leaves. Indigenous plant protection practice (IPPP) assembles awareness and understanding of various facts related to pest management that farmers have developed over a long period and continue to expand and diffuse across communities. Since the inception of agricultural practices, there has been a constant struggle between mankind and pests for better crop yield and survival, depicting the ethnic groups of Tripura that adopted various indigenous methods to manage insect pests.

1. You can combine the grain with sand, wood, and cow dung ash. These have the effect of filling the spaces between the granules, which limits the movement of the insects.
2. To prevent paddy blast and bacterial blight, a solution of cow dung made by combining 3 kg of cow dung with 3 L of water was sprayed on the paddy in the study area.
3. Neem, Karanja, or castor cakes were used as control measures for termites and soil-borne diseases like collar rot, root rot, and so on.
4. Kerosene was also frequently used in the field to combat termite infestations.

8.8 OPPORTUNITIES FOR INNOVATION THROUGH THE FUSION OF TRADITIONAL AND MODERN WISDOM

Combining traditional and modern knowledge opens up new possibilities for creative solutions that strike a balance between history and modern demands.

Sl. No.	Name of ITK	Details of ITK
1.	Using smoke from burning dry chilies and rice husks, rodent control	Rabid rodents severely damage paddy in jhum fields. The rats' burrow is filled with smoke by burning the paddy husks and dry chilies of the landrace variety
2	Turmeric powder can be used to get rid of red ants	One common way to keep red ants away from spinach is to scatter turmeric dust on it
3	Conventional ecotechnologies for disease control in a conventional land use system	Traditional technologies such as the bamboo drip method of irrigation of terrace rice
4	Control of nematodes in ginger, chilies, tomato, and turmeric by intercropping with *Chrysanthemum coronarium*	Intercropping with *Chrysanthemum coronarium*, *Tagetes erecta*, or growing *Tagetes erecta* as border crops Incorporate the leaves of these trap crops into the soil to enhance the effectiveness and nutrient enrichment of crops

Complementing traditional methods with contemporary technology promotes efficient and sustainable methods for resource management and agriculture. Communities can create integrated, flexible plans for environmental preservation and communal well-being by utilizing both traditional wisdom and modern innovations. In addition to preserving cultural diversity, this synergy opens up fresh approaches to solving difficult problems in a world that is changing quickly. Therefore, the conversation with the informants demonstrated that the integration of innovative, cutting-edge technology with traditional methods would strengthen the insect pest management approach, making it socially, economically, and environmentally acceptable [12].

8.9 FUTURE DIRECTIONS

8.9.1 THE POTENTIAL FOR A HYBRID APPROACH TO PLANT PROTECTION

The possibility of a hybrid approach to plant protection, combining the best elements of both worlds for workable and durable solutions, is presented by the integration of traditional and modern strategies. Plant protection systems that are both comprehensive and long-lasting can be created by fusing modern agricultural innovations, such as biopesticides and precision agriculture, with traditional farming methods, such as companion planting and local knowledge of plant-pest interactions. This hybridization can promote biodiversity preservation, reduce reliance on synthetic chemicals, and boost resistance to pests. By implementing an all-encompassing and adaptable approach that combines traditional wisdom with cutting-edge technologies, we can efficiently address pest management challenges and foster long-term environmental and agricultural sustainability.

8.9.2 RESEARCH GAPS AND AREAS FOR FURTHER EXPLORATION

- **Quantitative evaluation:** To systematically gauge the efficacy of indigenous wisdom and traditional knowledge (IWTK) in plant protection through multitrophic dynamics, more quantitative research is required. A deeper comprehension of the dependability and scalability of conventional methods will be possible through the establishment of precise metrics and the execution of thorough assessments.
- **Integration with contemporary science:** Future research should concentrate on establishing a connection between IWTK and contemporary scientific approaches. Our knowledge of multi-trophic interactions can be enhanced by collaborative research that blends traditional wisdom with modern scientific methods, providing thorough insights for sustainable pest management.
- **Impacts of climate change:** It is imperative to investigate how multitrophic dynamics are impacted by climate change within the framework of IWTK. Developing resilient plant protection strategies in the face of global climate shifts requires an understanding of how adaptable traditional knowledge is to changing environmental conditions.
- Research ought to tackle ethical considerations concerning the use of IWTK, such as rights to intellectual property, cultural sensitivity, and fair cooperation with indigenous communities. To build trust and encourage ethical engagement in the field, research must respect and benefit local communities.

8.10 CONCLUSION

The integration of traditional and modern strategies offers the potential for a hybrid approach to plant protection, combining the best features of both worlds for practical and long-lasting solutions. Comprehensive and durable plant protection systems can be produced by combining contemporary advancements like precision agriculture and biopesticides with traditional agricultural techniques like companion planting and indigenous knowledge of plant–pest relationships. This hybridization can increase resistance to pests, lessen dependency on artificial chemicals, and encourage the preservation of biodiversity. Through the adoption of a comprehensive and flexible strategy that integrates conventional knowledge with state-of-the-art technologies, we can effectively tackle pest management issues while promoting enduring environmental and agricultural sustainability. Researchers such as Patel et al. [11] and Gyawali et al. [7] have claimed that plant protection practices based on indigenous technology should be beneficial and a critical tool for sustainable crop production with assured food safety [10, 12]. The use of traditional methods for the management of insect pests serves as a better option when ecology and economy are both concerned [6].

KEYWORDS

- **plant protection**
- **biodiversity**
- **agricultural practices**
- **ecosystems**
- **biodiversity**

REFERENCES

1. Adul Ghany, S., Hena, H., Ardalan, S. Insecticidal effects of some aqueous plant extracts on the control of Khapra Trogoderma granarium, evert. *In International Conference on Chemicals, Biological, and Environmental Sciences, 201*, 55–70.
2. Changkija, S. (2014). *Biodiversity of Nagaland.* Department Forest, Ecology, Environment, and Wildlife: Kohima.
3. Chagnon, M., Kreutzweiser, D., Mitchell, E. A. D., Morrissey, C. A., Noome, D. A., Van, D. S. J. P. (2014). Risks of large-scale use of systemic insecticides to ecosystem functioning and services. *Environmental Science and Pollution Research, 22,* 119–134.
4. Chhetry, G. K. N., Belbahri, L. (2009). Indigenous pest and disease management practices in traditional farming systems in north east India, A review. *Journal of Plant Breeding and Crop Science, 1*(3), 28–38.

5. Deka, M. K., Bhuyan, M., Hazarika, L. K. (2006). Traditional pest management practices of Assam. *Indian Journal of Traditional Knowledge*, *5*(1), 75–78.
6. Grzywacz, D., Stevenson, P. C., Mushobozi, W. L., Belmain, S., Wilson, K. (2014). The use of indigenous ecological resources for pest control in Africa. *Food Security*, *6*(1), 71–86.
7. Gyawali, P., Khanal, S., Joshi, J. R. (2021). Crop protection practices in traditional agriculture in mid-hills of Western Nepal: A case of Palpa and Gulmi District. *International Journal of Applied Sciences and Biotechnology*, *9*(2), 138–151.
8. FAO. (2010). *Global Forest Resources Assessment.* FAO Forestry Paper No. 163. Food and Agriculture Organization: Rome.
9. Li, Y., Li, Z., Ye, M. (2008). The chemical compositions and their bioactivities in the different parts of Eupatorium adenophorum Spreng. *Journal of Yunnan Agricultural University*, *23*(1), 42–72.
10. Pal, S., Chakravarthy, A. K. (2020). *Advances in Pest Management in Commercial Flowers.* CRC Press.
11. Patel, S. K., Sharma A, Singh, G. S. (2020). Traditional agricultural practices in India: An approach for environmental sustainability and food security. *Energy, Ecology and Environment*, *5*(4), 253–271.
12. Uprety, Y., Asselin, H., Bergeron, Y., Doyon, F., Boucher, J. F. (2012). Contribution of traditional knowledge to ecological restoration: Practices and applications. *Ecoscience*, *19*(3), 225–237.

CHAPTER 9

Integrated Pest and Disease Management (IPDM) in Multitrophic Agroforestry

PARTHASARATHY SEETHAPATHY,[1*] JAGADEESH BATHULA,[2] MOHAMMAD ABDUL WASEEM,[3] and SIVARAJ PARAMASIVAM[4]

[1]*Agrotechnology Division, CSIR - Institute of Himalayan Bioresource Technology, Palampur, Himachal Pradesh, India*

[2]*Department of Natural Resource Management and Conservation, Forest College and Research Institute, Mulugu, Telangana, India*

[3]*Department of Basic and Social Sciences, Forest College and Research Institute, Mulugu, Telangana, India*

[4]*Department of Agricultural Extension, Amrita School of Agricultural Sciences, Amrita Vishwa Vidyapeetham, Coimbatore, Tamil Nadu, India*

*Corresponding author

ABSTRACT

In multitrophic agroforestry systems, pests and diseases can significantly affect productivity, and their impacts are often intensified in managed or plantation forests. Global plant transport, coupled with the unintentional introduction of invasive pests and pathogens, along with human-driven environmental changes, has further altered natural ecosystems. To address these challenges, Integrated Pest and Disease Management (IPDM) offers a holistic approach that combines cultural, biological, mechanical, and judicious chemical methods to reduce pest damage. Within agroforestry, IPDM emphasizes the conservation and enhancement of beneficial organisms

Multitrophic Interactions for Sustainable Agroforestry: Harnessing the Power of Beneficial Insects and Ecological Dynamics. Hanuman Singh Jatav, Vishnu D. Rajput, Ipsita Samal, and Tanmaya Kumar Bhoi (Eds).

DOI: 10.1201/9781779641250-9

such as predators, parasitoids, and pollinators, which contribute to natural pest regulation. Research has demonstrated that tree species diversification influences pest, pathogen, and beneficial organism dynamics within ecosystems. Habitat manipulation practices, including intercropping, relay cropping, and multitier cropping, have been shown to improve pest and disease management while enhancing system resilience. Recent advances also highlight the importance of plant diversity and plant–microbiome interactions in strengthening natural defenses. The phytobiome—comprising plants, soil microbes, and associated organisms—plays a critical role in disease suppression and pest resistance. A deeper understanding of these complex interactions holds promise for developing innovative, sustainable strategies to prevent pest outbreaks and improve the long-term resilience of agroforestry systems.

9.1 INTRODUCTION

Forests are very varied landscapes that store carbon, provide households for many species, and provide places for people to enjoy nature. Besides that, they provide wood and other valuable materials. At the same time, they face problems, including climate change that is getting worse, emerging diseases and pests, and other human-made problems. This chapter explores using scales from the molecular to the environment to understand trees' interaction with their living and nonliving environments, especially numerous biotic and abiotic pressures. The global agricultural sector experiences an approximate annual loss of 40% in crop yields due to pests and infestations despite the widespread utilization of pesticides [1]. Insect pests are responsible for an estimated 8% of the annual decline in foliage biomass production on a global scale [2]. Significant reductions in production can result from even minor but enduring chronic damage to long-lasting vegetation, such as forest trees [3]. Agroforestry principles, such as labor-driven intensification, sustainable growth, and agroecological transformations, have been employed to delineate the defining characteristics encompassing successful agroecological changes. The present discourse shall inevitably delve into the realm of exemplary agroforestry practices, commonly known as agroforestry, which pertains to the implementation of biophysical techniques aimed at augmenting the average lifespan of sustainable agricultural systems while simultaneously enhancing the soil ecosystems and biodiversity within the immediate vicinity, thereby reducing pest and disease will be sustainable.

Indigenous diseases, pests, droughts, storms, and fires sometimes affect forests or tree types, resulting in deceased or debilitated trees. These threats can have regional and widespread consequences, but they do not result in species extinction. Conversely, exotic invasive pests have the potential to endanger the sustainability of a species. The worldwide dissemination of tree pests and pathogens, affecting both native and cultivated forests, is rapidly increasing as an epidemic [4]. The bark and wood-drilling pests, primarily classified under the orders Coleoptera, Dictyoptera, Hymenoptera, Isoptera, and Lepidoptera, penetrate the wood to find nutrition or seek protection. The defoliators, skeletonizers, and sapsuckers of leaves belong to Coleoptera, Hemiptera, Lepidoptera, and Thysanoptera [5, 6].

In the dynamic landscape of modern agriculture, the pursuit of sustainable and resilient farming practices has become imperative. Integrating diverse elements within agroecosystems has emerged as a promising avenue to address the challenges associated with conventional farming systems. Among these innovative approaches, multitrophic agroforestry (MAF) has garnered significant attention for its potential to harmonize ecological processes while enhancing agricultural productivity [7]. This chapter delves into the intricate interplay between integrated pest and disease management (IPDM) strategies and MAF, exploring how the fusion of these two concepts can revolutionize the paradigm of sustainable farming.

9.2 RECENT OUTBREAKS OF PESTS AND DISEASES

Invasive alien pathogens and pests (IAPPs) severely harm or destroy many global forests every 10 years [8]. IAPPs severely threaten MAF environments worldwide, significantly impacting forest productivity, sustainability, and resilience (Table 9.1). Pests and other forest-related problems have significant impacts on the structure of forests through their interactions. The consequences of these relationships are expected to intensify in light of ongoing global change [9]. The Dutch elm disease (DED), butternut canker, and chestnut blight outbreaks were devastating tragedies that significantly impacted the ecological landscape and agricultural practices in the United States during the 20th century [10]. Between 1800 and 2009, more than 100 non-native tree diseases were introduced in Europe [11]. Over 400 woodland insect pests were identified in the United States between 1860 and 2000 [12]. These pest and pathogen invasions can profoundly impact the functioning of forest ecosystems and the services associated with landscapes dominated by trees [13]. Exploring the primary origins of plant diseases that may arise from forests needs to be improved. Fungi and diseases resembling fungi have historically been the predominant agents of plant-mimetic infectious diseases. From 1995 to 2010, the number of disease notices for plant-infecting fungi published by the Program for Monitoring Emerging Diseases increased 13-fold. Fungal and fungal-like pathogens are the leading causes of EIDs in forest trees. DED and chestnut blight are a few notable virulent infestations brought on by fungi that have become anthropogenically established [11]. The Ascomycota fungus is the predominant group responsible for developing forest diseases, accounting for 70% of European occurrences. Oomycota are the second most common community, accounting for 21% of cases. The introduction of ascomycetes has steadily increased in recent decades, while the rate of oomycete introduction has significantly accelerated in Europe since the 1990s [11]. *Phytophthora* diseases exhibit high pathogenicity and recombination rates and represent the pathogen category that will bring the most significant biosecurity issues. Additionally, there is a broader distribution of root-infecting diseases across various nations and host plant families than foliar-infecting diseases [14]. Viruses with a wide range of hosts pose a greater risk to plant health and safety. Two characteristics associated with generalist behavior that could serve as reliable indicators of danger are quick growth rates at their optimal temperature and prolonged longevity of latent spores.

TABLE 9.1 Global Invasive Pests and Diseases Affecting Agroforestry [10, 11, 14–16]

S. No.	Pest/Disease	Scientific Name	Host Tree
1.	Emerald ash borer	*Agrilus planipennis* Fairmaire	Ash and white fringetree
2.	Balsam woolly adelgid	*Adelges piceae* Ratzeburg	Spruce-fir
3.	Hemlock woolly adelgid	*Adelges tsugae* Annand	Carolina hemlock
4.	Spongy moth	*Lymantria dispar dispar* Linnaeus	Oak
5.	Asian long-horned beetle	*Anoplophora glabripennis* Motschulsky	Elm, maple, poplar, and willow
6.	Sirex woodwasp	*Sirex noctilio* Fabricius	Pine
7.	Southern pine beetle	*Dendroctonus frontalis* Zimmermann	Pine

TABLE 9.1 *(Continued)*

S. No.	Pest/Disease	Scientific Name	Host Tree
8.	Winter moth	*Operophtera brumata* Linnaeus	Cherries, maples, and oaks
9.	Oriental chestnut gall wasp	*Dryocosmus kuriphilus* Yasumatsu	Sweet chestnut trees
10.	Beech bark disease	Scale insects: *Xylococculus betulae* Pergande and *Cryptococcus fagisuga* Lindinger; Fungi: *Neonectria faginata* M.L. Lohman, A.M.J. Watson & Ayers and *Neonectria ditissima* (Tulasne & C. Tulasne) Samuels & Rossman	Beech trees
11.	Beech leaf disease	Nematode: *Litylenchus crenatae mccannii* Carta	Beech trees
12.	Chestnut blight	*Cryphonectria parasitica* (Murrill) Barr	American chestnut and American chinquapin
13.	Sudden oak death (SOD)	*Phytophthora ramorum* Werres, De Cock & Man in't Veld	Coast live oaks and Californian black oaks
14.	Dutch elm disease (DED)	*Ophiostoma himal-ulmi* Brasier & Mehrotra, *Ophiostoma novo-ulmi* Brasier, *Ophiostoma ulmi* (Buisman) Nannfeldt	Elm trees
15.	Thousand cankers disease (TCD)	*Geosmithia morbida* Kolarík, Freeland, Utley & Tisserat	Walnut
16.	Oak wilt	*Bretziella fagacearum* (Bretz) Z.W. de Beer, Marincowitz, T.A. Duong & M.J. Wingfield	*Quercus* spp.
17.	Butternut canker	*Ophiognomonia clavigignenti-juglandacearum* (Nair, Kostichka, & Kuntz) Broders & Boland	*Juglans cinerea*
18.	Laurel wilt	*Raffaelea lauricola* (T.C. Harr., Fraedrich & Aghayeva) Z.W. de Beer & M. Procter	Avocado
19.	Alder decline	*Phytophthora x alni* (Brasier & S.A. Kirk) Husson, Ioos & Marçais	Alder
20.	Pine wilt disease	*Bursaphelenchus xylophilus* (Steiner & Bührer) Nickle	Scots, Austrian, jack, mugo, and red pines

The monetary and biological hazards of nonindigenous timber pests and diseases are constantly growing due to the globalization of commerce and transportation. Therefore, the demand for adequate biosecurity measures to minimize these risks is more crucial than earlier. Recent data indicates that China and European Union countries have conducted more studies on forest insect pests and disease monitoring. Pine wilt disease was the most extensively researched disease, whereas bark beetles were the most prevalent pest incidence. The conifers most frequently examined in studies were pine, European spruce, and fir, whereas the most pervasive hardwoods were eucalypts [17]. Climate change can increase insect and disease numbers or alter trophic connections, modulating epidemic intensity [18]. Some of the latest bark beetle and defoliating insect epidemics are likely caused by climate change and affect landscapes and forest ecosystems that harbor insects and pathogens [19]. Notable recent instances include the exceptional destruction of ash trees in native and urban forests in America caused by the non-native phloem-feeding beetle emerald ash borer, which destroys the vascular system of trees [20]. The ash dieback phenomenon covered affected an area of 17,000 ha at 2008, has arrived in Europe from Asia, again reemerging as widespread new threat by *Hymenoscyphus fraxineus* (T. Kowalski) Baral, Queloz & Hosoya, which demise of European ash, poses a serious risk to its existence in countryside [21], and the influx of both native and invasive pests and pathogens that are undermining the productivity of eucalyptus plantations in areas where they are not naturally found [22].

9.3 PRINCIPLES OF PESTS AND DISEASE MANAGEMENT

Trees serve many ecological functions on land and are grown on managed farms and plantations to produce timber, paper, fiber, and food. Bacteria, fungi, viruses, and insect herbivores are the main pests and diseases that affect plants. Previous studies on tree diseases and pests focused on economically significant trees. However, several outbreaks and invasions have killed important trees in ecosystems worldwide. These have damaged critical ecosystems and highlighted the widespread effects of tree diseases and insect pests [23]. Critics claim it is fruitful to examine the way diseases and parasites impact the ecological functions of forests. Tree pests and diseases can affect not only the quality of forest and orchard products but also the capacity of forests to sequester carbon, mitigate flood risks, and purify water. They have the potential to impact the biodiversity of trees, the aesthetic appeal of woodlands, and their cultural significance. The absence of an assigned cost for these advantages and the fact that various groups perceive them raise concerns regarding accountability for safeguarding the welfare of trees. The content emphasizes alterations in the spread of diseases and insects, genetic combinations that generate unidentified disease risks, and global warming. Pests in a MAF system can impact the system's components, such as trees, crops, and other species, directly and indirectly. Understanding the principles of pests and disease development and management in these systems requires careful consideration of the ecological interactions and linkages within the ecosystem [24].

To implement IPDM in agroforestry, the following steps will be practical:

1. Monitoring and prevention of invasive, harmful pests and pathogens
2. Choose and rehabilitate eco-suitable trees that are resistant to pests and pathogens
3. Practice crop rotation between host trees and nonhost trees to prevent infestations
4. Employ noninvasive biological methods to control pest and pathogen populations
5. Enhance the population of pollinators by cultivating suitable intercrops and border crops
6. Adopt silviculture practices that do not amplify light levels in agroforestry systems.

9.4 INTEGRATED PESTS AND DISEASE MANAGEMENT IN MULTITROPHIC AGROFORESTRY

MAF, by definition, involves the intentional integration of trees, crops, and livestock within a single agricultural system. This synergistic arrangement mimics natural ecosystems, fostering biodiversity and promoting ecological balance. The inherent complexity of MAF provides a myriad of niches and habitats, creating a dynamic environment that can be harnessed to manage pests and diseases through natural and sustainable means [25]. On the other hand, IPDM is a holistic approach that transcends traditional pest control methods [26]. Instead of relying solely on chemical interventions, IPDM incorporates a spectrum of techniques, including biological control, ecological measures, cultural practices, host plant resistance, physical measures, and judicious use of chemicals when necessary [27]. This strategy aims not only to suppress pest and disease populations but also to minimize environmental impact and safeguard the long-term health of agroecosystems (Figure 9.1).

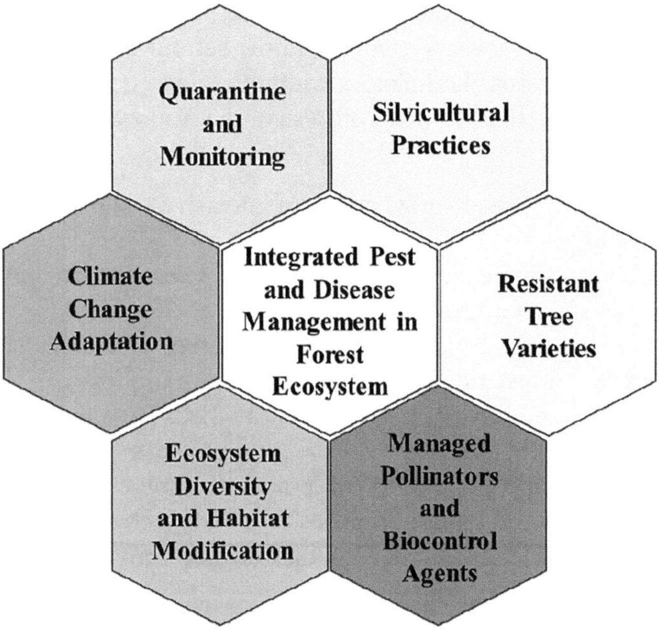

FIGURE 9.1 Integrated pest and disease management in forest ecosystem.

A mosaic of biotic and abiotic factors, including agroforestry, defines heterogeneous land use systems [28]. This mosaic results from the spatial arrangement of various plant species and their impacts on soil conditions, detritus, microclimate, and other elements. As a result, spatial patterns in the occurrence of pests and diseases within agroforestry systems may be apparent within agroforestry plots; therefore, farmers must be informed of these practices to detect pests and diseases in areas where they are most likely to manifest [29]. "Tree-crop interface" refers to the demarcation line between agricultural land and forested regions; in this context, significant localized variations in pests and diseases are frequently discernible [30]. Several researchers conducted experiments to examine the growth and productivity of crops at various distances from the boundary between trees and crops. Their goal was to measure the impact of competition between the shoots and roots of the plants. However, they found that these effects were mixed up with variations in the occurrence of pests and diseases. These interactions might begin early on when planting the crop. Establishing an annual yield along the tree-crop interface was frequently more challenging than further away from the trees due to an elevated level of seed predation[31].

9.4.1 MONITORING AND SURVEILLANCE

Implementing efficient methodologies for evaluating and surveilling forest pests and diseases is paramount in IPDM. These techniques, when accessible, offer valuable insights to natural resource supervisors, aiding them in the prioritization of management systems. In an MAF system, strategic decisions regarding the timing of harvesting, implementation of sanitation measures, and crop intensification efforts in neighboring forests are crucial for effective pest management. These decisions are most effectively made by relying on survey data predictions, specifically tree damage or mortality estimates. By utilizing such examinations, forest supervisors can proactively address potential outbreaks and mitigate the impact of pests and diseases in agroforestry. However, sampling insects and pathogens in MAF environments poses distinct challenges not typically encountered in agricultural systems or urban forests, where damage can be more readily observed and measured. Furthermore, the enigmatic characteristics exhibited by numerous insect species, coupled with their inhabitation of tree sections that pose sampling difficulties (such as the uppermost regions of tree crowns), contribute to the inherent challenges encountered during agroforest surveys.

Regular surveillance estimates pest density, allowing to plan and deploy IPM in advance. Monitoring depends on the pest and disease incidence assessment and evaluation of tree health parameters. A study correlated light interception and insect population in rice. Tree species with a scant canopy, allowing sunshine penetration, are ideal for tree-rice agroforestry. Rice bug infestation was lowest in the Albida-rice association, followed by the Jhau-rice association, and highest in the Akashmoni-rice association. The control plot, without tree species, had the highest light intensity and the least insect infestation compared to the other relationships [32]. Major attributes to evaluate in the context of disease development encompass the growth rate of the pathogen at the optimal microclimate. The progression of interdisciplinary and international cooperation is vital in the enhancement of horizon-scanning systems based on traits. This would require the establishment of integrated repositories that consolidate data pertaining to the transmission, characteristics, and lineage of pathogens [14].

9.4.2 SILVICULTURAL PRACTICES

Silvicultural techniques that promote vegetation diversity, such as mulching, stumping, organic supplements, mixed-species plantations, and selective logging, can be employed to enhance forest resilience. These practices aim to create a more balanced and resilient forest structure, reducing the risk of pest and disease outbreaks. Organic mulch enhances soil physical qualities and provides vital elements like carbon, phosphorus, and nitrogen to the soil, hence improving nutrient absorption by the roots of plants. Mulches made of organic matter produce a beneficial soil condition that promotes the plant's vigor while reducing the likelihood of infection and disease formation. Specifically, mulches can regulate soil temperature and moisture, decrease the presence of aphids and fungal spores dispersed by rainwater, and potentially lower the occurrence of certain diseases. In a study, it was found that using mulches made from a single species substantially impacted the extent of *Phytophthora* root rot lesions in horse chestnut trees [33]. These mulches also promoted the growth and overall health of the trees. Although diversification strategies, including agroforestry and barrier cropping (also known as strip cropping), have been examined less, they have demonstrated some potential in mitigating the spread of the cocoa swollen shoot virus disease (CSSVD). According to the findings, intentionally selected and effectively managed species may exhibit advantageous impacts on the growth and well-being of cocoa trees, thereby aiding in their resistance to CSSVD [34].

The efficacy of stumping as a forest management technique in mitigating the occurrence of *Armillaria* root rot in young trees has been established, and influences the soil bacterial population. Implementing stump removal, in conjunction with cultivating tree species resistant to *Armillaria* root rot alongside those vulnerable, can be a productive management strategy for reducing the disease. The integration of these methods has a mutually reinforcing impact on fostering a favorable bacterial microbiome, which has the potential to mitigate the effects of *Armillaria* root rot and enhance tree production [35]. Infestation rates of the coffee berry borer (CBB), coffee leaf miner (CLM), and the brown twig beetle were notably higher, by up to 7%, in young coffee farms that employed more intensive and conventional practices compared to those that used organic methods at the same or lower input level. The abundance of the entomopathogenic fungus *Beauveria bassiana* (Balsamo) Vuillemin was more significant in the traditional intensive plots, perhaps due to the higher prevalence of pest infestation. The presence of brown twig beetles was more prevalent in areas without shade. However, the absence of shade slightly decreased the severity of anthracnose, but by less than 2%. There is a trade-off between the damage caused by pests and diseases compared to shaded and unshaded circumstances. However, the decrease in pest infestation due to shade was higher than the rise in fungal infections [36]. They also documented that the organic farming practices with suitable shade trees are effective in the suppression of pests such as *Hypothenemus hampei* (Ferrari), *Xylosandrus morigerus* (Blandford), *Leucoptera coffeella* (Guérin-Méneville), and pathogens like *Colletotrichum* spp., *Cercospora coffeicola* Berkeley & Curtis, *Pellicularia koleroga* (Cooke) von Höhnel, and *Phoma* spp.

It is encouraged to introduce intercropping techniques in tea plantations by cultivating climate-resilient crops such as cowpea, sorghum, soybean, and aromatic plants, including ginger, marigold, flemingia, and semen cassiae. These suggested cover crop species effectively lure a wide range of natural predators and inhibit the growth of significant tea pests (such as mirid bugs, tea aphids, tea green leaf hopper, and white flies) and diseases (blister blight and grey blight) [6].

9.4.3 CROPPING PATTERNS

The pest and disease risk mitigation potential of agroforestry cropping systems has been extensively studied under various intercropping techniques such as mixed planting crops, row strip intercropping, intercropping, and relay intercropping [37]. Through the deliberate integration of multiple crops within a single field, agroforestry systems generate an environment characterized by increased complexity and diversity, potentially leading to disturbances in pathogen and disease cycles. The arrangement of these patterns may hinder the mobility of pests, thereby restricting their capacity to increase and inflict harm. Furthermore, the coexistence of numerous crops in close vicinity may impede the ability of pests to locate their preferred host plants due to the dilution effect.

Over time, the increased diversity of crop-tree canopy in the agroforestry disrupts the life cycle of soil-borne pests and diseases through below-ground mechanisms, resulting in nonhost impacts. The primary objective is to decrease the number of disease-causing organisms or the population that can survive from one season to the next owing to the demise of the host plant. However, the effectiveness of these preventive measures can vary depending on the range of host plants or the presence of specific survival mechanisms. Monoculture ecosystems typically exhibit an elevated incidence of biotic stress compared to MAF systems [38]. Enhancing the health of trees within agroforestry systems can be achieved by implementing management strategies that effectively augment light availability. Concurrently, establishing a sense of interconnectedness between these systems and the surrounding forest can improve tree protection. Agroforestry systems are fortified by incorporating germplasms derived from species with significant nutritional and commercial worth into native agroforests. The integration of germplasms in agricultural practices could raise concerns about reducing genetic diversity within the cultivated species. Incorporating arboreal vegetation within agricultural landscapes has substantially enhanced insect management strategies. The presence of trees can function as a physical impediment, impeding the locomotion of insects and thereby diminishing the likelihood of infection in both trees and crops.

Alley farming can have either a beneficial or detrimental impact on the progression of crop diseases. Earlier research indicated that alley cropping with common sesban decelerated the growth of corn rust [39]. In monocrop corn, there was a considerably higher proportion of diseased leaves per plant, more significant amounts of uredial pustules per plant, and a higher area under disease progress curve than in alley-cropped corn. The rust and late leaf spots in groundnuts were much higher in the crop rows heavily sheltered by the trees than in further areas in the characterized alley farming system [40]. The presence of hedgerows in this location significantly diminished potential evapotranspiration and caused groundnut leaves to dry later in the morning compared to regions with less shadow. The localized microclimatic effect countered the overall decrease in the occurrence of foliar diseases in the MAF. The green mulch likely caused this effect, as it was observed in plots with mulch but no hedgerows [41].

9.4.4 ECOLOGICAL SUCCESSION

The potential benefits of redirecting pests using specific plants in agroforestry include the enhancement of sustainability and ecological practices [42]. The variety of vegetation chosen would be contingent upon the pests and commodities under consideration. By integrating specific trees into agroforestry ecosystems, the detrimental effects of pests and diseases can be mitigated through a multitude of strategies.

The following methods encompass the agroecosystem-based succession of biotic stressors:

- Employing visible and volatile signals to suppress pests, complexing resources, and redirecting pests with inhibitors;
- Interfering with the geographic pattern of pests through nonhost plants;
- Eliminating host plants in order to reduce pest populations and interfere with the temporal rhythm;
- Implementing allelopathic reactions within the environment to suppress the soil-borne pests and pathogens;
- Promoting antagonists and biocontrol agents of pests and diseases or augmenting the suppressive capacity of the soil;
- Enhancing crop sustenance to bolster physiological resistance;
- Releasing predators or restoring natural enemies for managing pests that affect crops from the environment; and
- Developing natural barriers and influencing the microclimate to indirectly affect pests.

Vegetation diversity contributes to ecological succession, a natural process of change in the composition of plant species over time [42]. Increased vegetational diversification and the resulting biodiversity can sometimes lead to more efficient natural pest and disease control in tropical agroforestry. Different stages of succession may attract different pests and diseases. Managing forests to mimic natural succession can help control specific pests by altering the habitat conditions. The presence of uniform tree genetic material has been identified as a primary factor contributing to the rise of insect and diseases in monoculture fields [43]. Many studies have demonstrated that cocoa agroforestry can increase the productivity of cocoa plantations by reducing the incidence of diseases and pests [44]. Several trials conducted under varying ecosystems have demonstrated a decrease in insect activity when diversified vegetation is present, as opposed to monoculture. Cocoa cultivation in full-sun monocultures in the tropics, aimed at maximizing short-term productivity up to 45%, has led to issues like erosion, loss of biodiversity, and elevated vulnerability to pests and diseases. In contrast, long practiced in Latin American countries, dynamic agroforestry systems mimic natural forests, providing benefits such as enhanced fertility, reduced pest pressure, and erosion control [38]. The pests, especially the experts, show more extended periods of employment and tremendous success in eating and reproduction. Agroforestry enhances plant diversity within a given area by including various plant species over time and across different spatial dimensions. Monoculture ecosystems typically exhibit an elevated incidence of biotic stress compared to MAF systems. Enhancing the health of trees within agroforestry systems can be achieved by implementing management strategies that effectively augment light availability. Concurrently, establishing a sense of interconnectedness between these systems and the surrounding forest can further improve tree protection. Complex agroforestry systems closely resemble but are not identical to natural tree communities seen in stable ecosystems or biological successions [24]. In the latter case, the characteristics and arrangement of plant life are determined by natural factors. The plant communities that arise through the process of selection by nature possess a certain level of inherent resistance against insect and disease infestation.

Shade trees are crucial to coffee leaf rust (CLR) regulation in India. These trees provide controlled shade, which is vital for CLR management. Not all tree shade patterns can be experimentally justified for CLR [45]. This finding supports the idea that a mixed shade distribution and canopy light regulation are best for management in the environment under consideration. Subsequent investigations have shown that the impact of shade on coffee diseases is uncertain due to its dependence on the season and climate [46]. A piecewise structural equation model analysis was conducted on CLR disease as a case study to examine the connections between the environment and shading systems and their impact on disease intensity. The study evaluated the relationships between seasonal microclimate and CLR incidence variations at various altitudes and shading systems. Most studies investigating the correlation between diversity and disease risk have discovered data supporting the dilution effects. As the number of different species in an ecosystem rises, the spread of disease-causing agents and the incidence of diseases decline. Amplification effects can arise in communities when there is a rise in variety, leading to a higher risk of infection. This can happen when the spread of diseases is not dependent on the number of hosts, such as in the case of common diseases. Furthermore, combining different tree species can modify the regional weather and moisture levels, consequently affecting the environment for pathogens [47].

9.4.5 RESILIENCE TO OUTBREAKS

Introducing plant diseases to new areas is becoming increasingly frequent due to the increasing global connection. Assessing the potential danger, they present to the health of plants can be challenging without a comprehensive understanding of their behavior, distribution, and propagation rate in tropical agroforestry systems [18]. The coexistence of diverse agricultural practices, such as polyculture and agroforestry, can be likened to distinct manifestations of agricultural diversity. In diverse forests, the impact of a particular pest or disease may be limited to specific tree species, preventing widespread devastation. If one species is affected, others may remain unaffected, maintaining overall forest health and productivity [14]. A deeper understanding of the complex interplay between cultural practices and their environmental consequences can be achieved by conducting a comprehensive examination of the intricate dynamics at play within agricultural systems. A review paper revealed that the negative impact caused by monophagous insects, which feed exclusively on tree varieties within a specific genus, was reduced by 42%. Similarly, damage caused by oligophagous insects, which feed on various genera within specific relatives, was reduced by 15%. However, there was no substantial decrease in injury caused by polyphagous insects, which infest multiple tree systems [48].

The integration of arboreal components within the vicinity of vineyards, employing agroforestry methodologies such as alley cropping, wind screens, hedgerows, and multistory planting, holds promise in establishing vineyard landscapes characterized by enhanced diversity. This approach exhibits the potential to mitigate the prevalence of pests and diseases, curtail the reliance on pesticides, and bolster financial resilience [49]. The presence of host tree types is less concentrated or diluted among other related species of trees, resulting in lower availability of food supplies for specialized insect pests. As a result, these insect pests have limited potential to increase in number and cause damage rapidly [48]. In coffee and tea plantations, aromatic crops secrete secondary volatile compounds. These compounds efficiently impede the colonization of significant plantation pests. Integrating aromatic crops and legumes into the system provides an improved means of managing arthropod pests and diseases in plantations. Adopting this methodology diminishes the dependence on chemical pesticides, mitigating the adverse ecological impacts. Intercropping cover crops within Arabica and Canephora coffee and cocoa plantations is widespread across many nations. Although cover crops play a crucial role in soil health, there needs to be more research on the diversity and function of fungal communities in perennial cropping systems. The fungal genera *Alternaria*, *Aspergillus*, *Acremonium*, *Armillaria*, *Cylindrocarpon*, *Cladosporium*, *Microdochium*, *Phaeoacremonium*, *Penicillium*, *Phialophora*, and *Rosellinia* exhibited a positive correlation with elevated soil organic matter content and the breakdown of intricate polymers in olive orchards that were intercropped with grass cover crops. These fungi play a beneficial role in promoting soil health [50]. This is because cover crops comprised of crotalarias, jack bean, mucunas, and pigeon pea offer numerous advantages, including reduced reliance on chemical inputs in coffee cultivation, effective vegetation management, and soil conservation [51]. A recent study using sunflower and showy rattlebox as interrow cover crops effectively eliminated root-knot nematodes on coffee plantations in Brazil [52]. The pigeon pea-intercropped planting exhibited a lower incidence of CLM on coffee foliage. Furthermore, more significant indications of predation by wasps on CLM were detected in the plantings shaded by Leucaena [53].

9.4.6 NATURAL PREDATORS, ENEMIES AND BIOCONTROL CONTROL

By restoring the provision of ecosystem services, agroecoforestry practices can improve both biotic and abiotic factors that support the conservation of biodiversity. The research assessed the influence of various management approaches, agroforestry versus full-sun cultivation, on the efficacy of entomopathogenic fungi in insect control within Brazilian coffee plantations [54]. By combining survival analyses with the insect lure technique, the researchers determined the frequency of entomopathogenic fungal insect mortality in soil specimens from both systems. Researchers observed that the mortality rate of insects was significantly higher in agroforestry soils than in full-sun soils. *Metarhizium* was the fungus most frequently isolated from the lure insects, with *Beauveria* being the most commonly discovered subsequently. In the meantime, *Fusarium* was frequently attributed as the primary or secondary cause of different diseases. The authors hypothesized that the diverse insect survival rates indicated that entomopathogenic fungi can regulate insect populations in agroforestry. Furthermore, the efficacy of the insect bait method in suppressing pests was evaluated through the observation of lure insect mortality, a process known as the "bait survival technique."

The CBB is a widely distributed parasite currently in all countries that produce coffee, except Australia and Nepal. It is widely accepted as the most devastating insect to coffee. CLM populations are prevalent across the American continent. In general, a higher level of diversity in agroforestry systems is linked to enhanced management of insects that affect coffee crops. Introducing cover crops alongside coffee is a viable strategy to improve the availability of food and shelter and create an ideal microenvironment for predators and parasitoids. In exchange for the food that extrafloral nectarines provide, natural enemies protect against herbivory. Inga trees with extrafloral nectarines supply associational resistance to coffee plants in agroforestry systems. The additional nectar-producing glands of the inga plant serve to lure predators and parasitoids that prey on the CLM and CBB insects [55].

Consequently, this enhances their ability to survive and produce, increasing their effectiveness in controlling pests. Moreover, optimizing the selection of tree species for biological control can be accomplished by considering their interactions with pests and natural adversaries. Planting sunn hemp and buckwheat alongside coffee led to an augmentation in the rates of predation and parasitism, respectively, caused by wasps targeting the CLM. Compared to growing only coffee, incorporating sunn hemp into coffee cultivation increased the number of predatory mites from the Phytoseiidae family and decreased the number of herbivorous Tetranychidae [56]. Even though the presence of a variety of shade trees in coffee MAF of South India does not appear to have a significant impact on prey–predator interaction [57].

The study employed reductionist methods, including single and poly-species cultures of bacteria, and emergentist methods, such as bacterial applications in an oak infection with the inclusion of larval insects to understand the transcripts patterns and symptom outbreaks in the triadic relations between the host, microbiota, and insects in the acute oak decline (AOD) pathobiome. AOD is a degenerative disease caused by a bacterial complex of *Brenneria goodwinii* Denman, *Gibbsiella quercinecans* Brady et al., and *Rahnella victoriana* Brady et al. that includes contributing abiotic conditions, internal lesions caused by a bacteria pathogenome and microbiome, and larvae chambers of the bark-boring insect [58]. Forest ecosystems exhibit disease-suppressive soil development when trees are grown in different species, unlike a single plantation [35]. The cultivation of black pepper and vanilla together resulted in a decreased presence of *Fusarium oxysporum* Schlechtendal in the soil surrounding the roots of vanilla plants. Conversely, antagonistic fungal groups like *Trichoderma* spp. showed an increased presence in comparison. Intercropping offers advantages that extend beyond tropical agroforestry systems [59].

Biological control is essential for long-term management after an IAPP organism colonizes a forest. In some instances, natural treatment has restored sustainable pest population dynamics by keeping invading species densities below damage thresholds. Parasite *Torymus sinensis* Kamijo controls the Asian chestnut gall wasp *Dryocosmus kuriphilus*. This species effectively reduces insect populations worldwide, making it autonomous. Biological methods based on *Phlebiopsis gigantea* (Fries) Jülich effectively eliminate the root and butt rot fungus *Heterobasidion annosum* (Fries) Brefeld in European conifers [8]. Commercial formulations of this biocontrol agent are currently available.

9.4.7 ECOSYSTEM SERVICES

Diverse forest ecosystems offer an extensive array of services, including recycling nutrients, water purification, and soil stabilization. These functions collectively enhance the overall health and resilience of MAF. In forest production landscapes, the primary goal of management decisions is to optimize productivity, with less emphasis placed on the conservation of ecological services. Nevertheless, it is critical that ecological services substantially influence productivity. Establishing coffee plantations in tropical regions has significant environmental and economic ramifications, as it causes the clearance of numerous tropical evergreen forests. The predation levels by produced caterpillars were similar in both the diversified native-trees shaded and monodominant oak-shaded plantations of coffee in Kodagu [57]. Agronomic research indicates that decreasing or eliminating shade in coffee cultivation could improve coffee flowering and yield. Nonetheless, the diversity of shade trees is vital for providing essential pollination services and natural insect control [60]. A healthy ecosystem is better able to withstand and recover from pest and disease impacts. The three vital functions that coffee plants require from biodiversity within plantations of coffee are natural shade, pollination processes, and pest control. The quantity of shade and the presence of diverse plant species are crucial factors that affect infestations of pests, natural pest management, pollination, and crop output. The presence of shade trees in the understory layer of plantations plays a vital role in reducing pests in coffee by promoting the abundance of tiny birds. The continuous shading provided by trees in perennial crops may reduce insect populations in some instances [61]. The presence of more trees, causing enhanced shading, resulted in a reduction in both the quantity and intensity of pests on cocoa pods [62]. There is a negative correlation between the level of shade in coffee agroforests in Peru and the rates of pest infestations in berries [63].

H. annosum sensu stricto (Fr.) is responsible for a severe form of root rot, considered the most significant disease in coniferous forests in several regions of Europe and America. The fungus depends on tree wounds or root connections to spread by spore infestations. The pathogen movement is restricted in free soil due to the presence of adversaries. The disease is more widespread in managed forests than in natural forests due to the numerous opportunities for infection resulting from cut stumps and harvesting damage [64]. Extensive research has been conducted on the silvicultural control of root rot by mixed stands, as the biology of the fungus indicates that broader spacing between susceptible tree species and the creation of soil modifications with antagonistic characteristics can reduce the occurrence of root rot [65].

9.4.8 DISEASE RESISTANCE AND TOLERANCE

Mixed tree species have more tolerance and resilience to abiotic stress, disruptions, pests, and diseases than stands consisting of a single species.

The various levels of variety in trees at the stand level are achieved by selectively eliminating more particular species from the MAF habitat. Conducting studies that simulate species extinction caused by certain pests or diseases in the past has the potential to produce highly intriguing outcomes. A survey on insect defoliation in a network of 208 forest plots across Europe. The sites were selected along two gradients: one representing rising species of tree diversity and the other representing latitudes [47]. The richness of tree species has a beneficial impact on resistance to invasive *D. kuriphilus,* a univoltine monophagous pest of the European chestnut tree, regardless of the environment. Different tree species exhibit varying levels of resistance or tolerance to specific diseases [66]. A diverse forest is likely to contain species with natural resistance to various pathogens. In the examined area, the presence of the ink disease or chestnut blight, which is carried by *D. kuriphilus*, was similarly diminished. By planting a mix of tree species, forest managers can reduce the risk of widespread diseases affecting the entire forest. A recent review revealed that the resistance to pests in mixed forests grows as the population of pests becomes larger. It suggests that increasing the range of forests is a promising strategy for reducing harm caused by pests. Mixed-species hedgerows can restrict the ability of insect pests to access host trees and provide a habitat for natural predators. Ultimately, the selection of forest diversification options should be determined by a comprehensive analysis that considers economic, technical, and sociocultural limitations [67]. However, most investigations have only examined associational resistance to focal species of trees, focal species of pests, or both. They need to assess the harm caused by insect pests at the population and community level.

9.4.9 CLIMATE CHANGE ADAPTATION

Climate change affects forest pests for several species and guilds, although species-specific responses are generalized as challenging. It can alter the distribution and behavior of pests and diseases. Diverse forests are better equipped to adapt to changing environmental conditions, including those influenced by climate change. Global warming affects insect vectors that spread plant pathogens, resulting in more severe disease outbreaks caused by alterations in their reproductive and dispersion patterns. An instance of this is the spread of the mountain pine beetle, which is connected to the blue stain fungus that harms pine trees in western Canada was probably due to climate change [68]. This phenomenon has been attributed to climate change. Furthermore, the western corn rootworm, which transmits the maize chlorotic mottle virus, has shown a trend of spreading in Europe, potentially due to shorter and less severe winters. Plant infections can respond to external variations by undergoing forest host shifts, which occur when new pathogens arise after the spread of new host plants to other geographical sites [69]. Effective forest land management is crucial for the nation to meet international obligations about land degradation neutrality, nationally determined contributions, and sustainable development goals. Such goals include restoring degraded forests, biodiversity conservation, and adaptation and mitigation to climate change.

9.4.10 EXTENSION AND IMPLEMENTATION

Forest extension should empower rural populations, agricultural organizations, and farm foresters to access information and technologies. Alliances between service providers, farmers, and the commercial sector; collaborations between research and extension organizations, farmer's groups, and forest communities; and service provider partnerships are meant to facilitate these processes. Multipurpose and multifunctional forestry is growing among large and small producers. Disseminating agricultural technologies and information, such as IPM, to producers in a prompt and efficient manner is a problem for agricultural or forest extension workers. An activity incorporated within the Ecosystem Services Improvement Project in India, such as fortification of the Van Vigyan Kendra. The extension services are empowered to optimize the yield of preexisting MAF land, implementing integrated systems need IPM and integrated crop management, focusing on reduced chemical application and enhancing biodiversity [5]. Consequently, growers possessed a restricted comprehension of the underlying concepts that govern various management strategies [70]. By delivering IPM solutions tailored to the unique requirements of commercial tree nurseries, ensuring efficient management and regulation of biotic populations. Provide support in training and producing biopesticides isolated from native plants (*Trichoderma* and *Beauveria*) that are intended for use in nurseries to control soil pests and improve the quality of tree seedlings. However, research collaboration is the most important alliance for forestry extension to remain technologically and economically relevant to farmers worldwide.

9.5 AGROFORESTRY IN BIOTIC STRESS MITIGATION

The meta-analytical investigation of agroforestry systems found a significant reduction in weed abundance, coupled with an increase in the population of natural predators. Furthermore, it has been noted that the use of agroforestry systems in cultivating perennial crops has resulted in a significant reduction in insect populations. Agroforestry approaches have effectively reduced plant damage caused by pests and diseases [71]. To mitigate biotic stress and improve productivity, forest managers in several regions worldwide have imported non-native tree species due to their accelerated growth, increased volume production, or reduced vulnerability to pests and diseases. The reviewed literatures offer empirical evidence that supports the central hypothesis that greater habitat diversity in agroforestry systems leads to improved pest control results (Figure 9.2).

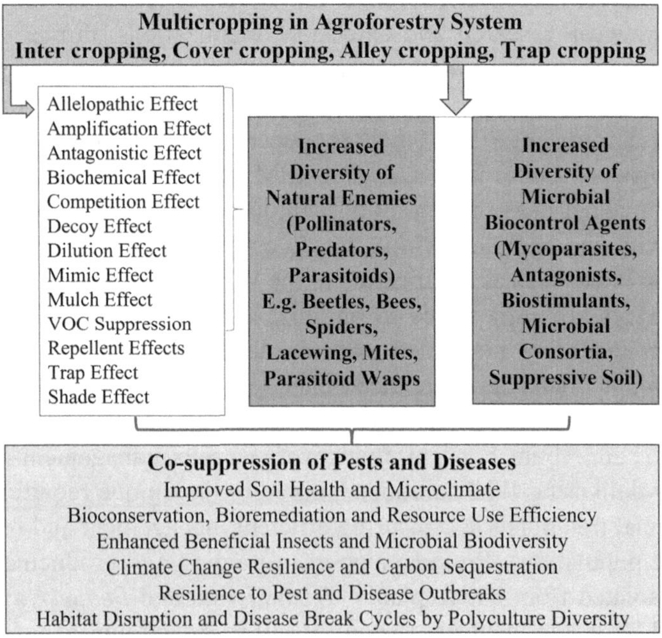

FIGURE 9.2 Sustainable ecosystem services in agroforestry on IPDM and biotic stress mitigation.

9.6 ORGANIZATIONS GOVERNING FOREST HEALTH

The International Union of Forest Research Organizations comprises multiple branches and units that focus on researching various aspects of forest health, such as bioprotection, invasive alien species, forest extension units, and the effects of pollutants in the atmosphere and global warming on forests and their ecosystems. The North American Forest Commission formed a task force on forest insects and diseases more than 40 years ago. Recently, the group has been expanded to deal with the issue of invasive plant species.

The regional organizations focused on forest pests, specifically forest invasive species, consist of the Asia-Pacific Forest Invasive Species Network, Commonwealth Agricultural Bureaux International, European and Mediterranean Plant Protection Organization, European Union Notification System for Plant Health Interceptions, European Alien Species Information Network, Forest Invasive Species Network for Africa, Global Invasive Species Programme, International Forestry Quarantine Research Group, and Near East Network on Forest Health and Invasive Species [4]. Periodically, new information may be incorporated into the portals of international agencies to establish a collaborative forecasting service for insect pests and diseases in agroforestry with other agencies.

9.7 FUTURE DIRECTIONS

- More investigation is required to elucidate the kinetics of pathogens and parasites within MAF systems. Insect and pathogen havocs that are unique to a particular agroforestry system can vary considerably from those encountered in other systems. Consequently, each system must be evaluated individually.
- The primary step in initiating epidemiological inquiries involves the monitoring of insect pests and disease outside an agroforestry system. In the vast array of harmful species inhabiting the ecosystem, it is essential to discern the perilous organisms by ascertaining their precise classification and evaluating the magnitude of the damage they have inflicted. The execution of this exercise is imperative for every distinct tree species encompassed within each constituent of the system. The resulting analysis will yield a comprehensive overview of the entirety of the pest and disease complex within the system.

- Identifying the pivotal species within a pest and disease complex is imperative to facilitate subsequent comprehensive investigations. Research on the life cycle of these pests should encompass various aspects such as their seasonal occurrence, peak activity period, method of attack, susceptible stages, predators, parasites, and antagonists specific to each insect species and pathogens, among other factors. The investigation encompasses a comprehensive approach incorporating field and laboratory research methodologies. An investigation into the factors influencing population growth necessitates carefully examining predisposing elements, such as temperature, RH, and light. These kinds of studies would result in the acquisition of valuable data that can be utilized for pest forecasting.
- Quantifying the detrimental impact inflicted by pests and diseases on individual tree species and the entire ecological arrangement necessitates accurate estimations. In contrast to the established techniques utilized for estimating crop losses, novel approaches may need to be developed to assess the impact of IAPPs on MAF. The valuation of component plants in MAF systems can vary, leading to differences in the methodology for estimating losses in each system design.
- An identification compendium at the global level can efficiently convey a full list of agroforestry insect pests and diseases. This compendium would include several figures and photos to help identify these pests. It would also reveal their injury ways, host range, life cycle, natural enemies, geographical distribution, and management policies.
- Biocontrol agents can be introduced and used in MAF systems. Biological control studies can move forward after collecting data on pest predators and infection competitors. Identifying various arboreal and shrubby plants that provide pollen and nectar to indigenous predators can also begin. Identifying barriers to natural predators is crucial. Established these compounds provide cost-effective, long-term host insect resistance.
- To comprehend the intricacies of insect and pathogen activity within agroforestry systems, a thorough comprehension of the mechanisms that govern the densities of these organisms is imperative. The management aspects pertain to examining the intercomponent interactions influence the behavior and progression of pathogens

and insects. Looking at the proportional abundances of pathogens and insects on a particular plant species across agroforestry and monoculture systems will provide explicit knowledge regarding the effects of co-cultivation on the dynamics of these organisms.

- In assessing a woody plant species for agroforestry, insects and pathogens from its natural habitat must be considered. If pests and pathogens migrate to the new area without natural predators, knowledge about pests, diseases, and predators will be invaluable.
- It is advisable to commence investigating for multipest and pathogen testing services that specialize in analyzing pathogen complexes and minor pests. Prioritize the cultivation of these plant species in MAF to mitigate the risk of invasion. One of the primary goals of tree development programs should be to incorporate insect and pathogen resistance. Long-term, if high-yielding species or provenances exhibit greater vulnerability to insect and pathogen infestations, the result may be lower compared to species with lower productivity but more excellent resistance to these threats.
- It is imperative to assess the allelopathic characteristics of the plant species being evaluated for potential use in agroforestry and their ability to attract or repel significant insect–pest and pathogen species present in the system. Encouraging the integration of species that demonstrate resistance to the parasites and pathogens present in other constituent species is recommended. Identifying plant species that function as collateral hosts or refuges for crop parasites or as vectors of plant diseases is essential. Possible superior alternatives might be employed as substitutes for these species.
- Further study should be conducted to ascertain the effectiveness of diverse tree-crop assemblages within different agroforestry frameworks in mitigating insect and pathogen populations. A comprehensive analysis of the incidence and severity of damage caused by insect pests and diseases on a particular host species relative to other plant species will elucidate the amalgamation of factors that impede the proliferation of these pests and diseases.
- In an MAF system, it is vital to research the effects of various spatial and temporal arrangements of woody perennials and cereals on insect predators and biocontrol agents that effectively mitigate the populations of pathogens and parasites. By analyzing various agricultural systems, it is possible to determine which

methodologies are most suitable for the sustainable control of pathogens and parasites. Evaluating designs across multiple geographic regions is crucial for determining their suitability for a specific agroforestry system.

9.8 CONCLUSION

The investigation of pest and disease management in MAF is a critical field of study in ecology and diversity. The integration of multiple crops and tree species in agroforestry systems promotes a dynamic and diverse environment that enhances natural pest and disease regulation. This approach aligns with principles of ecological balance and sustainable agriculture by reducing the reliance on chemical inputs and fostering a resilient farming system. Agroforestry systems are fortified by incorporating germplasms derived from species with significant nutritional and commercial worth into native agroforests. The integration of germplasms in agricultural practices could raise concerns about the reduction of genetic diversity within the cultivated species. Incorporating arboreal vegetation within agricultural landscapes has substantially enhanced IPDM strategies. The presence of trees can function as a physical impediment, impeding the locomotion of insects and thereby diminishing the likelihood of infection in both trees and crops. An optimal IPDM program for agroforestry should incorporate physical, cultural, and biological control measures to maintain pest and pathogen populations below threshold levels with minimal or no use of pesticides. It is recommended that pertinent data regarding insect pests, pathogens, and their inherent adversaries be compiled in a database and accessible to professionals engaged in agroforestry.

KEYWORDS

- **integrated pest and disease management**
- **multitrophic agroforestry**
- **invasive alien pathogens and pests**
- **agroecosystems**
- **cocoa swollen shoot virus disease**

REFERENCES

1. Ollinaho, O. I., Kröger, M. (2021). Agroforestry transitions: The good, the bad and the ugly. *Journal of Rural Studies*, *82*, 210–221. doi: 10.1016/j.jrurstud.2021.01.016.
2. Kozlov, M. V., Lanta, V., Zverev, V., Zvereva, E. L. (2015). Global patterns in background losses of woody plant foliage to insects. *Global Ecology and Biogeography*, *24*(10), 1126–1135. doi: 10.1111/geb.12347.
3. Zvereva, E. L., Zverev, V., Kozlov, M. V. (2012). Little strokes fell great oaks: minor but chronic herbivory substantially reduces birch growth. *Oikos*, *121*(12), 2036–2043. doi: 10.1111/j.1600-0706.2012.20688.x.
4. J., M., Brockerhoff, E. G., Wingfield, B. D., Slippers, B. (2015). Planted forest health: The need for a global strategy. *Science (80-.)*, *349*(6250), 832–836. doi: 10.1126/science.aac6674.
5. Angon, P. B., Mondal, S., Jahan, I., Datto, M., Antu, U. B., Ayshi, F. J., Shafiul Islam, M. (2023). Integrated Pest Management (IPM) in agriculture and its role in maintaining ecological balance and biodiversity. *Advances in Agriculture*, *2023*, 1–19. doi: 10.1155/2023/5546373.
6. Pokharel, S. S., Yu, H., Fang, W., Parajulee, M. N., Chen, F. (2023). Intercropping cover crops for a vital ecosystem service: a review of the biocontrol of insect pests in tea agroecosystems. *Plants*, *12*(12), 2361. doi: 10.3390/plants12122361.
7. Sheppard, J. P., Chamberlain, J., Agúndez, D., Bhattacharya, P., Chirwa, P. W., Gontcharov, A., Sagona, W. C. J., Shen, H. L., Tadesse, W., Mutke, S. (2020). Sustainable forest management beyond the timber-oriented status quo: Transitioning to co-production of timber and non-wood forest products—A global perspective. *Current Forestry Reports*, *6*(1), 26–40. doi: 10.1007/s40725-019-00107-1.
8. Panzavolta, T., Bracalini, M., Benigno, A., Moricca, S. (2021). Alien invasive pathogens and pests harming trees, forests, and plantations: Pathways, global consequences and management. *Forests*, *12*(10), 1364. doi: 10.3390/f12101364.
9. Canelles, Q., Aquilué, N., James, P. M. A., Lawler, J., Brotons, L. (2021). Global review on interactions between insect pests and other forest disturbances. *Landscape Ecology*, *36*(4), 945–972. doi: 10.1007/s10980-021-01209-7.
10. Parthasarathy, S., Thiribhuvanamala, G., Muthulakshmi, P., Angappan, K. (2021). *Diseases of Forest Trees and Their Management*. CRC Press. doi: 10.1201/9781003173861.
11. Santini, A., Ghelardini, L., De Pace, C., Desprez-Loustau, M. L., Capretti, P., Chandelier, A., Cech, T., Chira, D., Diamandis, S., Gaitniekis, T., et al. (2013). Biogeographical patterns and determinants of invasion by forest pathogens in Europe. *New Phytologist*, *197*(1), 238–250. doi: 10.1111/j.1469-8137.2012.04364.x.
12. Aukema, J. E., McCullough, D. G., Holle, B. V., Liebhold, A. M., Britton, K., Frankel, S. J. (2010). Historical accumulation of nonindigenous forest pests in the continental United States. *Bioscience*, *60*(11), 886–897. doi: 10.1525/bio.2010.60.11.5.
13. Ghelardini, L., Luchi, N., Pecori, F., Pepori, A. L., Danti, R., Della Rocca, G., Capretti, P., Tsopelas, P., Santini, A. (2017). Ecology of invasive forest pathogens. *Biological Invasions*, *19*(11), 3183–3200. doi: 10.1007/s10530-017-1487-0.

14. Barwell, L. J., Perez-Sierra, A., Henricot, B., Harris, A., Burgess, T. I., Hardy, G., Scott, P., Williams, N., Cooke, D. E. L., Green, S., et al. (2021). Evolutionary trait-based approaches for predicting future global impacts of plant pathogens in the genus Phytophthora. *Journal of Applied Ecology*, *58*(4), 718–730. doi: 10.1111/1365-2664.13820.
15. Grünig, M., Mazzi, D., Calanca, P., Karger, D. N., Pellissier, L. (2020). Crop and forest pest metawebs shift towards increased linkage and suitability overlap under climate change. *Communications Biology*, *3*(1), 1–10. doi: 10.1038/s42003-020-0962-9.
16. Williams, G. M., Ginzel, M. D., Ma, Z., Adams, D. C., Campbell, F., Lovett, G. M., Pildain, M. B., Raffa, K. F., Gandhi, K. J. K., Santini, A., et al. (2023). The global forest health crisis: A public-good social dilemma in need of international collective action. *Annual Review of Phytopathology*, *61*, 377–401. doi: 10.1146/annurev-phyto-021722-024626.
17. Duarte, A., Borralho, N., Cabral, P., Caetano, M. (2022). Recent advances in forest insect pests and diseases monitoring using UAV-based data: A systematic review. *Forests*, *13*(6), 911. doi: 10.3390/f13060911.
18. Guégan, J.-F., de Thoisy, B., Gomez-Gallego, M., Jactel, H. (2023). World forests, global change, and emerging pests and pathogens. *Current Opinion in Environmental Sustainability*, *61*, 101266. doi: 10.1016/j.cosust.2023.101266.
19. Pureswaran, D. S., Roques, A., Battisti, A. (2018). Forest insects and climate change. *Current Forestry Reports*, *4*(2), 35–50. doi: 10.1007/s40725-018-0075-6.
20. Sadof, C. S., McCullough, D. G., Ginzel, M. D. (2023). Urban ash management and emerald ash borer (Coleoptera: Buprestidae): Facts, myths, and an operational synthesis. *Journal Integrated Pest Management*, *14*(1). doi: 10.1093/jipm/pmad012.
21. Pacia, A., Borowik, P., Hsiang, T., Marozau, A., Matić, S., Oszako, T. (2023). Ash dieback in forests and rural areas—History and predictions. *Forests*, *14*(11), 2151. doi: 10.3390/f14112151.
22. Teshome, D. T., Zharare, G. E., Naidoo, S. (2020). The threat of the combined effect of biotic and abiotic stress factors in forestry under a changing climate. *Frontiers in Plant Science*, *11*(November). doi: 10.3389/fpls.2020.601009.
23. Boyd, I. L., Freer-Smith, P. H., Gilligan, C. A., Godfray, H. C. J. (2013). The consequence of tree pests and diseases for ecosystem services. *Science (80)*, *342*(6160) doi: 10.1126/science.1235773.
24. Deguine, J. P., Aubertot, J. N., Bellon, S., Côte, F., Lauri, P. E., Lescourret, F., Ratnadass, A., Scopel, E., Andrieu, N., Bàrberi, P., et al. (2023). Agroecological crop protection for sustainable agriculture. *Advances in Agronomy*, *178*, 1–59. doi: 10.1016/bs.agron.2022.11.002.
25. Keenan, R. J. (2015). Climate change impacts and adaptation in forest management: A review. *Annals of Forest Science*, *72*(2), 145–167. doi: 10.1007/s13595-014-0446-5.
26. Moricca, S., Panzavolta, T. (2021). Recent advances in the monitoring, assessment and management of forest pathogens and pests. *Forests*, *12*(12), 10–15. doi: 10.3390/f12121623.
27. Herms, D. A., McCullough, D. G. (2014). Emerald ash borer invasion of north america: history, biology, ecology, impacts, and management. *Annual Review of Entomology*, *59*, 13–30. doi: 10.1146/annurev-ento-011613-162051.

28. Zhang, T., Ding, G., Zhang, J., Qi, Y. (2022). Contributions of biotic and abiotic factors to the spatial heterogeneity of aboveground biomass in subtropical forests: A case study of Guizhou province. *Sustainable*, *14*(17), 10771. doi: 10.3390/su141710771.
29. Jaimes-Suárez, Y. Y., Carvajal-Rivera, A. S., Galvis-Neira, D. A., Carvalho, F. E. L., Rojas-Molina, J. (2022). Cacao agroforestry systems beyond the stigmas: Biotic and abiotic stress incidence impact. *Frontiers in Plant Science*, *13*(July), 1–14. doi: 10.3389/fpls.2022.921469.
30. Gonçalves, B., Morais, M. C., Pereira, S., Mosquera-Losada, M. R., Santos, M. (2021). Tree–crop ecological and physiological interactions within climate change contexts: A mini-review. *Frontiers in Ecology and Evolution*, *9*. doi: 10.3389/fevo.2021.661978.
31. Foxx, A. J., Fort, F. (2019). Root and shoot competition lead to contrasting competitive outcomes under water stress: A systematic review and meta-analysis. *PLoS One*, *14*(12), 1–17. doi: 10.1371/journal.pone.0220674.
32. Islam, K. K., Rahman, G. M. M., Rafiqul Hoque, A. T. M. (2006). Infestation of insect pests in tree-rice agroforestry system. *Journal of Forestry Research*, *17*(1), 44–46. doi: 10.1007/s11676-006-0010-8.
33. Percival, G. C. (2013). Influence of pure mulches on suppressing phytophthora root rot pathogens. *Journal of Environmental Horticulture*, *31*(4), 221–226. doi: 10.24266/0738-2898.31.4.221.
34. Andres, C., Blaser, W. J., Dzahini-Obiatey, H. K., Ameyaw, G. A., Domfeh, O. K., Awiagah, M. A., Gattinger, A., Schneider, M., Offei, S. K., Six, J. (2018). Agroforestry systems can mitigate the severity of cocoa swollen shoot virus disease. *Agriculture Ecosystems and Environment*, *252*(June 2017), 83–92. doi: 10.1016/j.agee.2017.09.031.
35. Modi, D., Simard, S., Lavkulich, L., Hamelin, R. C., Grayston, S. J. (2021). Stump removal and tree species composition promote a bacterial microbiome that may be beneficial in the suppression of root disease. *FEMS Microbiology Ecology*, *97*(1), 1–13. doi: 10.1093/femsec/fiaa213.
36. Piato, K., Subía, C., Pico, J., Calderón, D., Norgrove, L., Lefort, F. (2021). Organic farming practices and shade trees reduce pest infestations in robusta coffee systems in amazonia. *Life*, *11*(5), 413. doi: 10.3390/life11050413.
37. He, H. M., Liu, L. N., Munir, S., Bashir, N. H., Wang, Y., Yang, J., Li, C. Y. (2019). Crop diversity and pest management in sustainable agriculture. *Journal of Integrative Agriculture*, *18*(9), 1945–1952. doi: 10.1016/S2095-3119(19)62689-4.
38. Andres, C., Comoé, H., Beerli, A., Schneider, M., Rist, S., Jacobi, J. (2016). Cocoa in monoculture and dynamic agroforestry. In: *Sustainable Agriculture Reviews* (pp. 121–153). Springer International Publishing. doi: 10.1007/978-3-319-26777-7_3.
39. Yamoah, C. F., Burleigh, J. R. (1990). Alley cropping sesbania sesban (L) merill with food crops in the highland region of rwanda. *Systems*, *10*(L), 169–181.
40. Schroth, G., Krauss, U., Gasparotto, L., Duarte Aguilar, J. A., Vohland, K. (2000). Pests and diseases in agroforestry systems of the humid tropics. *Agroforestry Systems*, *50*(3), 199–241. doi: 10.1023/A:1006468103914.

41. Biffi, S., Chapman, P. J., Grayson, R. P., Ziv, G. (2022). Soil carbon sequestration potential of planting hedgerows in agricultural landscapes. *Journal of Environmental Management*, *307*(June 2021), 114484. doi: 10.1016/j.jenvman.2022.114484.
42. Ratnadass, A., Fernandes, P., Avelino, J., Habib, R. (2012). *Plant Species Diversity for Sustainable Management of Crop Pests and Diseases in Agroecosystems: A Review*, *32*, 273–303. doi: 10.1007/s13593-011-0022-4.
43. Tooker, J. F., Frank, S. D. (2012). Genotypically diverse cultivar mixtures for insect pest management and increased crop yields. *Journal of Applied Ecology*, *49*(5), 974–985. doi: 10.1111/j.1365-2664.2012.02173.x.
44. Kouassi, J. L., Kouassi, A., Bene, Y., Konan, D., Tondoh, E. J., Kouame, C. (2021). Exploring barriers to agroforestry adoption by cocoa farmers in south-western côte d'Ivoire. *Sustainable*, *13*(23), 1–16. doi: 10.3390/su132313075.
45. Narayana, M. R. (2013). Do coffee varieties and shade trees matter for management of leaf rust disease in India evidence for household farmers. *Forests Trees Livelihoods*, *22*(4), 275–288. doi: 10.1080/14728028.2013.846563.
46. Liebig, T., Ribeyre, F., Läderach, P., Poehling, H. M., van Asten, P., Avelino, J. (2019). Interactive effects of altitude, microclimate and shading system on coffee leaf rust. *Journal of Plant Interactions*, *14*(1), 407–415. doi: 10.1080/17429145.2019.1643934.
47. Baeten, L., Verheyen, K., Wirth, C., Bruelheide, H., Bussotti, F., Finér, L., Jaroszewicz, B., Selvi, F., Valladares, F., Allan, E., et al. (2013). A novel comparative research platform designed to determine the functional significance of tree species diversity in European forests. *Perspectives in Plant Ecology, Evolution and Systematics*, *15*(5), 281–291. doi: 10.1016/j.ppees.2013.07.002.
48. Castagneyrol, B., Jactel, H., Vacher, C., Brockerhoff, E. G., Koricheva, J. (2014). Effects of plant phylogenetic diversity on herbivory depend on herbivore specialization. *Journal of Applied Ecology*, *51*(1), 134–141. doi: 10.1111/1365-2664.12175.
49. Favor, K., Gold, M., Halsey, S., Hall, M., Vallone, R. (2023). Agroforestry for enhanced arthropod pest management in vineyards. *Agroforestry Systems*, *98*, 213–227. No. 0123456789. doi: 10.1007/s10457-023-00900-9.
50. Castellano-Hinojosa, A., Strauss, S. L. (2020). Impact of cover crops on the soil microbiome of tree crops. *Microorganisms*, *8*(3), 328. doi: 10.3390/microorganisms8030328.
51. Baligar, V. C., Elson, M. K., He, Z., Li, Y., De, A., Almeida, A. A. F., Ahnert, D. (2020). Light intensity effects on the growth, physiological and nutritional parameters of tropical perennial legume cover crops. *Agronomy*, *10*(10), 1–14. doi: 10.3390/agronomy10101515.
52. Bessi, R., Massayuki Inomoto, M. (2022). Host status of cover crops for the management of pratylenchus jaehni on coffee. *Journal of Nematology*, *54*(1), 1–5. doi: 10.2478/jofnem-2022-0052.
53. Hoshino, A. T., Bortolotto, O. C., Hata, F. T., Ventura, M. U., Júnior, A. de O. M. (2018). Effect of pigeon pea intercropping or shading with leucaena plants on the occurrence of the coffee leaf miner and on its predation by wasps in organic coffee plantings. *Ciência Rural*, *48*(3), 1–7. doi: 10.1590/0103-8478CR20160863.
54. Moreira, C. C., Celestino, D., Guerra Sobrinho, T., Cardoso, I. M., Elliot, S. L. (2019). Agroforestry coffee soils increase the insect-suppressive potential offered

by entomopathogenic fungi over full-sun soils: A case proposing a "Bait Survival Technique." *Ecology & Evolution*, *9*(18), 10777–10787. doi: 10.1002/ece3.5598.
55. Rezende, M. Q., Venzon, M., dos Santos, P. S., Cardoso, I. M., Janssen, A. (2021). Extrafloral nectary-bearing leguminous trees enhance pest control and increase fruit weight in associated coffee plants. *Agriculture Ecosystems and Environment*, *319*(May), 107538. doi: 10.1016/j.agee.2021.107538.
56. Venzon, M. (2021). Agro-ecological management of coffee pests in Brazil. *Frontiers in Sustainable Food Systems*, *5*(September), 1–13. doi: 10.3389/fsufs.2021.721117.
57. Sinu, P. A., Viswan, G., Fahira, P. P., Rajesh, T. P., Manoj, K., Hariraveendra, M., Jose, T. (2021). Shade tree diversity may not drive prey-predator interaction in coffee agroforests of the western Ghats biodiversity hotspot, India. *Biological Control*, *160*(May), 104674. doi: 10.1016/j.biocontrol.2021.104674.
58. Doonan, J. M., Broberg, M., Denman, S., McDonald, J. E. (2020). Host-microbiota-insect interactions drive emergent virulence in a complex tree disease. *Proceedings of the Royal Society B: Biological Sciences*, *287*(1933). doi: 10.1098/rspb.2020.0956.
59. Xiong, W., Zhao, Q., Xue, C., Xun, W., Zhao, J., Wu, H., Li, R., Shen, Q. (2016). Comparison of fungal community in black pepper-vanilla and vanilla monoculture systems associated with vanilla Fusarium wilt disease. *Frontiers in Microbiology*, *7*(FEB), 1–8. doi: 10.3389/fmicb.2016.00117.
60. Vaughan, M., Adamson, N., Macfarland, K. (2017). Using agroforestry practices to reduce pesticide risks to pollinators & other agriculturally beneficial insects. *Agroforestry Notes*, *35*(9), 1–8.
61. Philpott, S. M., Soong, O., Lowenstein, J. H., Pulido, A. L., Lopez, T., Flynn, D. F. B., Declerck, F., Lopez, D. T. (2014). Richness and ecosystem services: Functional bird predation in tropical on arthropods agroecosystems. *Ecological Applications*, *19*(7), 1858–1867.
62. Daghela Bisseleua, H. B., Fotio, D., Yede, Missoup, A. D., Vidal, S. (2013). Shade tree diversity, cocoa pest damage, yield compensating inputs and farmers' net returns in West Africa. *PLoS One*, *8*(3), e56115. doi: 10.1371/journal.pone.0056115.
63. Mariño, Y. A., Pérez, M. E., Gallardo, F., Trifilio, M., Cruz, M., Bayman, P. (2016). Sun vs. Shade affects infestation, total population and sex ratio of the coffee berry borer (Hypothenemus Hampei) in Puerto Rico. *Agriculture Ecosystems and Environment*, *222*, 258–266. doi: 10.1016/j.agee.2015.12.031.
64. Piri, T., Vainio, E. J., Nuorteva, H., Hantula, J. (2021). High seedling mortality of scots pine caused by heterobasidion annosum s.s. *Forests*, *12*(9), 1289. doi: 10.3390/f12091289.
65. Youssef, K., Dambrauskaite, M., Witzell, J., Rönnberg, J. (2023). Survey study reveals high prevalence of heterobasidion root rot infection in scots pine (Pinus Sylvestris) stands established on seemingly low-risk sites. *Forests*, *14*(5), 1–10. doi: 10.3390/f14051018.
66. Guyot, V., Castagneyro, B., Vialatte, A., Deconchat, M., Selvi, F., Bussotti, F., Jactel, H. (2015). Tree diversity limits the impact of an invasive forest pest. *PLoS One*, *10*(9), 1–16. doi: 10.1371/journal.pone.0136469.
67. Jactel, H., Moreira, X., Castagneyrol, B. (2021). Tree diversity and forest resistance to insect pests: patterns, mechanisms, and prospects. *Annual Review of Entomology*, *66*, 277–296. doi: 10.1146/annurev-ento-041720-075234.

68. Dhar, A., Parrott, L., Hawkins, C. D. B. (2016). Aftermath of mountain pine beetle outbreak in British Columbia: Stand dynamics, management response and ecosystem resilience. *Forests*, *7*(8), 1–19. doi: 10.3390/f7080171.
69. Lehmann, P., Ammunét, T., Barton, M., Battisti, A., Eigenbrode, S. D., Jepsen, J. U., Kalinkat, G., Neuvonen, S., Niemelä, P., Terblanche, J. S., et al. (2020). Complex responses of global insect pests to climate warming. *Frontiers in Ecology and the Environment*, *18*(3), 141–150. doi: 10.1002/fee.2160.
70. Peshin, R., Jayaratne, K. S. U., Sharma, R. (2014). *IPM Extension: A Global Overview*. doi: 10.1016/B978-0-12-398529-3.00026-9.
71. Pumariño, L., Sileshi, G. W., Gripenberg, S., Kaartinen, R., Barrios, E., Muchane, M. N., Midega, C., Jonsson, M. (2015). Effects of agroforestry on pest, disease and weed control: A meta-analysis. *Basic and Applied Ecology*, *16*(7), 573–582. doi: 10.1016/j.baae.2015.08.006.

CHAPTER 10

Symbiotic Relationships: Exploring the Role of Insects in Agroforestry Systems

DEEPAK KUMAR MAHANTA,[1*] J. KOMAL,[2] TANMAYA KUMAR BHOI,[3] and IPSITA SAMAL[4]

[1]*Forest Entomology Discipline, Forest Protection Division, Indian Council of Forestry Research and Education-Forest Research Institute (ICFRE-FRI), Dehradun, Uttarakhand, India*

[2]*Basic Seed Multiplication and Training Centre, Central Silk Board, Kharsawan, Jharkhand, India*

[3]*Forest Protection Division, Indian Council of Forestry Research and Education-Arid Forest Research Institute (ICFRE-AFRI), Jodhpur, Rajasthan, India*

[4]*ICAR-National Research Centre on Litchi, Mushahari, Ramna, Muzaffarpur, Bihar, India*

*Corresponding author

ABSTRACT

Agroforestry systems are dynamic ecosystems that are influenced by complex symbiotic connections between plants and insects, which, in turn, affect ecological processes and the sustainability of agriculture. This chapter explores how important insects are to agroforestry landscapes, outlining important discoveries and their implications for sustainable land management. In agroforestry systems, insects have a variety of roles to play, including pollination, controlling pests, and assessing the health of the ecosystem. Their symbiotic connections with plants increase

Multitrophic Interactions for Sustainable Agroforestry: Harnessing the Power of Beneficial Insects and Ecological Dynamics. Hanuman Singh Jatav, Vishnu D. Rajput, Ipsita Samal, and Tanmaya Kumar Bhoi (Eds).

DOI: 10.1201/9781779641250-10

agricultural output, improve ecosystem resilience, and promote biodiversity. In agroforestry contexts, integrated pest management tactics, conservation biological control techniques, and pollinator diversity promotion programs are crucial pathways for fostering these symbiotic relationships. Threats to insect populations and the sustainability of agroforestry ecosystems include habitat loss, pesticide use, climatic fluctuation, and the spread of invasive species. However, there are plenty of chances for cooperative research, multidisciplinary discussions, and policy interventions, providing avenues to lessen these difficulties and promote long-term agroecological practices. Our knowledge of symbiotic interactions and their implications for sustainable land management is being advanced via interdisciplinary research initiatives, stakeholder involvement, and policy frameworks adapted to agroforestry situations. Agroforestry practitioners may exploit the potential of symbiotic interactions to improve ecosystem resilience, promote biodiversity conservation, and ensure the long-term sustainability of agricultural landscapes by embracing ecological principles and promoting collaboration across sectors.

10.1 INTRODUCTION

Agroforestry is a dynamic, sustainable method of land management that combines cattle, crops, and/or trees on one plot of land [1]. This technique entails the intentional blending of trees or shrubs with livestock, agricultural products, and/or other plants in order to have positive social, economic, and environmental effects [2]. In contrast to traditional monoculture farming, which frequently results in degraded soil, decreased biodiversity, and heightened vulnerability to pests and diseases, agroforestry systems utilize ecological principles to boost output while protecting and replenishing natural resources [3]. Agroforestry systems come in a broad range of designs and compositions, from straightforward multistrata systems to intricate intercropping combinations [4]. Examples include silvopasture, which mixes trees with animal grazing, alley cropping, which plants trees in rows and grows crops in the alleyways between them, and windbreaks, which employ trees to reduce wind erosion and shield crops [5]. Agroforestry is a method that has its roots in traditional agriculture systems all over the world. In order to manage their lands and resources sustainably, indigenous peoples and traditional farmers have long understood the need of integrating trees with crops and animals [6]. Agroforestry has reportedly been used for thousands of years, as evidenced by the use of agroforestry techniques by ancient civilizations such as the Incas, Aztecs, and Mayans to sustain their cultures. Agroforestry came to light in more recent times in reaction to the environmental and socioeconomic problems that industrialized agriculture was causing. Food insecurity, deforestation, and soil degradation were the main causes for the emergence of the contemporary agroforestry movement in the middle of the 20th century [7]. Agroforestry has gained traction as a viable substitute for traditional agriculture, with researchers and practitioners highlighting its potential to increase production, preserve biodiversity, and improve incomes.

Agroforestry provides a number of social, economic, and environmental advantages, which makes it an essential tool for advancing sustainable agriculture (Figure 10.1) [8]. Compared with monoculture agriculture, agroforestry systems sustain higher levels of biodiversity by offering habitat to a wide range of plant and animal species [9]. Beneficial insects, pollinators, and soil microbes are just a few of the many creatures that may be found in the ecological niches that are created by the varied structure and composition of agroforestry landscapes [10]. In agroforestry systems, trees contribute to better soil structure, less erosion, and improved nutrient cycling [11]. Deep roots of trees help to improve water penetration and retention by rupturing compacted soil layers [12].

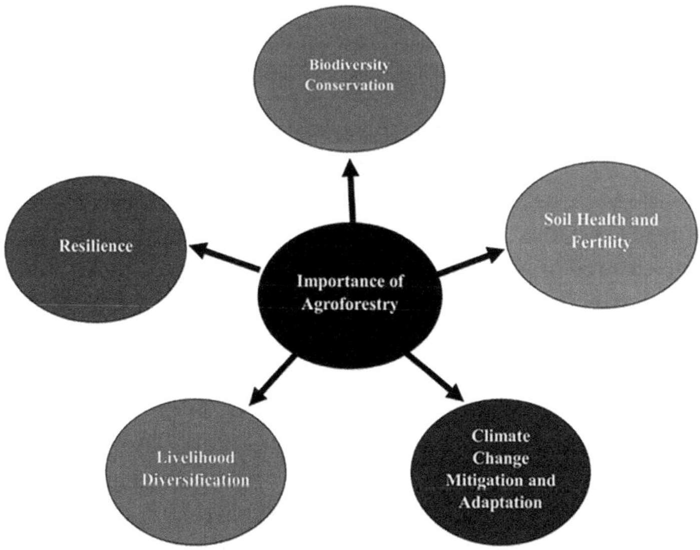

FIGURE 10.1 Representing importance of agroforestry in sustainable agriculture.

The organic matter from tree litter and root exudates contributes to soil fertility, enriching the soil with essential nutrients and enhancing its capacity to support plant growth [13]. By balancing greenhouse gas emissions, trees in agroforestry systems absorb carbon dioxide from the atmosphere and so slow down global warming. Agroforestry techniques such as windbreaks and alley cropping can also lessen the effects of extreme weather events such as floods and droughts by acting as natural barriers and regulating the microclimate [14]. Agroforestry helps farmers become less reliant on a particular crop or animal by providing them with a variety of revenue streams. Farmers can increase their income by selling lumber, fruits, nuts, and other tree products by combining trees with agricultural output [15]. Additionally, by distributing risks and acting as a buffer against environmental shocks and market volatility, agroforestry increases the resilience of farming systems [16]. In conclusion, agroforestry is a comprehensive method of farming that strikes a balance between social justice, economic feasibility, and ecological sustainability. Through the use of tree, crop, and animal synergies, agroforestry has a wealth of opportunities to tackle the complex issues confronting contemporary agriculture and foster a more robust and sustainable food system.

10.2 THE DIVERSITY OF INSECT COMMUNITIES IN AGROFORESTRY

In agroforestry systems, insects make up a large portion of the biodiversity because they participate in a variety of ecological activities such as pollination, pest control, and nutrient cycling [17–20]. Comprehending the taxonomy, distribution, and abundance of insect species (Figure 10.2) in agroforestry landscapes is crucial for both conservation and sustainable agricultural management [21–24]. In an agroforestry ecosystem, bees, butterflies, moths, beetles, and flies are some of the most frequent pollinators [25–28]. By spreading pollen from blossom to bloom, these insects help flowering plants reproduce and produce fruit and seeds [29]. Among the primary natural enemies of agricultural pests include flies, parasitic wasps, predatory beetles, lacewings, and ladybirds.

By feeding on or parasitizing pest species, these beneficial insects aid in the control of pest populations, hence lowering the demand for chemical pesticides [30–34]. Insects that consume plant tissues, such caterpillars,

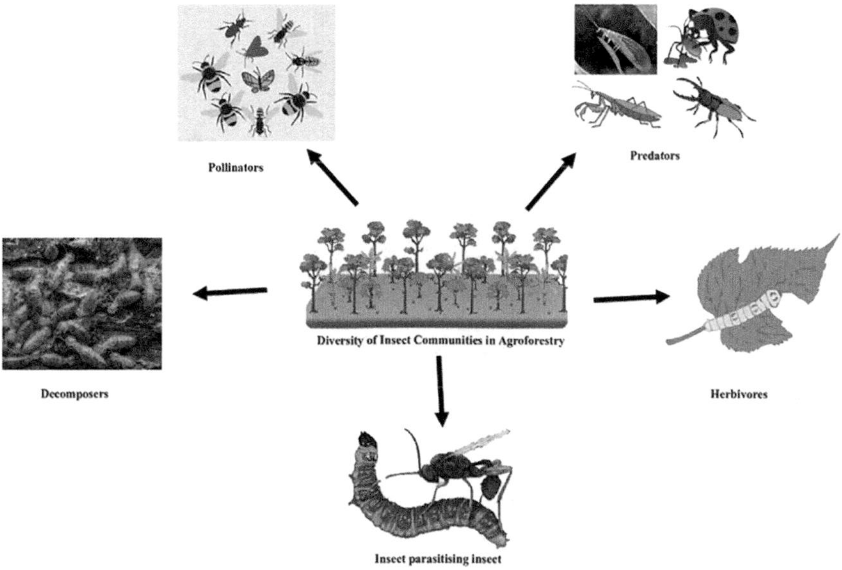

FIGURE 10.2 Representing diversity of insect communities in agroforestry.

aphids, grasshoppers, and beetles, can harm crops if their populations are not properly controlled [35–37]. By decomposing organic materials and recycling nutrients back into the ecosystem, soil-dwelling insects such as dung beetles, termites, and earthworms play important roles in nutrient cycling and soil health [38]. Agroforestry habitats' and insect populations' diversity and abundance are influenced by several variables (Figure 10.3).

A range of microhabitats, such as trees, shrubs, crops, and ground cover species, are usually available in agroforestry systems [39]. A vast variety of insect species with differing environmental requirements are supported by the many niches and resources this diversified terrain offers [40]. Agroforestry systems with a range of plant species increase floral diversity by giving insects host plants, pollen, and nectar throughout the growing season [41]. An ecosystem's resilience can be enhanced by a high plant variety, which can draw and harbor a wide range of insect pollinators and natural enemies [42]. Insects can find shelter and food in agroforestry systems with intricate vertical and horizontal plant structures, which increases the diversity and quantity of species [43]. In addition to acting as breeding grounds and feeding areas, trees and bushes offer protection from predators and bad weather. By modifying the shape of habitats,

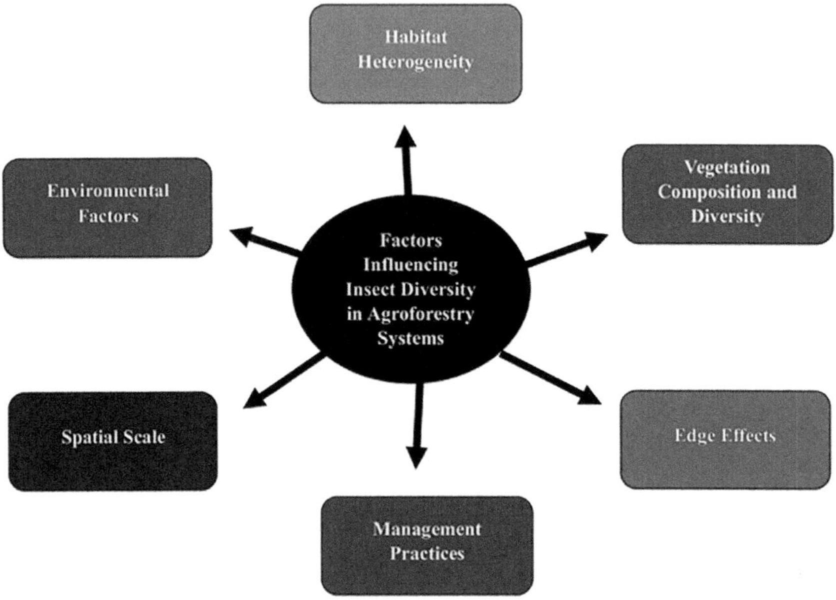

FIGURE 10.3 Depicting factors influencing insect diversity in agroforestry systems.

the availability of resources, and the dynamics of pest populations, land management techniques including mulching, pruning, intercropping, and adding organic soil amendments can have an impact on insect populations [44]. Biological, mechanical, and cultural control techniques are all part of integrated pest management (IPM) strategies, which can assist preserve insect variety while controlling pest numbers [45].

Around the world, a large number of researches have looked into the patterns of insect diversity in various agroforestry systems. Examples and case studies demonstrate the dynamic relationships that exist between agroforestry methods and insects, as well as the possible advantages of using insect-friendly management strategies. Compared with monoculture systems, research has demonstrated that alley cropping systems—which have alternating rows of crops and trees—can increase insect variety and abundance [46]. By giving beneficial insects a place to hide and find food, hedgerows and windbreaks enhance their ability to combat pests and facilitate pollination [47]. In silvopastoral systems, where grazing by livestock and trees coexists, insects are crucial to the decomposition, cycling of nutrients, and productivity of pastures [48].

Sustainable rangeland management can be aided by managed grazing techniques that emulate natural disturbance regimes [49]. These techniques can support a variety of insect species and encourage habitat heterogeneity. Coffee farms that use agroforestry, which includes shade trees and coffee bushes, are home to a wide variety of natural enemies and insect pollinators [50]. Shade-grown coffee systems support biodiversity conservation by offering habitat to native pollinators, butterflies, and birds. This improves pollination and pest control services.

10.3 INSECTS AS POLLINATORS IN AGROFORESTRY

As pollinators, insects are essential to agroforestry systems because they help blooming plants reproduce and produce fruits, nuts, seeds, and other agricultural goods [51]. Sustainable agricultural management and biodiversity conservation initiatives depend on an understanding of the role pollinators play in agroforestry, the effects of agroforestry practices on pollinator populations, and techniques for increasing pollinator variety and abundance [52]. In agroforestry systems, pollinators—which include bees, butterflies, moths, beetles, flies, and even some vertebrates such as birds and bats—are crucial for the sexual reproduction of several plant species [53]. By moving pollen from the male (anther) to the female (stigma) reproductive organs of flowers, these insects facilitate fertilization and the development of seeds and fruits [54]. Pollinators are essential to the reproductive success of crop plants and the trees that are included into the agroforestry environment [55]. Insect pollinators are necessary for cross-pollination and fruit set in fruit and nut trees, including apple, pear, almond, and walnut. In agroforestry systems, insect-mediated pollination helps several understory crops, herbs, and blooming shrubs provide seeds, berries, and other commercial goods [56]. In addition to enhancing agricultural yield, pollinators are essential for preserving the resilience and stability of ecosystems. Pollination services promote the total biodiversity of agroforestry ecosystems and improve plant adaptation to environmental changes by fostering genetic variation within plant populations [57].

Depending on how the system is specifically designed and managed, agroforestry techniques can affect pollinator populations in both beneficial and bad ways (Figure 10.4). Agroforestry systems with a variety of vegetation layers, such as ground cover plants, trees, and shrubs, offer a wealth of floral resources and places for pollinator nests [58]. Throughout the growing season, these environments can sustain a wide variety of insect species, which helps to keep pollinator populations healthy [59]. On pollinator health and abundance, the use of chemical pesticides in agroforestry can have negative consequences [50]. Pollinating insects may become acutely or chronically harmful due to pesticide contamination of floral supplies and nesting locations resulting from drift from nearby fields or direct application inside agroforestry plots [60]. Agroforestry systems that include a range of tree species and blooming crops can help to increase pollinator diversity by providing a steady stream of supplies of pollen and nectar [61]. Conversely, monoculture methods may reduce the amount of floral resources available and raise the possibility of nutritional stress for pollinators [62]. Pollinator diversity and abundance in agroforestry habitats can be influenced by the land use and composition of the surrounding environment [63]. While linked habitats and floral supplies can sustain bigger and more diversified pollinator populations, fragmented landscapes with little natural habitat may limit pollinator migration and foraging options [64].

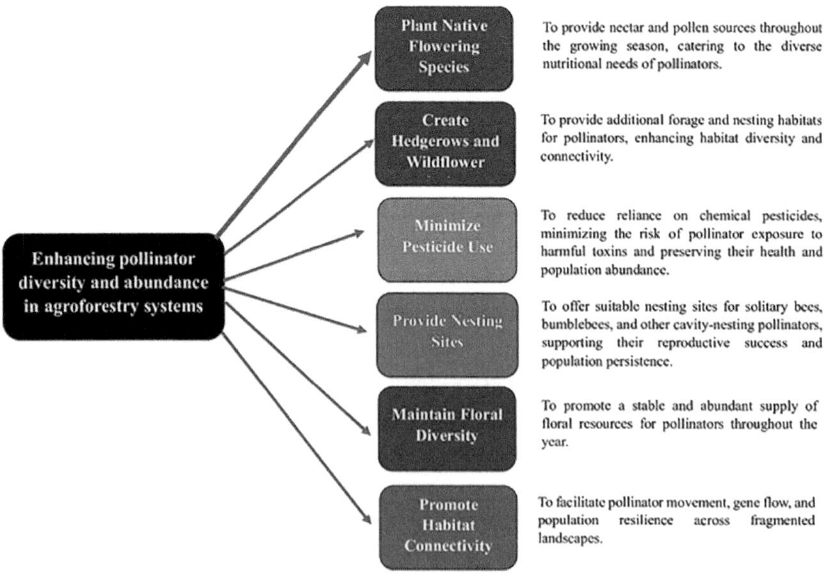

FIGURE 10.4 Strategies for enhancing pollinator diversity and abundance in agroforestry.

There are several ways that land managers and farmers may put into practice to improve pollinator variety and abundance in agroforestry systems. Agroforestry designs that use native blooming plants can draw a variety of native pollinators that are suited to the particular ecosystems in the area [65]. Protecting pollinator numbers and efficiently controlling pest demands can be achieved by implementing IPM techniques that reduce dependency on chemical pesticides [66]. For solitary bees and other cavity-nesting pollinators, installing bee hotels, nesting boxes, and other artificial structures can increase the number of available nesting sites [67]. Pollinators can have a steady supply of food to satisfy their nutritional demands if a variety of plant species continue to bloom throughout the growing season. Hedgerows, riparian buffers, and wildflower meadows are examples of nearby natural habitats that may be preserved and restored in order to improve landscape connectivity and encourage pollinator migration across habitat fragmentation.

10.4 INSECTS AS PEST REGULATORS IN AGROFORESTRY

In agroforestry systems, insects act as natural pest controllers, regulating pest insect populations and lowering the demand for chemical pesticides [68]. Sustainable pest management and agroecological resilience depend on an understanding of the principles of natural pest control, the dynamics of pest insect populations in agroforestry habitats, and the numerous tactics used by beneficial insects to regulate pest populations [69]. Natural pest management is the technique of suppressing pest populations without the use of synthetic pesticides by means of biological processes and ecological interactions [70]. By feeding on or parasitizing pest species, a wide range of beneficial insects—including diseases, parasitoids, and predators—help agroforestry systems naturally manage pest populations [71]. Beneficial insects include ladybirds, lacewings, predatory beetles, spiders, and predatory mites; these insects prey on nuisance insects including aphids, caterpillars, and leafhoppers [72]. By devouring individual prey, these predators help control pest populations by lowering their numbers and averting epidemics. Specialized insects called parasitoid wasps and flies deposit their eggs within or on the bodies of nuisance insects [73]. After developing within the host insect, the parasitoid larvae kill the insect and lower pest numbers. Biological control strategies frequently employ parasitoids as potent natural adversaries of agricultural pests. In agroforestry systems, microbial pathogens such as bacteria, fungi, and viruses have the ability to infect and kill pest insects, hence aiding in natural pest management [74]. *Beauveria bassiana* and *Metarhizium* spp. are two examples of entomopathogenic fungi that are frequently employed as biological control agents against a range of insect pests [75].

Agroforestry systems' pest insect populations are impacted by a range of biotic and abiotic variables such as the availability of host plants, weather patterns, natural enemies, and agricultural management techniques [76]. Seasonal and regional variations in pest population dynamics can be attributed to a variety of variables, including crop phenology, predator–prey interactions, and landscape structure [77]. Seasonal variations in pest bug populations are frequently seen as a result of shifting resource availability and environmental circumstances [78]. For instance, insect outbreaks could happen when crops are growing quickly or when the weather is nice; yet at times when pest numbers are at their highest, natural enemies might put more pressure on them [79]. In agroforestry settings, the spatial layout of trees, crops, and other landscape features can affect the dynamics of pest insects. The migration and distribution of pest and beneficial insects throughout the landscape can be influenced by edge effects, habitat fragmentation, and connectivity between various habitat types [80]. Crop rotation and a variety of plant species can upset the life cycles of pests and lower the likelihood of pest outbreaks in agroforestry systems. The implementation of mixed cropping and intercropping techniques can augment organic pest management by offering substitute hosts and diminishing the concentration of vulnerable plants [81]. In agroforestry systems, beneficial insects use a range of strategies to control pest populations. Pest insects are immediately consumed by predatory insects, which lowers their population and lessens agricultural damage. When parasitoid insects lay their eggs within or on top of pest insects, the host eventually perishes and pest numbers are reduced [82]. Certain helpful insects lower the fitness and reproductive success of pest species by competing with them for resources such as food, shelter, and mates. By changing pest behavior, decreasing pest fecundity, or preventing pest growth through chemical signals or physical interactions, beneficial insects can also indirectly affect pest populations. Agroforestry practitioners can lessen dependency on chemical pesticides and promote sustainable pest management strategies that uphold agroecological balance and resilience by increasing the diversity and abundance of beneficial insects and fostering ecological processes that support natural pest control.

10.5 THE INFLUENCE OF AGROFORESTRY PRACTICES ON INSECT HABITATS

Agroforestry methods shape the variety, structure, distribution, and microclimatic conditions of agroecosystems, which, in turn, has a significant impact on insect habitats [83]. It is crucial to comprehend how various agroforestry designs and management techniques affect insect habitats in order to enhance sustainable agricultural output, improve ecosystem services, and encourage biodiversity conservation [84]. Agroforestry systems offer a mosaic of habitats for various insect species, encompassing a broad spectrum of structural and functional diversity [85]: Multiple vegetation layers, such as trees, shrubs, herbaceous plants, and ground cover vegetation, are commonly seen in agroforestry systems [86]. Because of this vertical stratification, many insect species with differing habitat preferences and resource needs are supported in a range of microhabitats and ecological niches. Insect habitat quality and resilience are improved by the structural complexity of agroforestry landscapes, which are distinguished by a variety of plant species, canopy structures, and spatial heterogeneity [87]. Insects can find refuge, food sources, and nesting places in the vertical structure and canopy cover that trees and shrubs provide [88]. Meanwhile, understory and ground cover plants help control microclimate and supply resources. Agroforestry systems, which are impacted by crop phenology, seasonal variations, and management techniques, display dynamic temporal patterns in habitat structure and resource availability [89]. Throughout the growing season, the availability and diversity of floral resources—such as nectar, pollen, and extrafloral nectar—vary, which has an impact on insect activity, population dynamics, and community composition [90]. In agroforestry environments, edges are crucial in determining the distribution and quantity of insects. Ecotones, or the areas where multiple habitat types transition into one another, are hotspots for biological variety and interaction [91]. Foraging, dispersion, and migratory pathways for insects are provided by ecotones in agroforestry systems such as riparian zones, field margins, and forest edges [92]. This promotes habitat mobility and improves landscape connectedness. In agroforestry habitats, edge effects—which are defined by clear biological gradients from the core to the edge of habitat patches—have an impact on the distribution and behavior of insects [93]. In comparison to inner habitats, edge habitats may be subject to greater amounts of sun radiation, wind exposure, and temperature changes, which can have an impact on the microclimate, plant phenology, and insect activity patterns. Certain insect species have behaviors and ecological preferences connected with edges; they specialize in edge habitats and take use of resources that are exclusive to edge settings [94]. Edge-preferring butterflies, beetles, and grasshoppers are examples of edge-adapted species that may be more abundant and diverse close to agroecotones, field borders, and woodland margins.

In agroforestry systems, the microclimate has a significant influence on the biological processes and insect habitats. In agroforestry systems, canopy cover, tree density, and vegetation structure affect microclimatic parameters such as temperature, humidity, and airflow [95]. Agroecosystems' microclimatic gradients provide a variety of temperature and moisture regimes that have an impact on the physiology, behavior, and population dynamics of insects. Alley cropping, windbreaks, and riparian buffers are examples of agroforestry techniques that can alter soil moisture dynamics and water availability, resulting in microhabitat gradients that affect the distribution, reproduction, and survival of insects [96]. Water sources, such as ponds, wetlands, and streams, offer vital supplies to riparian and aquatic insect species, enhancing insect diversity overall and supporting ecosystem function [97]. In agroforestry environments, microclimatic variables influence the phenology, productivity, and quality of floral and foliar resources accessible to insects [98]. Insect foraging behavior, reproductive success, and population dynamics are shaped by temperature-driven phenological changes, precipitation patterns, and seasonal variations in resource supply [99]. These factors also impact community structure and ecosystem functioning. Farmers, land managers, and conservationists can develop and put into practice agroecological strategies that support the conservation of beneficial insect populations and biodiversity, foster habitat diversity, and improve ecological resilience by taking into account the impact of agroforestry practices on insect habitats and ecosystem processes.

10.6 MANAGEMENT STRATEGIES FOR ENHANCING SYMBIOTIC RELATIONSHIPS

Agroforestry systems depend heavily on the symbiotic interactions that exist between plants, insects, and other species for proper operation and productivity. Rigorous approaches to management that foster symbiotic relationships have the potential to augment ecosystem benefits, enhance pest control, and boost agricultural resilience [100]. Important strategies include conservation biological control techniques, IPM, and programs to support pollinator variety and health. A comprehensive strategy to pest control, IPM places an emphasis on using a variety of tactics to limit pest damage and lessen the need for chemical pesticides [101]. The following techniques are ways that IPM concepts can be used in agroforestry systems. By keeping a close eye on beneficial insect abundance, crop health, and pest populations, farmers may detect possible pest outbreaks early on and mitigate pest pressure [102]. Crop rotation, intercropping, mulching, and other cultural techniques can upset the life cycles of pests, lower their numbers, and improve the fertility and health of the soil [103]. Agroforestry techniques that increase habitat complexity and crop diversity can reduce the growth and reproduction of pests. Utilizing natural enemies such as parasitoids, microbial diseases, and predatory insects to manage pest populations is known as biological control [104]. In agroforestry settings, maintaining and improving beneficial insect habitats—such as hedgerows, cover crops, and flowering plants—can sustain natural enemies and improve biological pest management. Without using chemical pesticides, mechanical techniques such as hand-picking, trapping, and exclusion barriers can be utilized to physically remove or discourage pests from crops. Structures used in agroforestry, such as shade trees and windbreaks, can shield crops from insect pressure and bad weather.

The goal of conservation biological management in agroecosystems is to lower pest numbers and preserve ecological balance by protecting and boosting natural enemy populations. Natural enemies, such as predatory insects, parasitoids, and insectivorous birds, are provided with food, shelter, and nesting grounds when various habitats, such as hedgerows, insectary strips, and wildflower meadows, are established and maintained [105]. Reducing or completely eliminating the use of broad-spectrum insecticides lowers the possibility of damaging beneficial insects and upsetting populations of their natural enemies [106]. Effective pest population management may be achieved by using less toxic, targeted insecticides and applying pesticides selectively, both of which assist protect natural enemy communities [107]. Increased biological management of pests can be achieved by supplementing current populations with natural enemies such as predatory mites and parasitoid wasps [108]. In order to optimize their effect on the targeted pest species, biological control agents might be introduced during key periods that correspond with pest outbreaks. Planting floral hedgerows and buffer strips along the edges of fields gives beneficial insects, such as pollinators and natural enemies, a place to live and food supplies [109]. For improving conservation biological control, native plant species that sustain a variety of insect populations and offer constant bloom throughout the growing season are very beneficial.

Insects that pollinate crops and wild plants, such as bees, butterflies, moths, and other insects, are essential to food production, biodiversity preservation, and ecosystem resilience. The following tactics can be used in agroforestry landscapes to support pollinator health and variety. Adding blooming trees, shrubs, and cover crops to an agroforestry design gives pollinators vital supplies of nectar and pollen all during the growing season [110]. Promoting pollinator health and variety is especially beneficial for native plant species that sustain different insect groups and are acclimated to the local environment [111]. Reduced pesticide usage and the establishment of pesticide-free zones surrounding pollinator habitats are two ways to shield pollinators from dangerous chemical exposure [112]. The danger of pesticide-related mortality and sublethal impacts on pollinator populations is decreased by applying pesticides selectively and using bee-safe insecticides. For solitary bees, bumblebees, and other cavity-nesting pollinators, providing nesting and overwintering habitats—such as bee hotels, nesting boxes, and bare ground patches—is essential [113]. Diverse nesting substrates and cover alternatives improve the appropriateness of the habitat for a range of pollinator species. In agroforestry systems, maintaining a wide variety of crop species and flowering plants promotes floral diversity and pollinator-accessible feed. Adding more floral resources and bolstering pollinator populations can be achieved by planting mixed-species hedgerows, intercropping blooming herbs, and letting spontaneous vegetation flourish. Farmers, land managers, and conservationists can support sustainable agricultural production, foster ecological resilience, and improve symbiotic relationships between plants and insects while preserving biodiversity and ecosystem services by incorporating these management strategies into agroforestry practices.

10.7 CHALLENGES AND FUTURE DIRECTIONS

In order to advance sustainable farming practices, it is critical to resolve obstacles and map out future routes as agroforestry systems continue to acquire attention for their ecological and agricultural benefits. Insect population risks in agroforestry, research and partnership possibilities, policy consequences, and recommendations to assist agroforestry adoption and conservation efforts are all covered in this section. In agroforestry environments, the change of natural habitats to agricultural land and urban development is a major danger to insect populations [114]. Insect diversity and abundance fall as a result of habitat loss and fragmentation that also impair ecological linkages and make less appropriate habitat available [115]. Insect populations, particularly beneficial insects and pollinators, are seriously threatened by the careless application of chemical pesticides in agriculture [116]. Insect health, behavior, and reproduction can all be negatively impacted by pesticide pollution of soil, water, and plants, which can cause population decreases and ecological disturbance [117]. The phenomenon of climate change presents a range of difficulties for insect populations and their relationships with plants and other creatures, such as increased temperatures, modified precipitation patterns, and extreme weather events. In agroforestry systems, alterations in insect phenology, distribution, and behavior have the potential to upset ecological connections and modify ecosystem dynamics. The health of agroecosystems and native insect populations may suffer from the introduction of illnesses, pests, and invasive insect species. Invasive species have the potential to spread diseases, outcompete local species for resources, and harm the forestry and agriculture sectors financially.

The resilience and overall health of insect populations in agroforestry systems need to be evaluated through long-term ecological monitoring and research projects. In order to increase insect variety and abundance, collaborative research projects may be used to discover important ecological indicators, monitor population changes, and assess how well management strategies are working [118]. Understanding the intricate relationships between insects, plants, and human activity in agroforestry environments requires integrating several disciplines, including ecology, entomology, agronomy, and socioeconomics. Interdisciplinary research methodologies have the potential to produce comprehensive understandings of how ecosystems work, guide management choices, and advance sustainable farming methods. Promoting innovation and best practices in agroforestry requires facilitating information exchange and capacity building among researchers, practitioners, and policymakers. The implementation of sustainable agroecological principles may be improved through knowledge sharing, the development of technical proficiency, and the creation of collaborative networks, workshops, and training courses. Promoting participatory methods to agroforestry development and conservation requires involving stakeholders such as farmers, landowners, community groups, and government agencies. Building consensus, promoting social cohesion, and enabling local people to take charge of agroforestry efforts are all possible through cooperative decision-making procedures, stakeholder workshops, and participatory research projects.

To encourage farmers and landowners to embrace agroforestry methods, governments and politicians might introduce incentive programs such as grants, tax breaks, and subsidies specifically designed for agroforestry. Agroecosystem long-term sustainability may be supported, technical support can be given, and financial incentives can help defray the costs of initial investments. Maintaining insect populations and the health of ecosystems in agroforestry landscapes requires tightening laws and enforcement procedures to reduce the use of pesticides, save natural areas, and encourage biodiversity protection. Policies that place a high priority on ecosystem services—such as pest management, pollination, and soil preservation—can encourage sustainable land management techniques. By funding outreach, education, and extension initiatives, communities and farmers may become more aware of the environmental and financial advantages of agroforestry and be encouraged to embrace it. To promote the implementation and knowledge transfer of agroforestry, farmer field schools, demonstration plots, and extension services may offer hands-on training, technical support, and peer-to-peer learning opportunities. Funding and resources should be made available by governments, academic institutions, and international organizations to encourage new developments in the fields of agroforestry science and practice. Research on insect ecology, biodiversity preservation, climate change adaptation, and socioecological resilience in agroforestry systems ought to be prioritized for funding. Through the implementation of evidence-based policies and recommendations, addressing these obstacles, and embracing collaborative opportunities, stakeholders can jointly advance sustainable agroforestry development, preserve insect biodiversity, and improve ecosystem resilience for posterity.

10.8 CONCLUSION

Through the promotion of symbiotic interactions between plants, insects, and other species, agroforestry approaches provide viable solutions to solve ecological, economic, and social concerns in agricultural environments [52, 119–122]. The main conclusions and their implications for sustainability and agroforestry methods are outlined in this section. In agroforestry systems, insects are essential as pollinators, pest controllers, and health indicators for the ecosystem. Plant–insect symbiotic partnerships support agricultural production, biodiversity preservation, and ecosystem resilience. Improving symbiotic connections in agroforestry requires the application of IPM techniques, conservation biological control technologies, and pollinator health and variety promotion programs. In order to maintain beneficial insect populations and ecosystem services, these techniques place a strong emphasis on cultural practices, habitat variety, and lower pesticide usage. Threats to insect populations include invasive species, pesticide usage, habitat loss, and climate change. These factors highlight the importance of multidisciplinary research, stakeholder involvement, and collaborative research in agroforestry. Opportunities for study and cooperation, the consequences for policy, and suggestions emphasize how crucial it is to reward sustainable land management techniques and foster information exchange and capacity building.

Through the preservation of natural habitats, the promotion of biodiversity, and the support of ecological processes, agroforestry methods improve ecosystem resilience. Agroforestry systems may improve soil health and water quality while reducing the effects of pests, illnesses, and climate change by promoting symbiotic connections between plants and insects. By lowering reliance on chemical inputs, providing ecosystem services, and improving pest control, agroforestry techniques support sustainable agriculture. Agroforestry systems provide a variety of revenue sources, improve food security, and support social justice and cultural values in rural communities by combining trees, crops, and cattle. Agroforestry landscapes serve as a haven and home for a variety of plant and animal species such as birds, animals, insects, and microbes. Agroforestry techniques ensure the protection of biodiversity, genetic variety, and spatial connectivity by maintaining and reestablishing natural ecosystems. This supports the resilience and long-term health of ecosystems. To sum up, agroforestry's emphasis on symbiotic partnerships is critical to attaining sustainable land use and building robust, regenerative agricultural systems. Agroforestry practitioners have the potential to improve biodiversity, ecosystem services, and human well-being for future generations by embracing ecological principles, implementing creative management techniques, and encouraging stakeholder engagement.

KEYWORDS

- **agroforestry**
- **insects**
- **ecosystem**
- **pollination**
- **sustainability**

REFERENCES

1. Pantera, A., Mosquera-Losada, M. R., Herzog, F., Den Herder, M. (2021). Agroforestry and the environment. *Agroforestry Systems*, *95*(5), 767–774. doi: 10.1007/s10457-021-00640-8
2. Nair, P. R., Kumar, B. M., V. D. Nair, N., P. R., Kumar, B. M., V. D. Nair. (2021). Definition and concepts of agroforestry. In: *An Introduction to Agroforestry: Four Decades of Scientific Developments* (pp. 21–28).
3. Gupta, S. R., Dagar, J. C., Teketay, D. (2020). Agroforestry for rehabilitation of degraded landscapes: Achieving livelihood and environmental security. *Agroforestry for Degraded Landscapes*, *1*, 23–68.
4. Rafflegeau, S., Gosme, M., Barkaoui, K., Garcia, L., Allinne, C., Deheuvels, O., Grimaldi, J., Jagoret, P., Lauri, P., Merot, A., Metay, A., Reyes, F., Saj, S., Curry, G. N., Justes, E. (2023). The ESSU concept for designing, modeling and auditing ecosystem service provision in intercropping and agroforestry systems. A review. *Agronomy for Sustainable Development*, *43*(4). doi: 10.1007/s13593-023-00894-9
5. Dagar, J. C., Tewari, J. C. (2016). Agroforestry research developments: Anecdotal to modern science. In: *Agroforestry Research Developments* (pp. 1–45). Nova Press.
6. Altieri, M. A., Nicholls, C. I. (2017). The adaptation and mitigation potential of traditional agriculture in a changing climate. *Climatic Change*, *140*(1), 33–45. doi: 10.1007/s10584-013-0909-y
7. Tomar, J. M. S., Ahmed, A., Bhat, J. A., Kaushal, R., Shukla, G., Kumar, R. (2021). Potential and opportunities of agroforestry practices in combating land degradation. *Agroforestry-small Landholder's Tool for Climate Change Resiliency and Mitigation*. doi: 10.5772/intechopen.97843
8. Wilson, M. H., Lovell, S. T. (2016). Agroforestry—The next step in sustainable and resilient agriculture. *Sustainability*, *8*(6), 574. doi: 10.3390/su8060574
9. Udawatta, R. P., Rankoth, L. M., Jose, S. (2019). Agroforestry and biodiversity. *Sustainability*, *11*(10), 2879. doi: 10.3390/su11102879
10. Bentrup, G., Hopwood, J., Adamson, N. L., Vaughan, M. (2019). Temperate agroforestry systems and insect pollinators: A review. *Forests*, *10*(11), 981. doi: 10.3390/f10110981

11. Sileshi, G. W., Mafongoya, P. L., Nath, A. J. (2020). Agroforestry systems for improving nutrient recycling and soil fertility on degraded lands. *Agroforestry for Degraded Landscapes, 1*, 225–253.
12. Priori, S., Pellegrini, S., Vignozzi, N., Costantini, E. A. C. (2020). Soil physical-hydrological degradation in the root-zone of tree crops: Problems and solutions. *Agronomy, 11*(1), 68. doi: 10.3390/agronomy11010068
13. Lei, X., Shen, Y., Zhao, J., Huang, J., Wang, H., Yu, Y., Xiao, C. (2023). Root exudates mediate the processes of soil organic carbon input and efflux. *Plants, 12*(3), 630. doi: 10.3390/plants12030630
14. Biswas, P., Mondal, S., Maji, S., Mondal, A., Bandopadhyay, P. (2023). Microclimate modification in field crops: A way toward climate-resilience. In: *Climate-resilient Agriculture, 1: Crop Responses and Agroecological Perspectives* (pp. 647–666).
15. Kuyah, S., Sileshi, G. W., Luedeling, E., Akinnifesi, F. K., Whitney, C. W., Bayala, J., Kuntashula, E., Dimobe, K., Mafongoya, P. L. (2020). Potential of agroforestry to enhance livelihood security in Africa. *Agroforestry for Degraded Landscapes, 1*, 135–167.
16. Gitz, V., Meybeck, A. (2012). Risks, vulnerabilities and resilience in a context of climate change. *Building Resilience for Adaptation to Climate Change in the Agriculture Sector, 23*, 19.
17. Barrios, E., Valencia, V., Jonsson, M., Brauman, A., Hairiah, K., Mortimer, P. E., Okubo, S. (2018). Contribution of trees to the conservation of biodiversity and ecosystem services in agricultural landscapes. *International Journal of Biodiversity Science, Ecosystem Services and Management, 14*(1), 1–16. doi: 10.1080/21513732.2017.1399167
18. Bhoi, T. K., Samal, I., Majhi, P. K., Komal, J., Mahanta, D. K., Pradhan, A. K., Saini, V., Nikhil Raj, M., Ahmad, M. A., Behera, P. P., Ashwini, M. (2022). Insight into aphid mediated potato virus Y transmission: A molecular to bioinformatics prospective. *Frontiers in Microbiology, 13*, 1001454. doi: 10.3389/fmicb.2022.1001454
19. Mahanta, D. K., Jangra, S., Priti, A., Ghosh, A., Sharma, P. K., Iquebal, M. A., Jaiswal, S., Baranwal, V. K., Kalia, V. K., Chander, S. (2022). Groundnut bud necrosis virus modulates the expression of innate immune, endocytosis, and cuticle development-associated genes to circulate and propagate in its vector, Thrips palmi. *Frontiers in Microbiology, 13*, 773238. doi: 10.3389/fmicb.2022.773238
20. Bhatnagar, S., Mahanta, D. K., Vyas, V., Samal, I., Komal, J., Bhoi, T. K. (2023). Storage pest management with nanopesticides incorporating silicon nanoparticles: A novel approach for sustainable crop preservation and food security. *Silicon, 16*, 471–483.
21. Staton, T., Walters, R. J., Smith, J., Breeze, T. D., Girling, R. D. (2021). Evaluating a trait-based approach to compare natural enemy and pest communities in agroforestry vs. arable systems. *Ecological Applications, 31*(4), e02294. doi: 10.1002/eap.2294
22. Mahanta, D. K., Komal, J., Samal, I., Bhoi, T. K., Dubey, V. K., Pradhan, K., Nekkanti, A., Gouda, M. N. R., Saini, V., Negi, N., Bhateja, S., Jat, H. K., Jeengar, D. (2023a). Nutritional aspects and dietary benefits of “Silkworms”: Current scenario and future outlook. *Frontiers in Nutrition, 10*, 1121508. doi: 10.3389/fnut.2023.1121508

23. Samal, I., Bhoi, T. K., Vyas, V., Majhi, P. K., Mahanta, D. K., Komal, J., Singh, S., Kumar, P. V. D., Acharya, L. K. (2023a). Resistance to fungicides in entomopathogenic fungi: Underlying mechanisms, consequences, and opportunities for progress. *Tropical Plant Pathology*, *49*, 5–17. doi: 10.1007/s40858-023-00585-6
24. Mahanta, D. K., Bhoi, T. K., Komal, J., Samal, I., Nikhil, R. M., Paschapur, A. U., Singh, G., Kumar, P. V. D., Desai, H. R., Ahmad, M. A., Singh, P. P., Majhi, P. K., Mukherjee, U., Singh, P., Saini, V., Shahanaz, Srinivasa, N., Yele, Y. (2023b). Insect-pathogen crosstalk and the cellular-molecular mechanisms of insect immunity: Uncovering the underlying signaling pathways and immune regulatory function of non-coding RNAs. *Frontiers in Immunology*, *14*, 1169152. doi: 10.3389/fimmu.2023.1169152
25. Bentrup, G., Hopwood, J., Adamson, N. L., Powers, R., Vaughan, M. (2021). The role of temperate agroforestry practices in supporting pollinators. In: *Agroforestry and Ecosystem Services* (pp. 275–304).
26. Samal, I., Bhoi, T. K., Majhi, P. K., Murmu, S., Pradhan, A. K., Kumar, D., Saini, V., Paschapur, A. U., Raj, M. N., Ankur, Manik, S., Behera, P. P., Mahanta, D. K., Komal, J., Alam, P., Balawi, T. A. (2022). Combatting insects mediated biotic stress through plant associated endophytic entomopathogenic fungi in horticultural crops. *Frontiers in Plant Science*, *13*, 1098673. doi: 10.3389/fpls.2022.1098673
27. Komal, J., Desai, H. R., Samal, I., Mastinu, A., Patel, R. D., Kumar, P. V. D., Majhi, P. K., Mahanta, D. K., Bhoi, T. K. (2023). Unveiling the genetic symphony: Harnessing CRISPR-Cas genome editing for effective insect pest management. *Plants*, *12*(23), 3961. doi: 10.3390/plants12233961
28. Samal, I., Bhoi, T. K., Mahanta, D. K., Komal, J. (2023c). Establishing the role of silicon (Si) in plant resistance to insects: A bibliometric approach. *Silicon*, *16*, 2119–2128.
29. Dar, S. A., Hassan, G. I., Padder, B. A., Wani, A. R., Parey, S. H. (2017). Pollination and evolution of plant and insect interaction. *Journal of Pharmacognosy and Phytochemistry*, *6*(3), 304–311.
30. Ballal, C. R., Sreedevi, K., Salini, S., Gupta, A., Amala, U., Varshney, R. (2022). Biodiversity of agriculturally important insects: Status, issues, and challenges. In: *Biodiversity in India: Status, Issues and Challenges* (pp. 243–283). Springer Nature Singapore.
31. Majhi, P. K., Samal, I., Bhoi, T. K., Pattnaik, P., Pradhan, C., Gupta, A., Mahanta, D. K., Senapati, S. K. (2023). Climate-smart agriculture: An integrated approach to address climate change and food security. In: *Climate Change and Insect Biodiversity* (pp. 221–245). CRC Press.
32. Mahanta, D. K., Samal, I., Komal, J., Bhoi, T. K., Majhi, P. K., Ahmad, M. A. (2023c). Understanding anthropogenic climate change, its consequences on insect pests, and techniques in forecasting and monitoring pest dynamics: A contemporary scenario. In: *Climate Change and Insect Biodiversity* (pp. 44–67). CRC Press.
33. Bhoi, T. K., Samal, I., Mahanta, D. K., Komal, J., Jinger, D., Sahoo, M. R., Achary, G. C., Nayak, P., Sunani, S. K., Saini, V., Raghuraman, M., Singh, S. (2023). Understanding how silicon fertilization impacts chemical ecology and multitrophic interactions among plants, insects and beneficial arthropods. *Silicon*, *15*(6), 2529–2549. doi: 10.1007/s12633-022-02220-6

34. Sujatha, G. S., Singh, M. K., Mahanta, D. K., Bala, A. (2022). Pest management through ecological manipulation. *Vigyan Varta, 3*(5), 40–43.
35. Bhoi, T. K., Samal, I., Mahanta, D. K., Komal, J., Majhi, P. K., Ankur. (2024). Exploring melatonin's potential as an alternative strategy for protecting plants from biotic stresses. In: *Melatonin in Plants: A Pleiotropic Molecule for Abiotic Stresses and Pathogen Infection* (pp. 223–242). Springer Nature: Singapore.
36. Samal, I., Bhoi, T. K., Pradhan, A. K., Mahanta, D. K. (2023d). Plantix app: A success story of artificial intelligence in plant protection. *Van Sangyan, 24*.
37. Mahanta, D. K., Komal, J. (2022). Sericulture and its salient features. *Vigyan Varta, 3*, 105–110.
38. Sofo, A., Mininni, A. N., Ricciuti, P. (2020). Soil macrofauna: A key factor for increasing soil fertility and promoting sustainable soil use in fruit orchard agrosystems. *Agronomy, 10*(4), 456. doi: 10.3390/agronomy10040456
39. Pardon, P., Reheul, D., Mertens, J., Reubens, B., De Frenne, P., De Smedt, P., Proesmans, W., Van Vooren, L., Verheyen, K. (2019). Gradients in abundance and diversity of ground dwelling arthropods as a function of distance to tree rows in temperate arable agroforestry systems. *Agriculture, Ecosystems and Environment, 270–271*, 114–128. doi: 10.1016/j.agee.2018.10.017
40. Potapov, A. M., Beaulieu, F., Birkhofer, K., Bluhm, S. L., Degtyarev, M. I., Devetter, M., Goncharov, A. A., Gongalsky, K. B., Klarner, B., Korobushkin, D. I., Liebke, D. F., Maraun, M., Mc Donnell, R. J., Pollierer, M. M., Schaefer, I., Shrubovych, J., Semenyuk, I. I., Sendra, A., Tuma, J., Tůmová, M., Vassilieva, A. B., Chen, T.-W., Geisen, S., Schmidt, O., Tiunov, A. V., Scheu, S. (2022). Feeding habits and multifunctional classification of soil-associated consumers from protists to vertebrates. *Biological Reviews of the Cambridge Philosophical Society, 97*(3), 1057–1117. doi: 10.1111/brv.12832
41. Merle, I., Hipólito, J., Requier, F. (2022). Towards integrated pest and pollinator management in tropical crops. *Current Opinion in Insect Science, 50*, 100866. doi: 10.1016/j.cois.2021.12.006
42. Wyckhuys, K. A., Zhang, W., Colmenarez, Y. C., Simelton, E., Sander, B. O., Lu, Y. (2022). Tritrophic defenses as a central pivot of low-emission, pest-suppressive farming systems. *Current Opinion in Environmental Sustainability, 58*, 101208. doi: 10.1016/j.cosust.2022.101208
43. Parisi, F., Lombardi, F., Marziliano Pasquale, A., Russo, D., De Cristofaro, A., Marchetti, M., Tognetti, R. (2020). Diversity of saproxylic beetle communities in chestnut agroforestry systems. *iForest—Biogeosciences and Forestry, 13*(1), 456–465. doi: 10.3832/ifor3478-013
44. Las Casas, G., Ciaccia, C., Iovino, V., Ferlito, F., Torrisi, B., Lodolini, E. M., Giuffrida, A., Catania, R., Nicolosi, E., Bella, S. (2022). Effects of different inter-row soil management and intra-row living mulch on spontaneous flora, beneficial insects, and growth of young olive trees in Southern Italy. *Plants, 11*(4), 545. doi: 10.3390/plants11040545
45. Kaur, T., Kaur, M. (2020). Integrated pest management: A paradigm for modern age. *Pests, Weeds and Diseases in Agricultural Crop and Animal Husbandry Production.*
46. Tsonkova, P., Böhm, C., Quinkenstein, A., Freese, D. (2012). Ecological benefits provided by alley cropping systems for production of woody biomass in the

temperate region: A review. *Agroforestry Systems*, *85*(1), 133–152. doi: 10.1007/s10457-012-9494-8
47. Rourke, K. (2017). *Attracting Beneficial Insects to Your Farm: A Comparison of Habitat Modification Strategies*. Master's Projects and Capstones, 551.
48. Sarabia, L., Solorio, F. J., Ramírez, L., Ayala, A., Aguilar, C., Ku, J., Almeida, C., Cassador, R., Alves, B. J., Boddey, R. M. (2020). Improving the nitrogen cycling in livestock systems through silvopastoral systems. *Nutrient Dynamics for Sustainable Crop Production* (pp. 189–213).
49. Scheiter, S., Pfeiffer, M., Behn, K., Ayisi, K., Siebert, F., Linstädter, A. (2024). Managing southern African rangeland systems in the face of drought: A synthesis of observation, experimentation and modeling for policy and decision support. In: *Sustainability of Southern African Ecosystems Under Global Change Sciences for Management and Policy Interventions* (pp. 439–470). Springer International Publishing. doi: 10.1007/978-3-031-10948-5_16
50. Campera, M., Balestri, M., Manson, S., Hedger, K., Ahmad, N., Adinda, E., Nijman, V., Budiadi, B., Imron, M. A., Nekaris, K. A. I. (2021). Shade trees and agrochemical use affect butterfly assemblages in coffee home gardens. *Agriculture, Ecosystems and Environment*, *319*, 107547. doi: 10.1016/j.agee.2021.107547
51. Lee-Mäder, E., Vaughan, M., Goldenetz-Dollar, J. (2020). Agroforestry and cover cropping for pollinators. In: *Towards Sustainable Crop Pollination Services: Measures at Field, Farm and Landscape Scales* (pp. 105–126).
52. Sollen-Norrlin, M., Ghaley, B. B., Rintoul, N. L. J. (2020). Agroforestry benefits and challenges for adoption in Europe and beyond. *Sustainability*, *12*(17), 7001. doi: 10.3390/su12177001
53. Saha, H., Chatterjee, S., Paul, A. (2023). Role of pollinators in plant reproduction and food security: A concise review. *Research Journal of Agricultural Sciences*, *14*(1), 72–79.
54. Bhatla, S. C., Lal, M. A. (2023). Pollination, fertilization, and seed development. In: *Plant Physiology, Development and Metabolism* (pp. 583–605). Springer Nature Singapore.
55. Varah, A., Jones, H., Smith, J., Potts, S. G. (2020). Temperate agroforestry systems provide greater pollination service than monoculture. *Agriculture, Ecosystems and Environment*, *301*, 107031. doi: 10.1016/j.agee.2020.107031
56. Proctor, E., Nol, E., Burke, D., Crins, W. J. (2012). Responses of insect pollinators and understory plants to silviculture in northern hardwood forests. *Biodiversity and Conservation*, *21*(7), 1703–1740. doi: 10.1007/s10531-012-0272-8
57. Boreux, V., Kushalappa, C. G., Vaast, P., Ghazoul, J. (2013). Interactive effects among ecosystem services and management practices on crop production: Pollination in coffee agroforestry systems. *Proceedings of the National Academy of Sciences of the United States of America*, *110*(21), 8387–8392. doi: 10.1073/pnas.1210590110
58. Jose, S., Udawatta, R. P. (2021). Agroforestry for ecosystem services: An introduction. *Agroforestry and Ecosystem Services* (pp. 1–17).
59. Shivanna, K. R., Tandon, R., Koul, M. (2020). 'Global Pollinator Crisis' and its impact on crop productivity and sustenance of plant diversity. In: *Reproductive Ecology of Flowering Plants: Patterns and Processes* (pp. 395–413).

60. Wakgari, M., Yigezu, G. (2021). Honeybee keeping constraints and future prospects. *Cogent Food and Agriculture*, *7*(1), 1872192. doi: 10.1080/23311932.2021.1872192
61. Kay, S., Kühn, E., Albrecht, M., Sutter, L., Szerencsits, E., Herzog, F. (2020). Agroforestry can enhance foraging and nesting resources for pollinators with focus on solitary bees at the landscape scale. *Agroforestry Systems*, *94*(2), 379–387. doi: 10.1007/s10457-019-00400-9
62. Jones, J., Rader, R. (2022). Pollinator nutrition and its role in merging the dual objectives of pollinator health and optimal crop production. *Philosophical Transactions of the Royal Society of London. Series B, Biological Sciences*, *377*(1853), 20210170. doi: 10.1098/rstb.2021.0170
63. Millard, J., Outhwaite, C. L., Kinnersley, R., Freeman, R., Gregory, R. D., Adedoja, O., Gavini, S., Kioko, E., Kuhlmann, M., Ollerton, J., Ren, Z. X., Newbold, T. (2021). Global effects of land-use intensity on local pollinator biodiversity. *Nature Communications*, *12*(1), 2902. doi: 10.1038/s41467-021-23228-3
64. Fuccillo Battle, K., de Rivera, C. E., Cruzan, M. B. (2021). The role of functional diversity and facilitation in small-scale pollinator habitat. *Ecological Applications*, *31*(6), e02355. doi: 10.1002/eap.2355
65. Peters, V. E., Carroll, C. R., Cooper, R. J., Greenberg, R., Solis, M. (2013). The contribution of plant species with a steady-state flowering phenology to native bee conservation and bee pollination services. *Insect Conservation and Diversity*, *6*(1), 45–56. doi: 10.1111/j.1752-4598.2012.00189.x
66. Egan, P. A., Dicks, L. V., Hokkanen, H. M. T., Stenberg, J. A. (2020). Delivering integrated pest and pollinator management (IPPM). *Trends in Plant Science*, *25*(6), 577–589. doi: 10.1016/j.tplants.2020.01.006
67. Rahimi, E., Barghjelveh, S., Dong, P. (2021). How effective are artificial nests in attracting bees? A review. *Journal of Ecology and Environment*, *45*(1), 1–11.
68. Favor, K., Gold, M., Halsey, S., Hall, M., Vallone, R. (2023). Agroforestry for enhanced arthropod pest management in Vineyards. *Agroforestry Systems*, *98*, 213–227.
69. Brévault, T., Clouvel, P. (2019). Pest management: Reconciling farming practices and natural regulations. *Crop Protection*, *115*, 1–6. doi: 10.1016/j.cropro.2018.09.003
70. Dubey, N. K., Ravindra Shukla, R. S., Ashok Kumar, A. K., Priyanka Singh, P. S., Bhanu Prakash, B. P. (2011). Global scenario on the application of natural products in integrated pest management programmes. In: *Natural Products in Plant Pest Management* (pp. 1–20). CABI Publishing.
71. Ambele, C. F., Bisseleua, H. D., Djuideu, C. T., Akutse, K. S. (2023). Managing insect services and disservices in cocoa agroforestry systems. *Agroforestry Systems*, *97*, 965–984.
72. Alford, D. V. (2019). *Beneficial Insects*. CRC Press.
73. Fei, M., Gols, R., Harvey, J. A. (2023). The biology and ecology of parasitoid wasps of predatory arthropods. *Annual Review of Entomology*, *68*, 109–128. doi: 10.1146/annurev-ento-120120-111607
74. Balla, A., Silini, A., Cherif-Silini, H., Chenari Bouket, A., Moser, W. K., Nowakowska, J. A., Oszako, T., Benia, F., Belbahri, L. (2021). The threat of pests and pathogens and the potential for biological control in forest ecosystems. *Forests*, *12*(11), 1579. doi: 10.3390/f12111579

75. Meyling, N. V., Eilenberg, J. (2007). Ecology of the entomopathogenic fungi Beauveria bassiana and Metarhizium anisopliae in temperate agroecosystems: Potential for conservation biological control. *Biological Control*, *43*(2), 145–155. doi: 10.1016/j.biocontrol.2007.07.007
76. Harterreiten-Souza, É. S., Togni, P. H. B., Pires, C. S. S., Sujii, E. R. (2014). The role of integrating agroforestry and vegetable planting in structuring communities of herbivorous insects and their natural enemies in the neotropical region. *Agroforestry Systems*, *88*(2), 205–219. doi: 10.1007/s10457-013-9666-1
77. Welch, K. D., Harwood, J. D. (2014). Temporal dynamics of natural enemy–pest interactions in a changing environment. *Biological Control*, *75*, 18–27. doi: 10.1016/j.biocontrol.2014.01.004
78. Johnson, C. A., Coutinho, R. M., Berlin, E., Dolphin, K. E., Heyer, J., Kim, B., Leung, A., Sabellon, J. L., Amarasekare, P. (2016). Effects of temperature and resource variation on insect population dynamics: The bordered plant bug as a case study. *Functional Ecology*, *30*(7), 1122–1131. doi: 10.1111/1365-2435.12583
79. Lamichhane, J. R., Barzman, M., Booij, K., Boonekamp, P., Desneux, N., Huber, L., Kudsk, P., Langrell, S. R. H., Ratnadass, A., Ricci, P., Sarah, J., Messéan, A. (2015). Robust cropping systems to tackle pests under climate change. A review. *Agronomy for Sustainable Development*, *35*(2), 443–459. doi: 10.1007/s13593-014-0275-9
80. Tscharntke, T., Steffan-Dewenter, I., Kruess, A., Thies, C. (2002). Characteristics of insect populations on habitat fragments: A mini review. *Ecological Research*, *17*(2), 229–239. doi: 10.1046/j.1440-1703.2002.00482.x
81. Snyder, L. D., Gómez, M. I., Power, A. G. (2020). Crop varietal mixtures as a strategy to support insect pest control, yield, economic, and nutritional services. *Frontiers in Sustainable Food Systems*, *4*, 60. doi: 10.3389/fsufs.2020.00060
82. Tougeron, K., Brodeur, J., Le Lann, C., van Baaren, J. (2020). How climate change affects the seasonal ecology of insect parasitoids. *Ecological Entomology*, *45*(2), 167–181. doi: 10.1111/een.12792
83. Burgess, A. J., Correa Cano, M. E. C., Parkes, B. (2022). The deployment of intercropping and agroforestry as adaptation to climate change. *Crop and Environment*, *1*(2), 145–160. doi: 10.1016/j.crope.2022.05.001
84. Duru, M., Therond, O., Martin, G., Martin-Clouaire, R., Magne, M. A., Justes, E., Journet, E., Aubertot, J., Savary, S., Bergez, J., Sarthou, J. P. (2015). How to implement biodiversity-based agriculture to enhance ecosystem services: A review. *Agronomy for Sustainable Development*, *35*(4), 1259–1281. doi: 10.1007/s13593-015-0306-1
85. Montes-Londoño, I. (2017). Tropical dry forests in multi-functional landscapes: Agroforestry systems for conservation and livelihoods. In: *Integrating Landscapes: Agroforestry for Biodiversity Conservation and Food Sovereignty* (pp. 47–78). doi: 10.1007/978-3-319-69371-2_3
86. Bagella, S., Caria, M. C., Seddaiu, G., Leites, L., Roggero, P. P. (2020). Patchy landscapes support more plant diversity and ecosystem services than wood grasslands in Mediterranean silvopastoral agroforestry systems. *Agricultural Systems*, *185*, 102945. doi: 10.1016/j.agsy.2020.102945
87. Boesing, A. L., Nichols, E., Metzger, J. P. (2017). Effects of landscape structure on avian-mediated insect pest control services: A review. *Landscape Ecology*, *32*(5), 931–944. doi: 10.1007/s10980-017-0503-1

88. Ulyshen, M. D. (2011). Arthropod vertical stratification in temperate deciduous forests: Implications for conservation-oriented management. *Forest Ecology and Management, 261*(9), 1479–1489. doi: 10.1016/j.foreco.2011.01.033
89. Marsden, C., Martin-Chave, A., Cortet, J., Hedde, M., Capowiez, Y. (2020). How agroforestry systems influence soil fauna and their functions—A review. *Plant and Soil, 453*(1–2), 29–44. doi: 10.1007/s11104-019-04322-4
90. Tew, N. E., Baldock, K. C. R., Vaughan, I. P., Bird, S., Memmott, J. (2022). Turnover in floral composition explains species diversity and temporal stability in the nectar supply of urban residential gardens. *Journal of Applied Ecology, 59*(3), 801–811. doi: 10.1111/1365-2664.14094
91. Samways, M. J. (2007). Implementing ecological networks for conserving insect and other biodiversity. *Insect Conservation Biology*, 127–143. doi: 10.1079/9781845932541.0127
92. Schroth, G., da Fonseca, G. A., Harvey, C. A., Gascon, C., Vasconcelos, H. L., Izac, A. M. N. (Eds.). (2013). *Agroforestry and Biodiversity Conservation in Tropical Landscapes*. Island Press.
93. González, E., Salvo, A., Valladares, G. (2017). Natural vegetation cover in the landscape and edge effects: Differential responses of insect orders in a fragmented forest. *Insect Science, 24*(5), 891–901. doi: 10.1111/1744-7917.12377
94. Murphy, S. M., Battocletti, A. H., Tinghitella, R. M., Wimp, G. M., Ries, L. (2016). Complex community and evolutionary responses to habitat fragmentation and habitat edges: What can we learn from insect science? *Current Opinion in Insect Science, 14*, 61–65. doi: 10.1016/j.cois.2016.01.007
95. Kanzler, M., Böhm, C., Mirck, J., Schmitt, D., Veste, M. (2019). Microclimate effects on evaporation and winter wheat (Triticum aestivum L.) yield within a temperate agroforestry system. *Agroforestry Systems, 93*(5), 1821–1841. doi: 10.1007/s10457-018-0289-4
96. Jose, S., Holzmueller, E. J. (2021). Tree–crop interactions in temperate agroforestry. In: *North American Agroforestry* (pp. 67–87). doi: 10.1002/9780891183785.ch4
97. Schindler, D. E., Smits, A. P. (2017). Subsidies of aquatic resources in terrestrial ecosystems. *Ecosystems, 20*(1), 78–93. doi: 10.1007/s10021-016-0050-7
98. Gordo, O., Sanz, J. J., Lobo, J. M. (2010). Determining the environmental factors underlying the spatial variability of insect appearance phenology for the honey bee, Apis mellifera, and the small white, Pieris rapae. *Journal of Insect Science, 10*(1), 34. doi: 10.1673/031.010.3401
99. Ma, C. S., Ma, G., Pincebourde, S. (2021). Survive a warming climate: Insect responses to extreme high temperatures. *Annual Review of Entomology, 66*, 163–184. doi: 10.1146/annurev-ento-041520-074454
100. Trivedi, P., Mattupalli, C., Eversole, K., Leach, J. E. (2021). Enabling sustainable agriculture through understanding and enhancement of microbiomes. *New Phytologist, 230*(6), 2129–2147. doi: 10.1111/nph.17319
101. Ragunathan, V., Divakar, B. J. (2020). Integrated pest management strategies. In: *Molecular Biology of the Biological Control of Pests and Diseases of Plants* (pp. 173–196). CRC Press.
102. Deguine, J. P., Aubertot, J. N., Flor, R. J., Lescourret, F., Wyckhuys, K. A. G., Ratnadass, A. (2021). Integrated pest management: Good intentions, hard realities.

A review. *Agronomy for Sustainable Development*, *41*(3), 38. doi: 10.1007/s13593-021-00689-w

103. Ratnadass, A., Fernandes, P., Avelino, J., Habib, R. (2012). Plant species diversity for sustainable management of crop pests and diseases in agroecosystems: A review. *Agronomy for Sustainable Development*, *32*(1), 273–303. doi: 10.1007/s13593-011-0022-4
104. Jeffers, A., Chong, J. H. (2021). Biological control strategies in integrated pest management (IPM) programs. *Clemson University Cooperative, Land-grant Press by Clemson Extension, LGP, 1111*, 1–9.
105. Rodríguez-Gasol, N., Alins, G., Veronesi, E. R., Wratten, S. (2020). The ecology of predatory hoverflies as ecosystem-service providers in agricultural systems. *Biological Control*, *151*, 104405. doi: 10.1016/j.biocontrol.2020.104405
106. Khan, B. A., Nadeem, M. A., Nawaz, H., Amin, M. M., Abbasi, G. H., Nadeem, M., Ali, M., Ameen, M., Javaid, M. M., Maqbool, R., Ikram, M., Ayub, M. A. (2023). Pesticides: Impacts on agriculture productivity, environment, and management strategies. In: *Emerging Contaminants and Plants: Interactions, Adaptations and Remediation Technologies* (pp. 109–134). Springer International Publishing. doi: 10.1007/978-3-031-22269-6_5
107. Schmidt-Jeffris, R. A. (2023). Nontarget pesticide impacts on pest natural enemies: Progress and gaps in current knowledge. *Current Opinion in Insect Science*, 58, 101056. doi: 10.1016/j.cois.2023.101056
108. Bielza, P., Balanza, V., Cifuentes, D., Mendoza, J. E. (2020). Challenges facing arthropod biological control: Identifying traits for genetic improvement of predators in protected crops. *Pest Management Science*, *76*(11), 3517–3526. doi: 10.1002/ps.5857
109. Alignier, A., Uroy, L., Aviron, S. (2020). The role of hedgerows in supporting biodiversity and other ecosystem services in intensively managed agricultural landscapes. In: *Reconciling Agricultural Production with Biodiversity Conservation* (pp. 177–204). Burleigh Dodds Science Publishing. doi: 10.19103/AS.2020.0071.09
110. Kamala, I. M., Devanand, I. I. (2021). Pollination and ecological intensification: A way towards green revolution. In: *Ecological Intensification of Natural Resources for Sustainable Agriculture* (pp. 381–427).
111. Katumo, D. M., Liang, H., Ochola, A. C., Lv, M., Wang, Q. F., Yang, C. F. (2022). Pollinator diversity benefits natural and agricultural ecosystems, environmental health, and human welfare. *Plant Diversity*, *44*(5), 429–435. doi: 10.1016/j.pld.2022.01.005
112. A, M., Pandey, R., Mall, P. (2023). Protecting honeybees from pesticides: A call to action. *Biodiversity*, *24*(3), 117–123. doi: 10.1080/14888386.2023.2230181
113. Antoine, C. M., Forrest, J. R. K. (2021). Nesting habitat of ground-nesting bees: A review. *Ecological Entomology*, *46*(2), 143–159. doi: 10.1111/een.12986
114. Dagar, J. C., Gupta, S. R., Sileshi, G. W. (2023). Urban and peri-urban agroforestry to sustain livelihood and food security in the face of global environmental change and epidemic threats. In: *Agroforestry for Sustainable Intensification of Agriculture in Asia and Africa* (pp. 89–118). Springer Nature: Singapore. doi: 10.1007/978-981-19-4602-8_4

115. Dianzinga, N. T., Moutoussamy, M. L., Sadeyen, J., Ravaomanarivo, L. H. R., Frago, E. (2020). The interacting effect of habitat amount, habitat diversity and fragmentation on insect diversity along elevational gradients. *Journal of Biogeography*, *47*(11), 2377–2391. doi: 10.1111/jbi.13959
116. Zaller, J. G. (2020). Pesticide impacts on the environment and humans. *Daily Poison*, 127–221. doi: 10.1007/978-3-030-50530-1_2
117. Ali, S., Ullah, M. I., Sajjad, A., Shakeel, Q., Hussain, A. (2021). Environmental and health effects of pesticide residues. In: Inamuddin, Ahamed, M.I., Lichtfouse, E. (Eds.), *Sustainable Agriculture Reviews 48. Sustainable Agriculture Reviews*, Vol. 48, pp. 311–336. Springer: Cham. https://doi.org/10.1007/978-3-030-54719-6_8
118. Grames, E. M., Montgomery, G. A., Boyes, D. H., Dicks, L. V., Forister, M. L., Matson, T. A., Nakagawa, S., Prendergast, K. S., Taylor, N. G., Tingley, M. W., Wagner, D. L., White, T. E., Woodcock, P., Elphick, C. S. (2022). A framework and case study to systematically identify long-term insect abundance and diversity datasets. *Conservation Science and Practice*, *4*(6), e12687. doi: 10.1111/csp2.12687
119. Mahanta, D. K., Bhoi, T. K., Komal, J., Samal, I., Mastinu, A. (2024). Spatial, spectral and temporal insights: Harnessing high-resolution satellite remote sensing and artificial intelligence for early monitoring of wood boring pests in forests. *Plant Stress*, *11*, 100381.
120. Samal, I., Mahanta, D. K., Bhoi, T. K., Komal, J., Jatav, H. S., Jatav, S. S., Sathyanarayana, E. (2024). *Approaches of Biochar in Ecosystem Management: Current Scenario and Future Perspectives*. IntechOpen
121. Vyas, V., Bhoi, T. K., Samal, I., Singh, S., Mahanta, D. K. (2024). Revitalizing degraded soils with agroforestry interventions: Opportunities, challenges, and future direction. In: *Agroforestry to Combat Global Challenges: Current Prospects and Future Challenges* (pp. 529–549).
122. Samal, I., Bhoi, T. K., Mahanta, D. K., Komal, J., Singh, S. (2024). Biorational pest management: Potentials, unintended consequences, and future concerns. In: *Biorationals and Biopesticides: Pest Management* (pp. 47–75).

Index

A

Acute oak decline (AOD), 228
Agricultural productivity
  crop diversity and stability, 160
  economic value, 161
  enhancing crop yields, 160
  importance of pollinator, 159
  resilience to environmental changes, 161
  sustainable agriculture, 161
Agroecological phenomenon, 52
Agroecological plant protection, 194
Agroecosystems, 24, 68, 166, 213, 218
Agroforest habitats, 53
Agroforestry, 2
  advantages for pollinators, 6
  influence pesticide exposure, 7
  meteorological factors, 7
  sustainable agriculture, 245
  tree and shrub species, 6
  using biofertilizer, 36–38
  weather conditions, 7
  windbreaks, 7
  woody species, 6
Allelopathy, 126, 199, 202
Alley-cropping system, 67–68, 121–122
Amplified fragment length polymorphisms (AFLPs), 96
Amplified ribosomal DNA restriction analysis (ARDRA), 99
Angraecum sesquipedale (Madagascan orchid), 153
Anoplotrupes stercorosus, 11
Arbuscular mycorrhizal (AMF) fungi, 27, 36
Armillaria root rot, 220, 221

B

Bacillus thuringiensis, 57
Bacterial–plant association, 88
Bait survival technique, 227
Beauveria bassiana, 57, 221, 227, 252
Bees, 5, 153
Beneficial insects
  climate change impact, 17
  decomposers
    black soldier fly, 12–13
    dung beetle, 10–11
    scavenger, *see* Scavengers
  habitat fragmentation and loss, 17–18
  lack of awareness and education, 16–17
  natural enemies, 8–10
  nutrient cycling capacity, 4
  organic matter decomposition, 4
  pesticide usage and its impact, 17
  pollinators, 3, 5–7
Biodiversity conservation, 2, 18, 151, 165, 186, 249
Biofertilizers
  advancements, 39
  application challenges
    change in climate, 34–35
    complications in formulation, 34
    conditions of storage, 35
    global market, 35
    intellectual property, 36
    legal affairs, 36
    registration, 36
  definition, 25
  environmental benefits, 32–33
  KSB, 27–28
  methods of application
    direct application in soil, 31
    seed treatment, 32
    seedling root dipping, 31–32
  microorganisms, 25
  mode of action
    biological control of plant pathogens, 29
    improved soil health, 30
    improved tolerance to abiotic stress, 30
    ISR, 29
    nitrogen-fixing capability, 28–29

phytohormones, 29
solubilize nutrients, 28
N_2-fixers, 26
PGP microbes, 37
PSB, 26
VAM, 27
Bioinformatics, 103
Bioinoculants, 27, 35
Biological control, 61
benefits of, 137–138
challenges, 138–139
mechanisms, 135–136
multitrophic systems, 136–137
strategies for enhancing, 136
Biological nitrogen fixation (BNF), 87
Biotic stress mitigation, 232
Black soldier fly larvae (BSFL), 12
Bombus terrestris, 161, 165
Braconid wasps, 56
Bradyrhizobium strains, 88, 93
Bt generate proteins, 57
Burkholderia, 87, 89

C

Cacao agroforestry systems, 137
Capsicum annuum, 164
Carrion, 13, 14
Chemical control
emphasis in MAF, 140–141
reasons for minimizing, 140
strategies for minimizing, 141–142
Chemical fertilizers, 25, 32, 151
Chemical pesticides, 2, 3, 39, 52, 65, 71, 161, 195, 246
Cocoa swollen shoot virus disease (CSSVD), 220
Coffee leaf rust (CLR), 224
Coleomegilla maculata, 8
Coppicing, 134
Coprophagy, 10

D

Darwin's coevolutionary theory, 154
Denitrification, 26
Diazotrophs, 86, 88
Dimethyl trisulfide (DMTS), 11, 14
Dipterans, 8, 160
DNA–DNA hybridization, 91, 96, 100
Dryocosmus kuriphilus, 228
Dung beetle, 4
advantages of, 11
butyric acid, 11
types, 11
Dutch elm disease (DED), 213
Dwellers, 11

E

Ecological balance, 2, 4, 71, 187, 198, 255
Ecosystem services (ESs), 52, 122, 155, 228–229
Entomopathogenic fungi, 57, 73, 227
Ethnobotanical knowledge, 181, 201–202
Ethnobotanical studies
cultured-based plant materials, 185
ideas and concepts, 184
independent communities, 183
indigenous communities, 184–185
Indigenous medicine, 185
spiritual-cultured aspects, 184
Eucalyptus torelliana, 53, 56

F

Floristic diversity, 53
Forest gardening, 122
Forest health, 233
Forest pests, 219, 230, 233
Frankia, 88
Fusarium oxysporum, 228

G

Grassy strips, 10
Ground beetles, 55, 64
Ground-nesting bee, 6
Grubs, 11, 179

H

Haber–Bosch process, 86
Harvest festivals, 201
Heterobasidion annosum, 228
Honey bees, 7
Horizontal gene transfer (HGT), 93

Host-specific nodulation *(hsn)* genes, 90
Hoverflies, 55, 137
Hymenopterans, 8, 160
Hymenoscyphus fraxineus, 216

I

Ichneumonid wasps, 56
Indian Council of Agricultural Research (ICAR), 12
Indigenous agricultural practices
  beneficial plant associations, 182
  community-based conservation practices, 183
  cultural preservation, 186
  engaging with indigenous communities, 186
  fire management practices, 183
  planting for pest control, 182
  seed saving and exchange, 182
  traditional knowledge, 186
  tribes cultivation, 179
Indigenous knowledge, 176, 187, 193, 197, 209
Indigenous plant protection practice (IPPP), 206
Indigenous technical knowledge (ITK), 194, 204
Indigenous wisdom, 176, 192
  crop protection, 206
  innovation, 206–207
  modern agriculture, 203–206
  research gaps, 208
Indigenous wisdom and traditional knowledge (IWTK), 192, 208
Indole-3-acetic acid (IAA), 94, 95
Induced systemic resistance (ISR), 29
Insect communities
  challenges, 256–258
  factors influencing insect diversity, 248
  impact on habitats, 253–254
  management strategies, 254–256
  microhabitats, 247
  parasitizing pest species, 246
  as pest regulators, 251–252
  as pollinators, 249–251
Insect pollinators, 161, 249–251
  diversity of, 152–153
  impact on crop productivity, 154
  plant–insect relationships, 153–154
  services, 153
  sustainability and resilience, 155
  trophic interactions, 155
Insect scavengers, 4
Insect-friendly habitats, 4, 18
Integrated pest and disease management (IPDM), 213
  adaptive management, 124
  allelopathic characteristics, 235
  biocontrol agents, 234
  cultural and ethical considerations, 123–124
  detrimental impact, 234
  economic viability, 123
  enhanced ecosystem services, 124
  environmental sustainability, 123
  epidemiological inquiries, 233
  food safety, 124
  global level, 234
  human health, 124
  identifying the pivotal species, 234
  implement, 217
  investigation, 233
  MAF, 217–219
    biocontrol control, 226–228
    climate change adaptation, 230–231
    cropping patterns, 221–223
    disease resistance and tolerance, 229–230
    ecological succession, 223–225
    ecosystem services, 228–229
    extension and implementation, 231–232
    monitoring and surveillance, 219–220
    natural predators, 226–228
    resilience to outbreaks, 225–226
    silvicultural practices, 220–221
  pathogen activity, 234–235
  preservation of biodiversity, 123
  principle, 216–217
  resilience against pest outbreaks, 123
  woody plant species, 235
Integrated pest management (IPM), 54, 135, 193, 248
Intergenic spacer (IGS), 99

Intergovernmental Science-Policy Platform on Biodiversity and Ecosystem Services (IPBES) report, 5
Invasive alien pathogens and pests (IAPPs), 213

K

K-synthetic bacteria (KSB), 27–28

L

Lacewings, 55, 135, 161
Ladybird beetles, 53, 70
Land management system, 52
Legume–rhizobia association, 89–90
Legumes, 86–90, 137, 226
Leguminous trees, 87
List of Prokaryotic Names with Standing in Nomenclature (LPSN), 88
Lotus japonicus, 90

M

Medicago truncatula, 90
Memorandum of Understanding (MoU), 12
Metabolic fingerprinting, 92
Metarhizium species, 227, 252
Microbial biofertilizers, 32, 33
Microhabitats, 53, 154, 253
Modern agriculture, 39, 213
Monocropping, 62–63
Monoculture farming, 244, 248
Multilocus sequence analysis (MLSA), 92, 101
Multitrophic agroforestry (MAF)
  alley cropping, 121–122
  benefits of, 120–121
  biodiversity, 120
  definition, 120
  disease prevalent, 132
  features, 122
  forest gardening, 122
  historical roots, 122
  IPDM strategies
    biology, *see* Biological control
    chemical control, *see* Chemical control
    cultural control, 132–133
    silvicultural practices, *see* Silviculture
  Middle story
    coffee, 130
    guava, 130
    tea, 129–130
  mutual exchange, 177
  pest and disease suppression
    allelopathy, 126
    habitat manipulation, 125
    nutrient cycling, 125
    physical barriers, 124–125
    resource competition, 125
  silvopasture, 121
  Under story
    ginger, 131
    pineapple, 131
    turmeric, 131
  Upper story
    bamboo, 127
    casuarina, 128
    eucalyptus, 129
    gambhar, 128
    mango, 128–129
    poplar, 128
    teak, 127
Multitrophic dynamics, 177
  Africa, 181
  community-based approaches, 203
  crop varieties, 200
  cultural preservation and sustainability, 178
  ecological balance and resilience, 198
  ecosystem health, 178
  in ecosystems, 197
  indigenous knowledge, 178
  indigenous perspectives, 199
  interconnected relationships, 178
  North America, 180–181
  pest management
    harmony and balance, 201
    medicinal plants, 202
    plant-pest relationships, 202–203
    rituals and ceremonies, 200–201
    spiritual and cultural practices, 200
  pest resistance, 199
  plant protection strategies, 178, 192–193
    agricultural practices, 195
    biodiversity and, 198–199

cultural beliefs, 197
hybrid approach, 207–208
indigenous pest control methods, 195–196
plants/pests/ predators, 198
South and Central America, 181–182
Southeast Asia, 181
Multitrophic pollination, 151
diversity, 157–158, *see also* Pollinator diversity
natural plant communities, 158
sustaining agricultural productivity, 159, *see* also Agricultural productivity
ecological intensification (EL)
diverse farming approach, 157
plant–pollinator interaction, 156–157
insect pollinators
diversity of, 152–153
impact on crop productivity, 154
plant–insect relationships, 153–154
services, 153
sustainability and resilience, 155
trophic interactions, 155

N

Native populations, 179–180
Natural enemies (NE), 3, 8–10, 9
birds, 58
conservation measures, 60–62
abiotic factors, 63–64
agroforestry planting, 65
alley–cropping system, 67–68
avoidance of homogenization, 62–63
cropping patterns, 63
ecological engineering, 68–69
impact of pesticides, 69–71
planting density, 64–65
silvo-arable system, 66–67
vegetable planting, 65–66
ecological sustainability
biological control strategies, 71–72
botanicals and semiochemicals, 72
CBC, 72
habitat management, 72
technological innovations, 72
parasitic nematodes, 58
parasitoids
braconid wasps, 56
ichneumonid wasps, 56
Trichogramma wasps, 56
pathogens
Bacillus thuringiensis, 57
entomopathogenic fungi, 57
pest control, 54
biological control, 59
chemical inputs, 59
diversity and stability, 59
ecosystem resilience, 60
pollination services, 60
reduction, 59
seasonal, 59
predators
ants, 55
bugs, 56
ground beetles, 55
hoverflies, 55
lacewings, 55
ladybird, 55
mites, 56
rove beetles, 55
spider, 57
success rate, 72–74
Nodulating bacteria, 87
Nutrient cycling, 125, 134
Nutrient mineralization, 4

O

Oligolecty, 6

P

Parasite, 8, 53, 180, 216, 228, 235
Parasitoid wasps, 10, 16, 251
Parasitoids, 8, 56, 61, 255
Pathogens, 8, 29, 57
Phacelia tanacetifolia, 55
Phenotypic tests, 92
Phlebiopsis gigantea, 228
Phosphate-solubilizing bacteria (PSB), 26, 94
Phytohormones, 29, 95
Plant growth promoting (PGP) traits, 25, 39

Plant growth-promoting rhizobacteria (PGPR), 94–95
Plant–pollinator interactions, 151, 156
Planting ceremonies, 200
Pollarding, 134
Pollination services, 52, 153, 162, 249
Pollinator diversity
  agricultural output, 163–164
  antagonistic interactions, 166
  climate change, 161
  environmental indicators, 164
  natural plant communities, 162
  necessity to enhance biodiversity, 165
  pest and disease control, 164–165
  plant production and recruitment, 162
  pollen limitation, 163
Pollinators, 3, 5–7, 53, 66, 136, *see also* Insect pollinators
Polyacrylamide gel electrophoresis (PAGE), 96
Polymerase chain reaction (PCR), 96
Predatory flies, 16
Predatory insects, 8, 56, 155, 252, 255

R

Random amplified polymorphic DNA (RAPD), 96
Refuge, 68, 235
Restriction fragment length polymorphisms (RFLPs), 96, 98–99
Rhizobia, 87, 88
  characterization of
    biochemical characters, 93
    molecular, 95–97
    multilocus fingerprinting analysis, 97–98
    multilocus sequence analysis methods, 101
    PGPR, 94–95
    phenotypic characteristics, 92–93
    RFLP, 98–99
    ribosomal gene, 99–100
    16S rRNA sequence analysis, 101
    sym loci, 101–103
  legume-rhizobia association, 89–90
  RNB, *see* Root nodule bacteria (RNB)
  taxonomy of, 88–89
Rhizobium-legume symbiosis, 87
Rollers, 11
Root nodule bacteria (RNB), 89–91
Rove beetles, 55

S

Scavengers, 4
  insect activity, 15–16
  insect orders, 13–14
  process of decomposition, 14–15
  stages of decomposition, 14
Sequence analysis of symbiotic (SYM) loci, 101–103
Shade-grown coffee systems, 249
Silviculture
  biodiversity, 134
  competition for resources, 134
  coppicing, 134
  disease and pests, 135
  labor and knowledge, 135
  nutrient cycling, 134
  pollarding, 134
  tree pruning and thinning, 134
  tree selection and planting, 133–134
Silvo-arable system, 66–67
Silvopastoral systems, 248
Silvopasture, 121, 244
Sinorhizobium, 88
Smallholder farmers, 24
Soil inoculants, 37
Strip cropping, 220
Sustainable agriculture, 24
Sustainable land management, 2
Symbiosomes, 90
Symbiotic nitrogen fixation, 87
Synthetic fertilizers, 32

T

Tinawon, 201
Torymus sinensis, 228
Tritrophic relationship, 53
Tropical agroecosystems, 9
Tunnelers, 11

V

Volatile organic compounds (VOCs), 14, 29

W

Walnut tree-cereal agroforestry system, 10
Waste management, 4, 12, 18

For Product Safety Concerns and Information please contact our EU representative GPSR@taylorandfrancis.com Taylor & Francis Verlag GmbH, Kaufingerstraße 24, 80331 München, Germany

Batch number: 10725937

Printed by Printforce, the Netherlands